BRAIN TUMOR PATHOLOGY:
CURRENT DIAGNOSTIC HOTSPOTS AND PITFALLS

Brain Tumor Pathology: Current Diagnostic Hotspots and Pitfalls

by

DAVIDE SCHIFFER
University of Turin, Italy
Neuro-bio-oncology Center,
Policlinico di Monza Foundation

Springer

A C.I.P. Catalogue record for this book is available from the Library of Congress.

ISBN-10 1-4020-3997-2 (HB)
ISBN-13 978-1-4020-3997-3 (HB)
ISBN-10 1-4020-3998-0 (e-book)
ISBN-13 978-1-4020-3998-0 (e-book)

Published by Springer,
P.O. Box 17, 3300 AA Dordrecht, The Netherlands.

www.springer.com

Printed on acid-free paper

Printed in the Netherlands.

TABLE OF CONTENTS

INTRODUCTION

Since Bailey and Cushing (1926), all brain tumor classifications have been called histogenetic. The nosographic position that the tumor types progressively acquired in the classification systems derived from the resemblance of tumor cells to those of the cytogenesis, modified whenever new information became available from different biological research fields and especially from molecular genetics. Classically, on the basis of the rough correspondence between the mature/immature aspect of tumor cells and the benign/malignant biological behavior of the tumors, the histological labels contained a prognostic significance. The supposed origin of the tumors was thus a factor for prognosis. Later on, with the concept of anaplasia (Cox, 1933; Kernohan et al., 1949) new criteria were introduced for establishing the malignancy grades of tumors. Immunohistochemistry and later molecular genetics further refined the prognostic diagnoses, substantially increasing the opportunities to recognize the cell origin of tumors, beside revealing the pathogenetic mechanisms. Prognoses became more accurate, as required by the greater and more targeted possibilities of therapy.

Molecular genetics, on the one hand, gave us a deeper knowledge of tumorigenesis and tumor transformation, but on the other hand, it made things more complicated, demonstrating for example that phenotypically similar tumors may have different genetic assets and vice versa, with important implications for prognoses drawn from diagnoses. Recently, microarray gene profilings are demonstrating that genetically-based prognosis may be more reliable than histologically-based prognosis (Nutt et al., 2003). In the meantime, new variants and new tumor entities have been described (Cenacchi and Giangaspero, 2004) which should be added to the WHO book (Kleihues and Cavenee, 2000), showing the double relation to molecular genetics.

The prognosis drawn from histological diagnosis had to become more refined, also because new radiotherapeutic procedures and chemotherapeutic strategies became more demanding, as they were retailed on specific prognoses and diagnoses. Another requirement for more precise prognostic categorization of tumors derived from the introduction of the statistical-mathematical method, epitomized in multivariate analysis, used for the study of the outcome of tumors or of their time to progression (TTP), was assumed as a means for the evaluation of the efficacy of therapies.

In the last few years, the increased diagnostic-prognostic requirement was paralleled by a reduced quantity of tumor tissue available for examination. The clinical diagnostics by neuro-imaging techniques, with functional, diffusion and intraoperative MRI and spectroscopy (Rees, 2003), SPECT and PET procedures have been strongly facilitated to the point that often the tumor nature can be foreseen. This led to earlier discovery of tumors, at a stage when they are still of small dimensions. At the same time, surgical procedures also improved, for example with the introduction of the neuro-navigator, so that quite often a sheer diagnostic function is required from neurosurgery. Moreover, the availability of different modalities of radiotherapy, such as neutron-, proton- and ion-radiation with a better therapeutic planning and a conformational-three-dimensional implementation, the possibility to reach deeply located and irregularly shaped tumors, stereotactic radiotherapy and the γ knife, not to mention chemotherapy, use of stem cells, immunotherapies, biodegradable polymers, convection-enhanced drug delivery (Dunn and Black, 2003), oncolytic viruses (Jiang et al., 2003), etc., contributed enormously to modifying the collaboration between neurosurgeon and pathologist. The tumors as a consequence are currently recognized earlier and therefore they are smaller and surgical specimens are of reduced size, but at the same time a more precise prognosis from histology is being required. The introduction of stereotactic biopsy procedures was the first step in this direction. If the discovery of new variants and of new prognostic categories of brain tumors is added, a greater possibility of pathology error becomes comprehensible.

The advancement of neurobiological studies of the development of the nervous system and the recent emphasis given to stem cells progressively modified our conception of the cell composition of a tumor, since the morphological and antigenic aspect of its cells oscillates between that of progenitors to that of mature specific cells with the possibility of going in one direction or the other, the so-called de-differentiation a Shih and Holland, 2004). This may lead us to evaluate the origin of the tumor, which in turn may have an influence on the prognosis drawn from histology in practice.

The aim of this work is to discuss the practical importance, at the moment of a histological diagnosis, of some not yet resolved biological problems of brain tumors, the components of which may have influence on the diagnosis, and to emphasize the dilemmas that arise in attributing one significance or another to the different findings.

Chapter 1

THE ORIGIN OF GLIOMAS IN RELATION TO THE HISTOLOGICAL DIAGNOSIS

The old question of the origin of brain tumors becomes once again of great interest with the therapeutic application of the concept of stem cells. The parallelism between the morphological aspect of tumor cells and that of the cells during cytogenesis, on which Bailey and Cushing's and the so-called histogenetic classifications were based (Figure 1), has recently been shaken by a number of *in vivo* and *in vitro* observations. These can influence both the diagnostics, when it has to be performed in small samples, where the histological patterns of the tumors are frequently lacking, and the experimental therapeutic strategies.

In humans, it is still impossible to identify the cells which give rise to tumors prior to their transformation (Holland, 2001), because the earliest stages of glial tumor development are not known; and the first visible lesions, i.e. for example astrocytomas, are already organized as tumors when they are recognized.

Theoretically, it is accepted that astrocytomas derive from astrocytes and oligodendrogliomas from oligodendrocytes; and, since adult glia does not proliferate, whereas basically the development of a tumor requires that the transforming events affect proliferating cell populations, their original cells must be precursor cells or neural stem cells. It is also very well known that the proliferation is slow in low-grade gliomas and quick in high-grade gliomas, and that a transition from the former to the latter through anaplasia can occur. Our working concepts on tumor formation are based on the old multistage model which establishes that tumors develop in the three stages of initiation, promotion and progression.

Years of experimental studies on the oncogenetic effects of nitrosourea derivatives in rats have made it clear that the induced tumors arise from primitive neuroepithelial cells of the ventricular zone (VZ) or from its derivative subventricular zone (SVZ) or from cells of the so-called "renewal" of the adult that occurs in the remnants of SVZ or sub-ependymal layer, hippocampus, cerebellum, first cortical layer (Figure 2).

Classically, neurons and glia derive from primitive neuroepithelial cells or neural stem cells of the VZ and SVZ, characterized by self-renewal and multipotency (Figure 3). They differentiate along different pathways under extrinsic and intrinsic stimulations. Neurogenesis occurs first through organizing centres which generate signals inducing the expression of patterning genes encoding

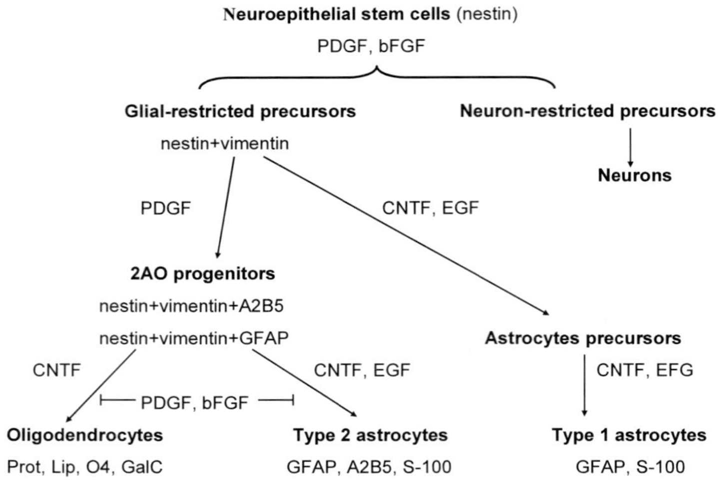

Figure 1. Cell differentiation in the course of cytogenesis

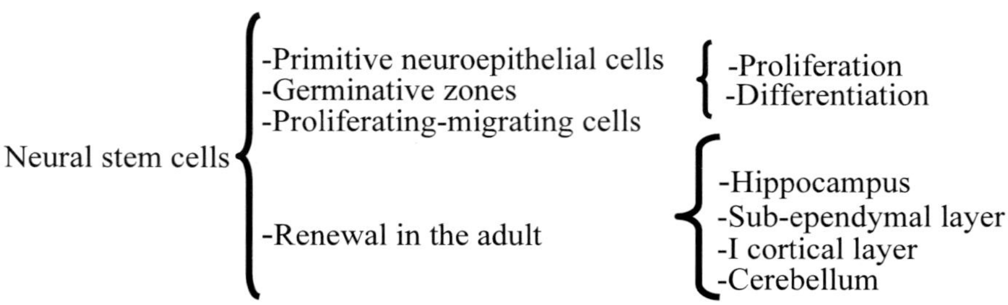

Figure 2. Cells of origin of gliomas

transcriptional factors and controlling neuronal subtypes in the adjacent neuroepithelial cells (Kobayashi et al., 2001).

Markers are in order nestin, vimentin, A2B5, GFAP and then O4, proteolipid protein, galactocerebroside, myelin basic protein and synaptophysin and neurofilaments for neurons. Growth factor signaling controls the passage from one stage to the other: PDGF and bFGF promote the passage from stem cells to precursors, the passage from precursors to O2A progenitors and to astrocyte

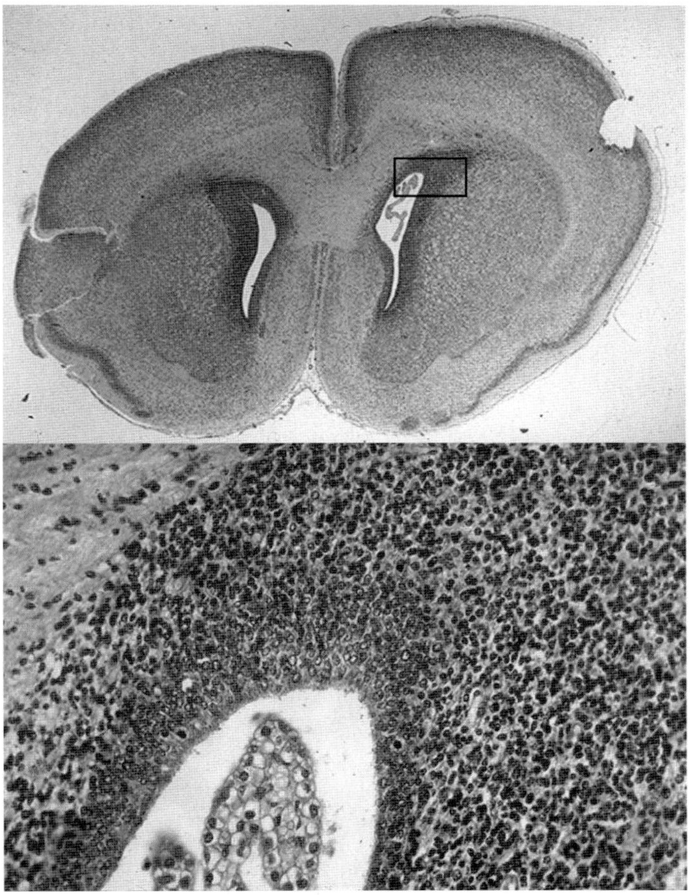

Figure 3. Germinative zone in the rat, H&E, x 25 and x 200

precursors is promoted by PDGF and CNTF, EGF, respectively. The passage from O2A progenitors to oligodendrocytes and type 2 astrocytes is promoted by CNTF and CNTF, EGF, respectively and inhibited by PDGF and bFGF (Goldman, 2000; Holland, 2001). During migration, glia cells continue to proliferate. In time, the VZ disappears and the SVZ decreases in size persisting in the adult as a sub-ependymal cell layer (Figure 4).

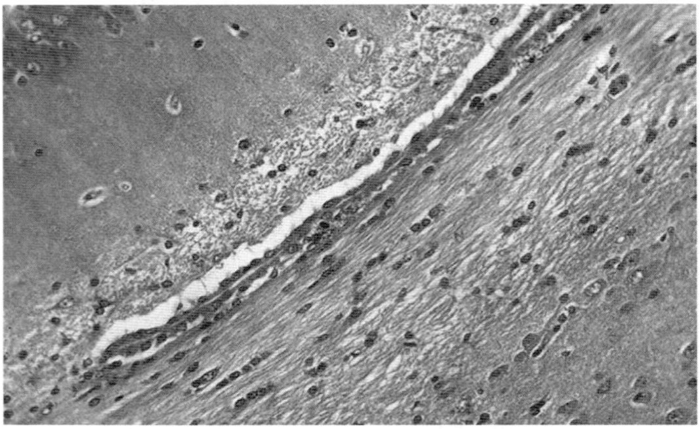

Figure 4. Sub-ependymal layer in adult rat, H&E, x 400

Radial glia that derives from neuroepithelial stem cells at the onset of neurogenesis is particularly important. The soma borders on the ventricle and the processes extend to the pial surface as scaffolding to migrating neurons (Figures 5, 6).

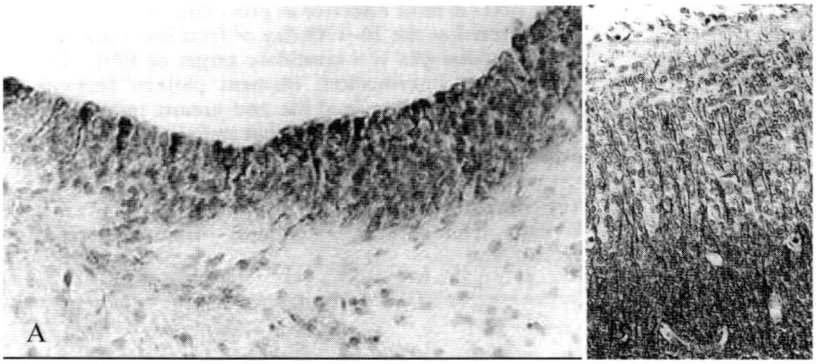

Figure 5. Radial glia in the rat. Vimentin, DAB. A. The soma borders on the ventricular cavity. B. Cell processes reach the meninges. x 200. From Giordana et al., 1990

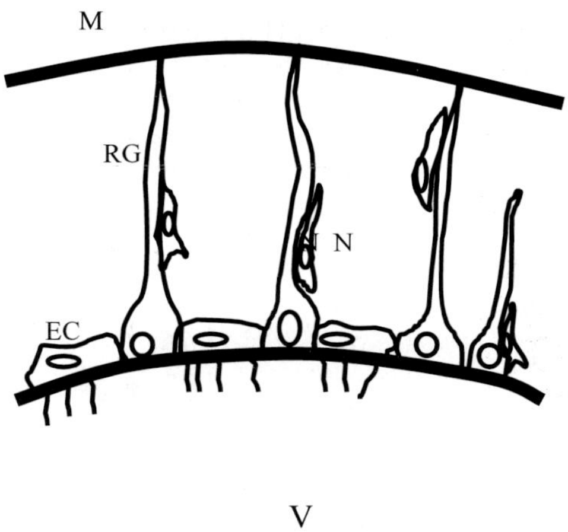

Figure 6. Radial glia with migrating neurons.

EC = Ependymal cells, N = Neurons, RG = Radial glia, V = Ventricle

Radial glia has astrocytic characteristics and most progenitors of VZ possess radial glia features. It expresses RC2, nestin, vimentin, GFAP, GLAST (glutamate-aspartate transporter) and it is neurogenetic (Ever and Gaiano, 2005). Radial glia state is maintained by Notch signaling through ligand Delta1, ErbB through Neuroregulin and FGFR through FGF (Ever and Gaiano, 2005). On the other hand, it is known that these are critical for glioma cell survival and proliferation (Purow et al., 2005). At the end of neuron migration radial glia transforms into astrocytes, as committed to the astroglial lineage (Schmechel and Rakic, 1979). It reveals stem cell characteristics as do the derived astrocytes either during development or in the adult (Laywell et al., 2000; Doetsch, 2003).

Neurogenesis and gliogenesis continue in the adult mammalian brain (Gage, 2000, 2002) from neural stem cells occurring in the sub-granular zone of the dentate gyrus of the hippocampus and as astrocytes and ependymal cells under the anterior lateral ventricular wall (Clarke, 2003) (Figure 7). The existence of neuroectodermic stem cells not only in embryonal life, but also in the adult, changed our interpretation of many processes occurring in the brain and it became a basic working concept. Particularly important are the recent interpretations already mentioned that neurospheres formed from SVZ cells, both in embryos and in the adult, are ultimately astrocytes and that radial glia also may represent stem cells or progenitors during embryonal development (Doetsch et al., 2003), confirming the hypothesis that stem cells are found within the lineage neuroepithelium – radial glia – astrocytes (Alvarez-Buylla et al., 2001). To this hypothesis belongs also the concept that in the adult, glia cells or a

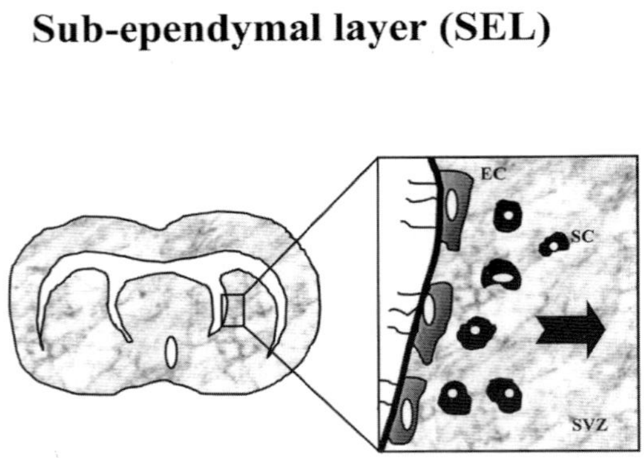

Figure 7. Scheme of cell migration from the sub-ependymal layer, EC = ependymal cells, SC = stem cells

subset of them may represent latent stem cells throughout the brain and there is evidence that cells from non-neurogenetic regions, if cultured with bFGF/EGF, become neurons (Doetsch et al., 2003).

Tumors can arise from stem cells either of VZ or SVZ during embryonal life or in adult life and most of our knowledge is based on experimental studies of glioma induction in the rat by transplacental Ethylnitrosourea (ENU). The tumors were observed to originate from the VZ and from the derived SVZ (Schiffer et al., 1978-80; Lantos and Pilkington, 1979). The target of Ethylnitrosourea administered transplacentally was the germinative zone. Here the alkylation of O6 of guanine, O2 of cytosine and O2 and O4 of thymidine took place with consequent coupling errors during transcription (Kleihues and Rajewsky, 1984) and the defective repair of DNA (Goth and Rajewesky, 1974). Short-term, as stop to mitoses and nuclear deaths in the germinative zones (Bosch, 1977), and long-term phenotypic effects, such as early lesions, microtumors and tumors in the brain hemispheres, starting in the periventricular white matter (Schiffer, 1997, 78), represented the consequences of nitrosourea derivative action (Figures 8, 9, 10, 11).

The vulnerability of neuroepithelial cells to neoplastic transformation resulted from the interaction of several factors, among which were the number of replicating cells at risk at a particular time, the length of time during which a cell population remains in cycle, the state of differentiation etc. It explained the different incidence of the

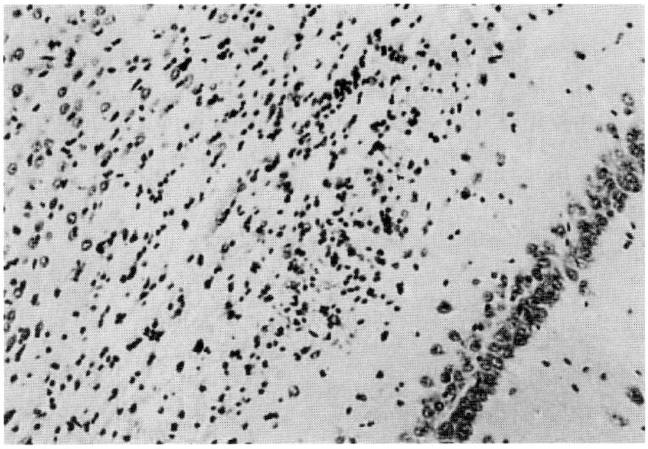

Figure 8. Transplacental ENU tumors in the rat. Early phase in the periventricular white matter, H&E, x 200. From Schiffer et al., 1978

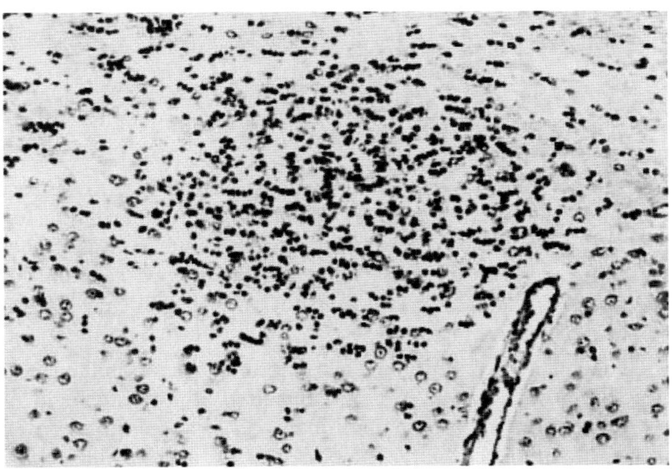

Figure 9. Transplacental ENU tumors in the rat. Periventricular oligodendroglial micro-tumor, H&E, x 200. From Schiffer, 1997

various tumor types and also the phenotype of tumors. On these observations was based the concept of "window of vulnerability" for each precursor cell type (Rubinstein, 1985, 1987). Another concept emerged as very important, i.e., that experimentally more genetic alterations are needed for tumor transformation as more advanced the differentiation status of progenitor cells becomes (Shih and Holland, 2004). The origin of tumors from transplacental ENU has been recently studied again and it has been demonstrated that either SVZ cells or cells of early lesions in the white matter were nestin-positive and so were cells cultivated from SVZ of exposed rats, confirming definitely that tumors arise from stem or precursor cells (Recht et al., 2003).

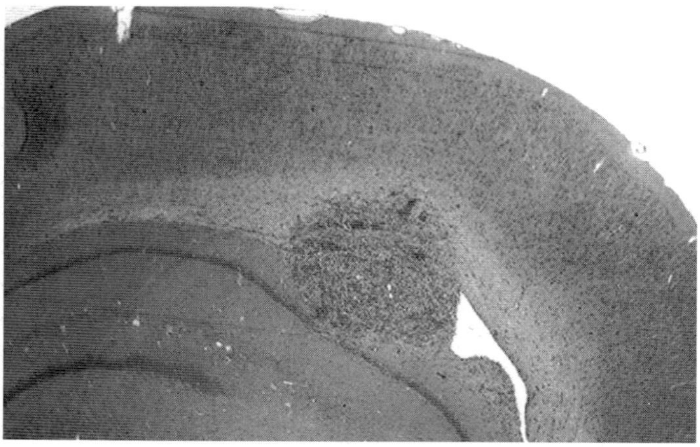

Figure 10. Transplacental ENU tumors in the rat. Micro-tumor developing into a tumor in periventricular white matter, H&E, x 100. From Schiffer, 1991c

Using as tumor inducer Methylnitrosourea, instead of Ethylnitrosourea, either intra-cerebrum or intra-peritoneum or subcutaneously, and administering it in the adult, when the germinative zone is no more present, it was demonstrated that tumors arose in the periventricular white matter, corpus callosum, hippocampus (Schiffer et al., 1970) and their origin could not have been different than that of the SVZ or the sub-ependymal layer (Figure 12).

The relationship between differentiation of neuroepithelial cells during cytogenesis and glioma formation points out firstly the existence of characteristics common to progenitor cells and malignant cells, represented by simplicity of the form, proliferation capacity, potentiality to differentiate and capacity of migration (Dai and Holland, 2003). Secondly, some of the genetic alterations that characterize malignant gliomas, such as those that activate signal transduction pathways and those disrupting the cell cycle arrest machinery and represent the molecular signature of these tumors, concern genes/proteins involved in the regulation of differentiation during cytogenesis and may have effects on the differentiation/de-differentiation status of the cells (Dai and Holland, 2003). In the differentiation process the main signals are: EGF, FGF, PDGF, CNTF, IGF, SHH,

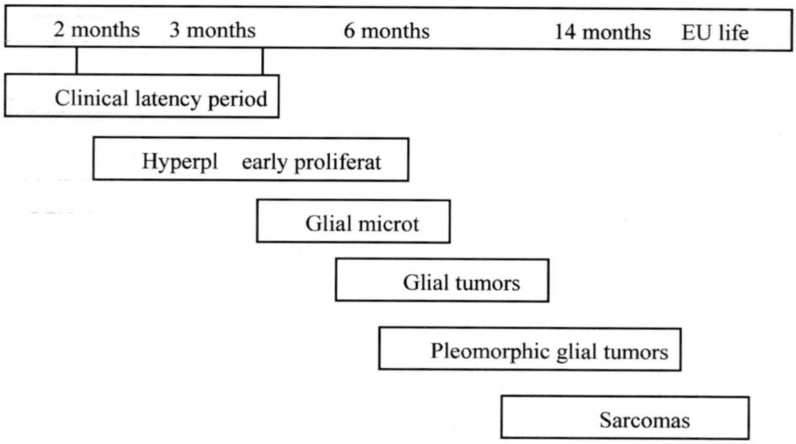

Figure 11. Transplacental ENU tumors. Scheme of developing tumors. From Schiffer et al., 1978

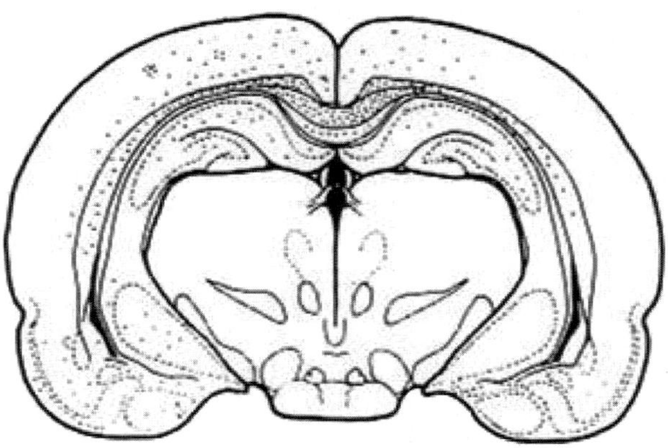

Figure 12. Tumors from intra-peritoneal MNU in the rat. Reprinted from J. Neurol. Sci 11, 559, 1970, Schiffer D, Fabiani A, Grossi-Paoletti, E., and Paoletti, P., (1970), Experimental brain tumors induced in rats by nitrosourea derivatives. Part I. Morphological aspects of methylnitrosourea tumours. Year permission 2005.

etc. and are at the same time mitogenic for cell proliferation. EGF/EGFR activation or inactivation produce increased numbers of astrocytes, for signaling to RAS/MAPK (Bergman et al., 2002) or apoptosis. FGF and PDGF are also mitogens, whereas CNTF promotes astrocytic differentiation (Bonni et al., 1997) that, on the contrary, can be prevented by CpG methylation of a STAT3 (Takizawa et al., 2001). Signaling through receptor Notch regulates astrocytic and oligodendrocytic differentiation, and PDGF is known to be involved in that of oligodendrocytes etc. (Shih and Holland, 2004).

In humans, the possibility of tracing back the origin of the tumors from the aspect of the cells is impossible. Only in animals can this be achieved through the "lineage tracing". In developed human tumors, the morphology of their elements does not correspond entirely to that of mature cells of a certain line, but it is rather composed of a mixture of cells either resembling the stages of cytogenesis, or even expressing markers of immature cell types (Dai and Holland, 2003). As a consequence, the tumor cells will appear de-differentiated if compared with mature cells; or, alternatively, this means that they originate from progenitor-like cells in the tumor.

The evidence that maximally makes the resemblance of tumor cells to cells of the cytogenesis less reliable for establishing the cell origin of the tumors shows that the differentiation of cells during cytogenesis may undergo environmental, epigenetic and genetic influences. This means that the histology of a tumor "would be more a reflection of the environment and time of initiation than the cell of origin" and this would decide whether a tumor ultimately becomes, for example, an astrocytoma or an oligoastrocytoma (Recht et al., 2003). This is demonstrated by some experiments. Introducing Akt and Kras in mouse brains by a retroviral technique it is easier to obtain tumors in nestin-expressing than in GFAP-expressing cells, especially if there is a loss of CDKN2A (Uhrbom et al., 2002), so that genetic deregulations would appear more important than the cell stage of origin. Epigenetic events could be responsible for dedifferentiation: in U 373 MG cell lines of glioblastoma, TGFα or other factors acting on TKRP can reduce GFAP mRNA and enhance nestin expression, whether or not affecting vimentin (Sultana et al., 1999; Zhou and Skalli, 2000). The transforming event could also block the differentiation of a neural stem cell (Pereira et al., 1998), according to the old concept of maturation arrest (Cairncross, 1987), and it could hit either a neural stem cell or a tumor cell which re-acquires properties of a neural stem cell. Other examples are available: in rat progenitor cells the over-expression of Akt or Ras produces tumors with the phenotype of human glioblastomas which, on the other hand, are known for showing an over-expression of them (Holland et al., 2000), whereas the over-expression of PDGFR.B gives tumors with the phenotype of oligodendrogliomas (Dai et al., 2001). It should also be taken into account that different phenotypes can be sustained by the same genetic basis (Kraus et al., 1995). In practice, the neoplastic transformation of glial precursors produces tumors with the phenotypes of astrocytoma and oligodendroglioma and this depends on the activation or inactivation of specific protein pathways. A deeper knowledge of the relationship between molecular pathways and tumor phenotype is very important for discovering the origins of gliomas that tentatively are at the moment only deduced from the phenotype of tumor cells.

The following properties are recognized in neural stem cells: undifferentiation, capacity to form neurospheres and to proliferate, self-maintenance and clonogenicity (Reynolds and Weiss, 1992). Neurospheres derive from one neural stem cell and can be recognizable because they express nestin and a new surface antigen, CD133, which is a 120 kDa five-transmembrane cell surface protein, once a haematopoietic stem cell marker (Uchida et al., 2000). In various experiments all these properties have been demonstrated in neurospheres formed from surgical specimens of glioblastoma. Therefore, the problem of the origin of gliomas needs to take on board not only the relationship of tumor cells to neural stem cells, but also the existence of tumor stem cells, since only a proportion of tumor cells are clonogenic when xenografted (Recht et al., 2003).

The hypothesis that in general cancer stem cells exist is based on the observation that tumor cells are heterogeneous and variably express differentiated antigens typical of the organ, but only a minority of them are self-renewing, multipotent, clonogenic and continuously replenishing mature cells (Reya et al., 2001). *In vitro,* cells from human gliomas generating clusters of clonally derived cells resembling neurospheres, self-renewing and proliferating, and capable of differentiation have been demonstrated (Singh et al., 2003), also in pediatric tumors (Hemmati et al., 2003). Brain tumors, therefore, besides arising from the transformation of neural stem cells, can contain tumor stem cells that once transplanted into mice reproduce tumors with the characteristics of glioblastoma (Galli et al., 2004). It has been shown that the $CD133^+$ cell sub-population from human tumors exhibits stem cell properties *in vitro* and show a capacity for self-renewal and reproduce tumors which can be serially transplanted. On the contrary, $CD133^-$ cells engrafted do not reproduce tumors (Singh et al., 2004). Whether these are real tumor stem cells deriving from the transformation of normal neural stem cells (Shih and Holland, 2004) or they represent the product of an extreme de-differentiation due to anaplasia, i.e. they correspond to the new clones developed by accumulation of mutations and are selected by competition, with a high proliferation rate and lacking any differentiation antigen, is really difficult to say, also because the latter may have acquired stem cell properties (Figure 13). Paraffin FISH studies on $CD133^+$ xenografts from a GBM demonstrated that tumor cells showed amplification of the EGFR gene, so that both $CD133^+$ and $CD133^-$ cells bear the same cytogenetic alterations and therefore they are clonally derived (Singh et al., 2004). A peculiar hypothesis has been put forward for gliomas, i.e. that tumor stem cells divide asymmetrically. One daughter cell remains as a cancer stem cell in the germinative zone and the other migrates away and proliferates. This would have therapeutic consequences (Berger et al., 2004).

The demonstration of stem cell markers in the adult brain and in brain tumors and of the occurrence of neural stem cells in the adult brain and at variance with tumor stem cells in brain tumors is becoming a very complicated and challenging matter. Uchida et al. (2004) propose three explanations for the occurrence of stem cell markers in brain tumors: positive cells are transformed neural stem cells expressing nestin and Musashi-1; they are cells re-expressing nestin (Dahlstrand et al., 1992;

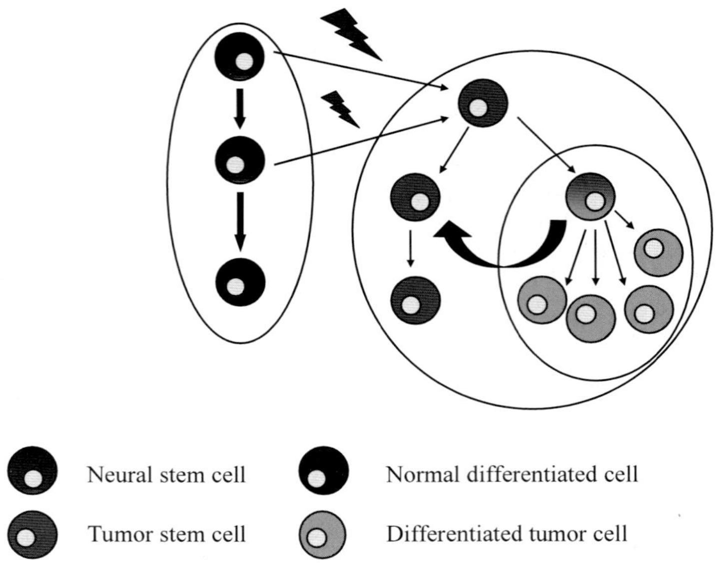

Figure 13. Neural and tumor stem cells

Tohyama et al., 1992) because de-differentiated; they are exogenous stem cells attracted to the tumor tissue (Park et al., 2002). In an infantile tumor the expression of nestin + GFAP and nestin + Tuj-1, similarly to what can be seen in the normal post-natal sub-ventricular zone, has been demonstrated (Uchida et al., 2004), even though this is not a proof that the former derives from the latter. An interesting observation was that the grafted tumor cells in the adult brain, but also *in vitro* after some time, became quiescent, similar to normal post-natal or adult neural stem cells *in situ* (Morshead et al., 1994).

In all this matter the role of nestin is not so clear-cut. Nestin is typical of neuroepithelial/progenitor cells in rats and humans and its expression precedes that of vimentin and GFAP, disappearing from the CNS in adult age with the exception of ependymal cells. This has been demonstrated in the progenitor cells during development and in the subependymal layer of adults (Hockfield and McKay, 1985; Lendhal et al., 1990) with no definite temporal relationship with vimentin and GFAP that should appear later in the development. There is a certain discrepancy among the various researchers concerning technical problems and specificity of the staining. Nestin expression, for example, is considered characteristic of progenitor cells, but since it characterizes reactive astrocytes in brain injury (Tohyama et al., 1992; Lin et al., 1995; Krum and Rosenstein, 1999), one wonders whether the latter cells de-differentiate in their cell reaction or whether nestin cannot be taken as specific of undifferentiation (Holland, 2001). In the first case, nestin could not be indicative of a stem cell status or there would be no possibility to distinguish in tumors between neural stem cells and tumor stem cells, unless the latter are capable of re-acquiring nestin expression even if they represent a transformed and not an undifferentiated phenotype.

In reactive astrocytes (Figure 16A) the expression of nestin would represent the embryonal regression of cytoskeleton connected with their morphological plasticity. In hippocampus, it decreases with the age of the subject (Abdel-Rahman et al., 2004). There are alternative interpretations, very complicated, concerning the up-regulation of multiple embryonic proteins in adult astrocytes following injury for re-enacting a microenvironment reminiscent of that during the embryonic period (Clark et al., 1994; Nakamura et al., 2003).

In some findings nestin and vimentin are not markers of stem cells (Singh et al., 2004). In pediatric brain tumors, nestin has been found to be expressed in PNET, anaplastic astrocytomas and ependymomas, but almost never in low-grade astrocytomas in one series (Tohyama et al., 1992) and not expressed in another series of gliomas (Dahlstrand et al., 1992). Using two different antibodies (Tohyama et al., 1992 and Grigelioniené et al., 1996), after antigen retrieval nestin was demonstrated in ependymal cells, cells of the germinal matrix and radial fibres of a human foetus, co-expressed with vimentin, but not with GFAP, and in tumor cells of pediatric ependymoma, PNET, glioblastoma and pilocytic astrocytoma associated with vimentin, GFAP and S-100. In our experience nestin is expressed in ependymomas, pilocytic astrocytomas and much more in glioblastomas than in astrocytomas (Figure 14). It is not expressed in oligodendrogliomas, with the exception of GFOC (Figure 15). Also endothelial cells were found to be positive (Almqvist et al., 2002) and

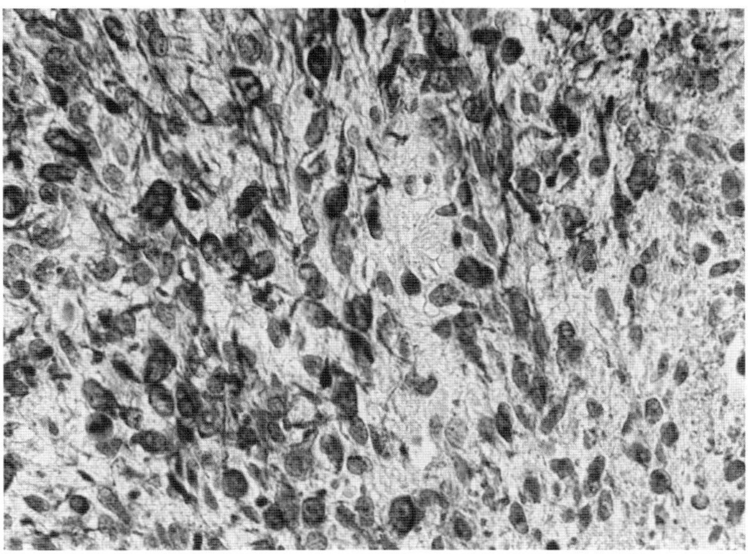

Figure 14. Glioblastoma. Nestin-positive cells, DAB, x 400

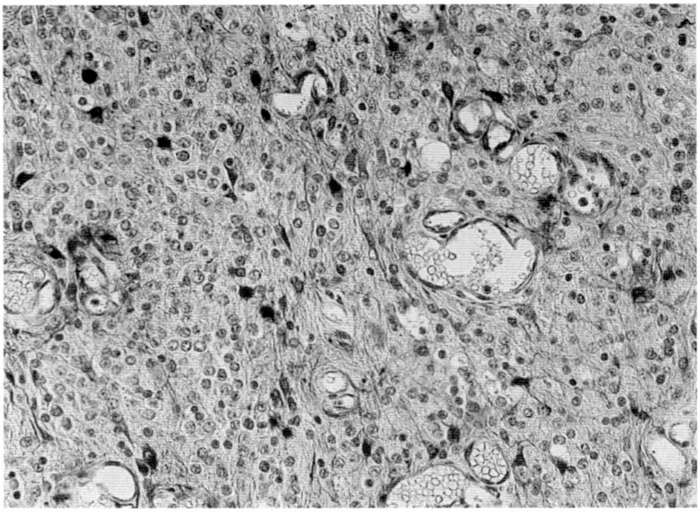

A B

Figure 15. Oligodendroglioma. Tumor cells negative for nestin and positive minigemistocytes, DAB, x 200

nestin has been considered to be a marker of rapidly growing endothelial cells (Sugawara et al., 2002). In our experience, it is expressed in micro-vascular proliferations of glioblastomas and oligodendrogliomas (Figure 16).

Vimentin was shown to be expressed in all the three cell lines from a malignant astrocytoma, whereas nestin was variably positive in the most motile and invasive cells (Rutka et al., 1999). In U-373 MG cells, TGFα was observed to reduce GFAP mRNA; however it did not modify vimentin but increased nestin (Zhou and Skalli, 2000).

The discussion about the relationship between progenitor cells and tumor development has just started and a fundamental question, concerning not only transplacental ENU tumors, but also human tumors, is the timing of cell migration from the subependymal layer, considering that its cells are normally capable of migrating and providing a continual source of parenchymal cells (Levison and Goldman, 1997). One wonders whether an established tumor derives from recently migrated or from the first migrated cells.

New hypotheses on the glioma origin concern astrocytes and radial glia as possible multipotent stem cells, both *in vitro* and *in vivo* (Steindler and Laywell, 2003). Astroglial lineage, from radial glia to astrocytes, might act as stem cells either in embryos and in adulthood (Doetsch, 2003). SVZ astrocytes have been

A

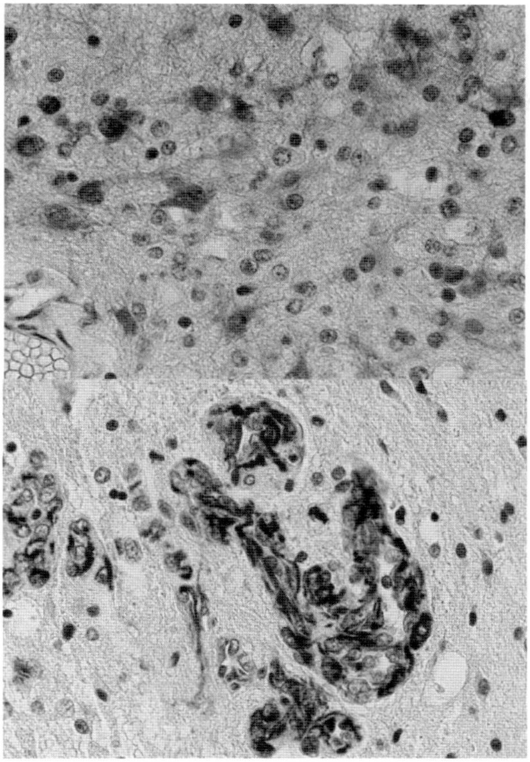

B

Figure 16. Nestin, DAB. A. Infiltrating glioblastoma: reactive astrocytes; B. microvascular proliferation in an oligodendroglioma, x 400

considered *in vivo* primary precursors and acting as stem cells *in vitro* (Doetsch et al., 1999) and forming neurospheres. These cells may undergo transformation.

There is also enough evidence, with the mechanism of the "lineage tracing" in the adult CNS, of a trans-differentiation from one cell type into another and that mutations occurring in gliomas may influence the differentiation or the trans-differentiation state. Cultured p16 and p14 –/– astrocytes maintain a diploid status, but shift to a rapidly proliferating status losing GFAP and acquiring nestin expression (Holland et al., 1998). This means that p16 and p14 keep astrocytes in a differentiated status, but in glioblastomas they are inactivated (Holland, 2001). This brings us back to the distinction between the undifferentiated or the transformed phenotype of tumor stem cells. All these changes must be interpreted in terms of new clones with the new phenotype. Another important demonstration of how a differentiation can be influenced is that gliomas may originate because of the inability of progenitor cells to differentiate, as happens with PDGF on precursors expressing GFAP which assume oligodendroglial morphologies (Dai et al., 2001).

There is no evidence that tumors can develop from the proliferating reactive glia; however, they might originate from radial glia that is capable of proliferation, into which differentiated astrocytes can regress under certain stimuli (Magavi et al., 2000). Today bone marrow stem cells must also be considered as a possible source of tumors, because of their capacity to differentiate along the neuroectodermal line (Mezey et al., 2000).

Chapter 2

MOLECULAR GENETICS OUTLINE OF BRAIN TUMORS

Theoreticallly, molecular genetics is well able to be of help at the moment of diagnosis, even of small samples, but actually its procedures are rather complicated and need time, so that they are feasible only in selected and specifically well equipped laboratories. The use of molecular genetics for the identification of the tumor type, sub-type or grade is still at present a challenge, also because a complete histological validation of molecular genetics data is still lacking and it is not known how far the correlation found between molecular genetics data and survival, statistically studied, can be of use in single cases. Treatises of the last ten years (Schiffer, 1997; Kleihues and Cavenee, 2000, Ironside et al., 2002; Burger et al., 2002) represent a good basis for codifying brain tumor diagnoses, whereas molecular approaches in this direction only recently started to acquire a practical importance (Louis et al., 2001). For example, by multidimensional scaling analysis of gene expression profiles, a good distinction of glioblastomas, anaplastic astrocytomas, oligodendrogliomas and anaplastic oligodendrogliomas has been achieved, including a small group of glioblastomas with extended survival, and it parallels morphological classification showing a good correlation with survival (Fuller et al., 2002). The problem, however, is still how far all this can be used in the single case at the moment of diagnosis.

Brain tumors are supposed to originate as monoclonal and, for genetic instability, to undergo a genetic heterogeneity, accompanied with increased mutation rate and proliferation capacity and followed by a phenotypic heterogeneity. New clones arise that better adapt themselves to the environment, show a greater proliferation potential and compete with the predecessors in a kind of selection by competition, losing the differentiating capacity. Inactivation of tumor suppressor genes and accumulation of mutations are the main genetic characteristics. The loss of differentiation characteristics of a given cytogenetic stage, with regression to those of preceding stages, is called anaplasia (Russell and Rubinstein, 1989). It can be realized also by an accelerated growth of already differentiating cells or by maturation arrest (Cairncross, 1987). In the different phases of tumor progression by anaplasia, genotypic alterations are associated with pathologic phenotypes, according to an established scheme (Louis, 1997) (Figure 1).

Tumor stage	Associate pathology	Genetics
Astrocytoma	Proliferation Apoptosis	TP53 mutations PDGFR over-expression 17p, 22q losses
Anaplastic Astrocytoma	Cell cycle Deregulation	CDKN2A/p16 deletion RB mutation CDK4 amplification 9q, 13q, 19q, 11p losses
Glioblastoma	Necroses Angiogenesis Clonal selection	EGFR amplification/truncation PTEN mutations pRb pathway alterations

Figure 1. Progressing anaplasia and genetic alterations

In the stage of astrocytoma grade II the most common finding is the absence of wt p53 and the existence of a non-functional p53 pathway. Roughly 60% of tumors show loss of alleles on 17p including TP53 locus, whereas the retained TP53 allele is mutated in most cases (Rasheed et al., 1994). Over-expression of PDGFR is another alteration of this stage where mechanisms allowing cells to evade apoptosis are supposed to occur (Louis et al., 1997) (Figure 2). In anaplastic astrocytomas, alterations of oncogenes/proteins regulating the cell cycle occur, in the so-called Rb-pathway.

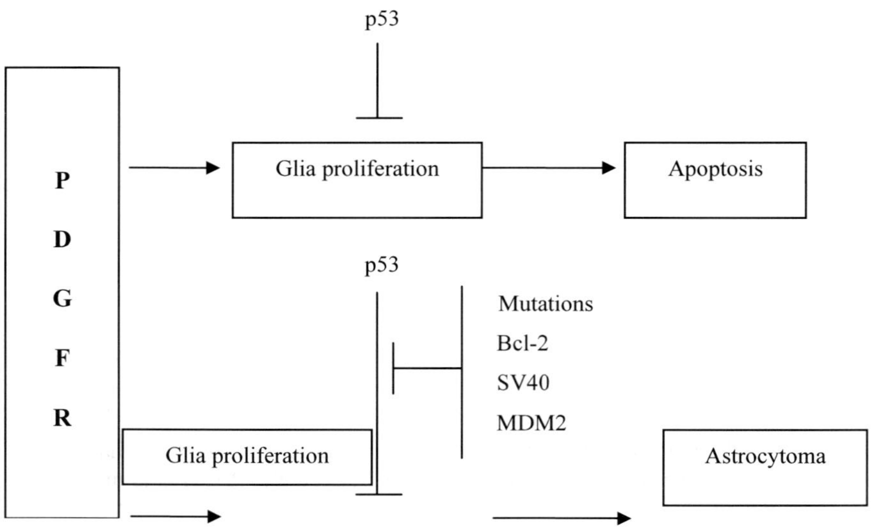

Figure 2. Role of PDGFR and apoptosis in glioma development

Normally, Rb1 sequesters E2F transcription factor, but when phosphorylated or when RB1 is mutated, E2F initiates the entry into S phase. Rb1 phosphorylation is produced by the complex cyclin D1/CDK4-6 heterodimer. Normally p16 binds CDK4 inhibiting the formation of CDK4/Cyclin D1 heterodimer, but CDKN2A/p16 homozygous deletion and CDK4 over-expression can activate the cycle (Figure 3).

All the events leading to deregulation of the cell cycle are mutually exclusive. In glioblastomas, alterations of many oncogenes serving different cell functions are found. Beside those of the PDGF/PDGFR and EGF/EGFR pathways, mitogenic through Ras/MAPK pathway, p53 pathway modulated by Mdm2 and p14 (Figures 2, 4), pRb pathway with involvement of cell cycle regulation through cyclins and kinases, and PTEN pathway are active.

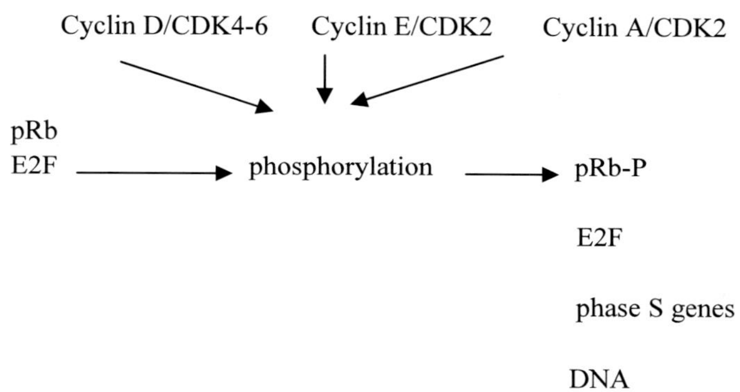

Figure 3. pRb pathway

The occurrence of highly amplified cells for EGFR at the tumor edges has a particular meaning (Okada et al., 2003), whereas of special interest is the frequent finding of truncated EGFR, ΔEGFR, with auto-phosphorylation of tyrosine kinase, continuous signaling and no internalization-degradation of the complex L/R (Figure 5). PTEN through its proteinphosphatase activity regulates cell migration and invasion and through its phophoinositolphosphatase via Akt/PKB regulates cell proliferation, survival and apoptosis (Figure 5) (Louis and Cavenee, 2001; Maher et al., 2001; Knobbe et al., 2002; Collins, 2004).

Particular importance is attributed today to the genetic changes of PI3K cascade. Its activation follows amplification, rearrangement or over-expression of EGFR (Collins, 2002), but one third of glioblastomas show mutations of PTEN and this gene inhibits PI3K activation of Akt (Knobbe et al., 2003). On the other hand, Akt is activated in the majority of glioblastomas (Choe et al., 2003) and it plays an important role in the development of these tumors (Holland et al., 2000; Sonoda et al., 2001). It has been found that carboxyl-terminal modulator protein (CTMP) inhibits Akt phosphorylation at threonine 308 and serine 473 (Maira et al., 2001), so that inactivation of this protein can abolish inactivation of Akt. In a series of glioblastomas and glioblastoma cell lines, CTMP did not show mutations or deletions. However, in 40% glioblastomas and 67% glioblastoma cell lines, reduced mRNA levels associated with hypermethylation of the CTMP promoter were found and became thus a common finding in glioblastoma (Knobbe et al., 2004).

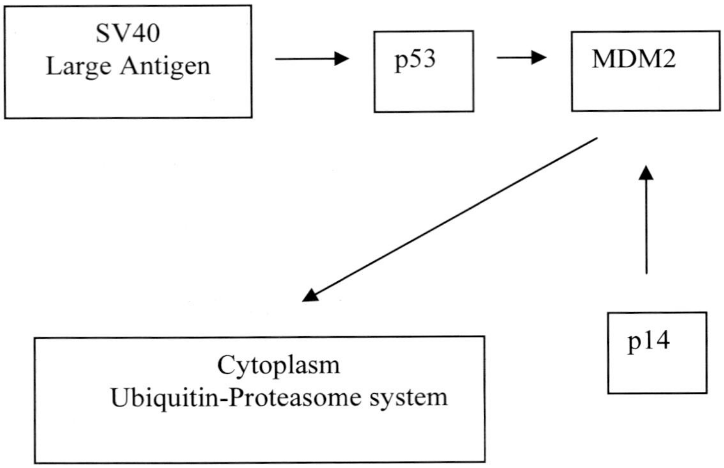

Figure 4. The role of MDM2

Many alterations accompany the formation and progression of gliomas, appearing also specific in some tumors, so that the clinician is provided with biological and pathogenetic information supplementary to those deriving from the biopathology of tumors and even more clarifying. In some tumors, prognostic subtypes, for example, can be defined only on their molecular features or better on these than on their sheer phenotype. In the same way, sensitivities to certain therapies can be discovered which could not be detected otherwise and this confers the new genetically integrated classification clinico-practical goals.

Gene expression profiling is substantially contributing to these new requirements (Louis et al., 2003). For example, the promoter hyper-methylation of MGMT in relation to TP53 mutations is emerging as an important factor for prognosis and therapy in astrocytic tumors. It is associated with a reduced period free survival (PFS) in diffuse astrocytomas to the point that it is considered as a

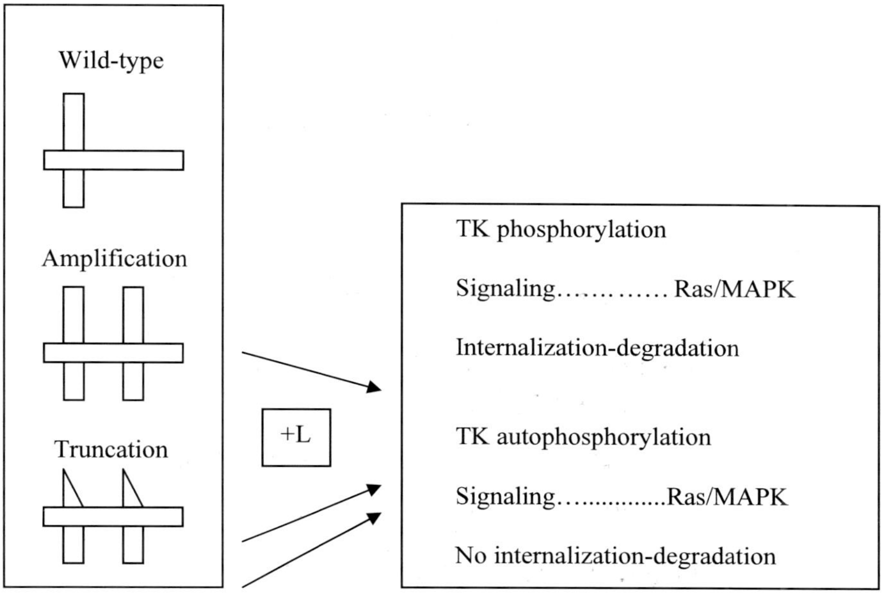

Figure 5. EGFR pathway

predictive factor in the clinical course, more than of malignant progression, and as indicating possible efficacy of chemotherapy (Komine et al., 2003).

There are special events that are considered as specific of malignancy and show very complicated phenotypic and genotypic characteristics which must be considered separately, i.e. cell migration and invasion, angiogenesis and apoptosis. The knowledge of their regulating molecular pathways and of their morphogenesis is believed to offer opportunities for identifying therapeutic approaches.

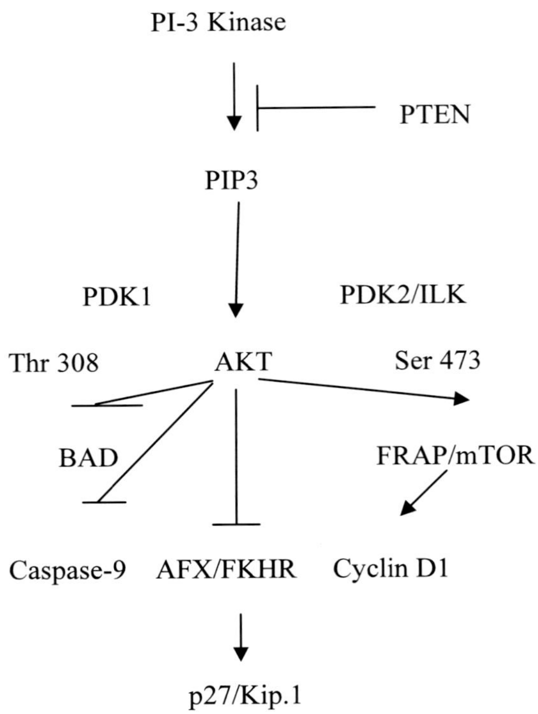

Figure 6. Akt pathway

Chapter 3

GENERAL REMARKS

The purpose of this work is to discuss hotspots in surgical pathology diagnostics on small samples of brain tumors in their important relevance to prognosis and to therapy, as they are encountered in everyday clinical practice. The hotspots may become diagnostic pitfalls. Excellent reviews are available on every aspect of brain tumors: general biology (Behin et al., 2003), epidemiology (Wrensch et al., 2002), genetics (Maher et al., 2001) etc., but our goal is to discuss the problems as they arise in the single case and at the very moment when clinical decisions have to be made on the basis of pathology.

Taking into account all the contributions produced in the different fields, especially in that of molecular genetics, in order to discover new criteria of classification of brain tumors, directed towards more refined prognoses and therapies, it can be summarized that tumors arise from precursor cells, that genetic events acting on them are responsible for the transformation and that the acquired phenotype depends on the genetic events, but also on environmental influences (Louis et al., 2001). The consequences are that the biological definition of tumor type and of tumor grade with the relevant prognosis has become over time more difficult, especially in small surgical specimens. One of the most frequent mistakes is, for example, to give a wrong prediction of survival of a tumor compared with the real outcome observed later and to underestimate the degree of malignancy (Glantz et al., 1991). The consequence is the missed treatment of a tumor which could have benefited, on the contrary, from radio- or chemotherapy. Another not infrequent mistake is to overestimate the malignant significance of pathologic characteristics which may lead to over-treatment of tumors with side effects that appear during an unexpectedly long post-operative survival. A study has been dedicated to avoid the possibility of over-treatments drawing attention to those features in an algorithm that require interdisciplinary review of clinical, radiographic and pathologic findings, before starting post-surgical therapy for a tumor (Burger et al., 1997). As a matter of fact, it may happen that the morphological assessment of a tumor is in contrast with its interpretation based on more recent neuro-imaging and clinical data and, therefore, this may be confusing. A recent study demonstrated, for example, that disagreements among neuropathologists on brain tumor diagnosis on surgical samples were potentially affecting treatment in 15% of cases, whereas in 7.9% of cases they were non-treatment altering, when two groups of neuropathologists were compared: one with access to all clinical, radiological and neuropathological information and the other with limited access to neuropathology only (Murphy et al., 2002). This means that either neuropathologists in their every day work can dispose of every kind of extra-neuropathological information or they must have trained in clinical and radiological fields during their training, or, alternatively, the decision on treatment and prognosis of a tumor becomes a matter of multidisciplinary discussion.

In stereotactic biopsies, the discrepancy with the diagnosis on resected specimens is 38% of cases. Since these procedures are associated with a 4% morbidity risk, some people even think that they might be unnecessary (Jackson et al., 2001).

The goal of this study is to identify and to discuss the challenging controversial points in the everyday diagnostics of tumors. The increasing sophistication of diagnostic and prognostic criteria keeps up with the tremendous amount of contributions from research and this can be confusing in turn. Pathologists are aware that their diagnosis reverberates on post-surgical therapeutic strategies and on the statistical analyses of the efficacy of therapeutic drugs. The study needs to be especially addressed to neuropathologists who participate in teams of clinical neuroscience who are taking care of patients with brain tumors.

Chapter 4

ASTROCYTIC TUMORS I

One of the frequent pitfalls in the prognostic diagnosis of gliomas is encountered in the most classical and easy tumor: diffuse astrocytoma. Classically, the diffuse astrocytoma, subdivided into fibrillary, protoplasmic and gemistocytic variant, does not show problems of recognition. The three variants when fully expressed do not leave doubts as to their identification. Difficulties arise when the surgical samples are too small, as in stereotactic biopsies, or they have not been removed from the full tumor, so that they do not contain cell distribution patterns which make the tumor recognizable. If the sample is taken from the tumor periphery, all normal cells of the nervous tissue will be present and it is common knowledge that one cell of astrocytoma, individually, cannot be distinguished from a normal or reactive astrocyte.

Many diagnostic difficulties can be encountered in anaplastic astrocytomas and glioblastomas, especially when the surgical sample is taken purposely from a non-necrotic area or there is a sampling error. Stereotactic biopsy is the main procedure for a pre-surgical histological diagnosis with low risk, accuracy and with minimally invasive nature. However, it has limitations and not infrequently it is useless up to the point that it has even been regarded as unnecessary (Vaquero et al., 2000; Jackson et al., 2001).

1. DIFFERENTIATION OF TUMOR VS NON-TUMOR TISSUE

Very frequently the small sample is removed from the very periphery of the tumor or from a lesion that is not a tumor, for example ischemic or demyelinating areas (Figure 1). This aspect can be dominated by apparently normal cells and the first question is to establish whether their number has increased, they are glial or neuronal or both, and whether at least some of them are neoplastic.

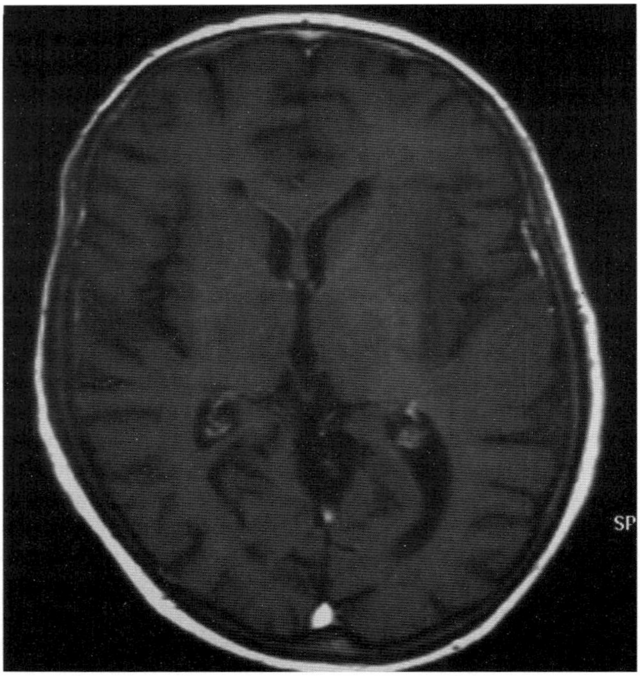

Figure 1. Hypodense ischemic insular lesion, MRI, T1. From the Neuroradiology Unit, Dpt Neuroscience, University of Turin

The increase in the number of cells cannot always be easily assessed, especially when it does not exceed 30%. Sometimes the number of nuclei seems to be normal, but their volume is increased (Figure 3A, B). Normal neurons are easily recognizable, once the existence of a neuronal tumor has been excluded. Inflammatory cells can be recognized by the small and dark nucleus and by their crowding around the vessels in addition to being scattered thoughout the tissue. Microglia cells are also easily recognizable by the irregular form of the nucleus. Specific immunohistochemical methods will be of help, such as those employing CD3, CD20, CD68 (Figure 2A, B).

In the cerebral cortex, the number of peri-neuronal satellites, either oligodendroglial or astroglial, may be increased and, letting aside frank aspects of satellitosis that may not be so easily attributed to astrocytomas or oligodendrogliomas, quite often it is even difficult to assess (Figure 4A).

A

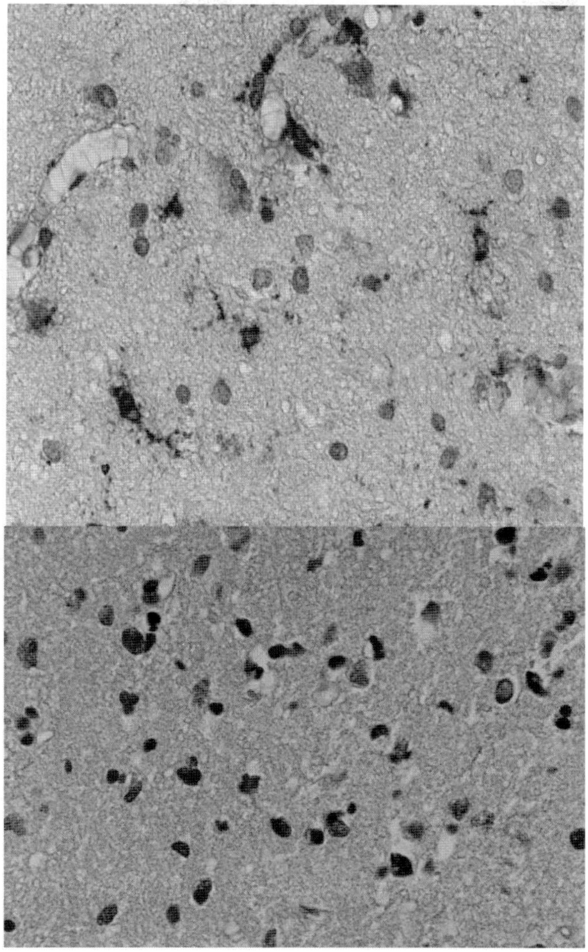

B

Figure 2. A. Microglia cells in the cortex, CD68, DAB, x 400; B. Inflammatory cells in the cortex, H&E, x 400

A

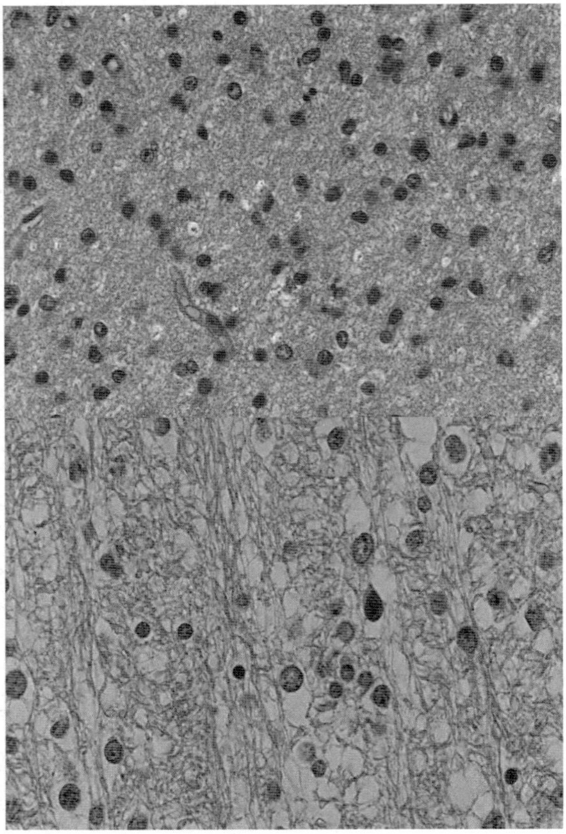

B

Figure 3. Stereotactic biopsies from a glioma periphery. A. Likely increase of nuclei in the white matter, H&E, x 400; B. Increase of the nuclear volumes, H&E, x 400

A

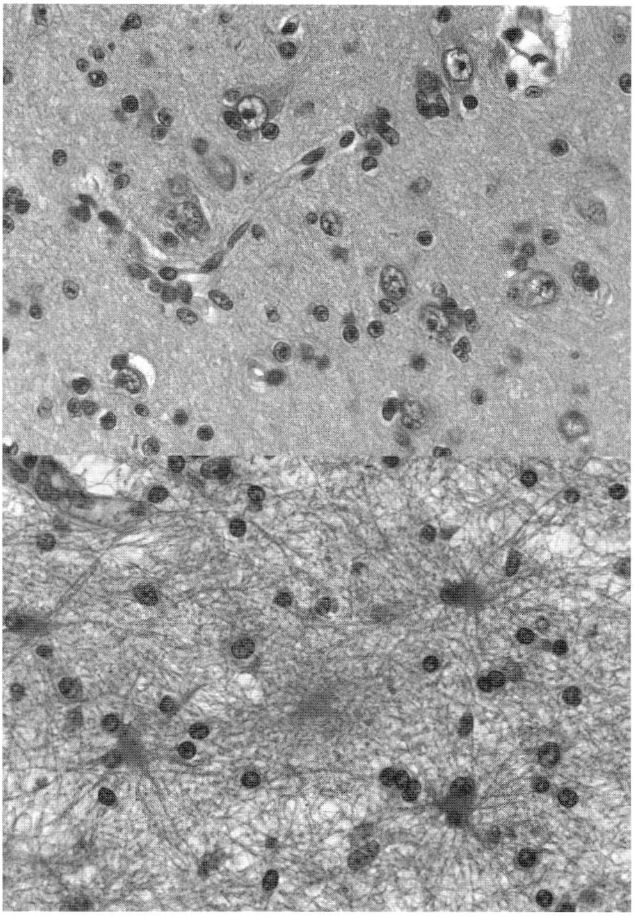

B

Figure 4 . A. Infiltrated cortex. Peri-neuronal satellitosis, H&E, x 400; B. Reactive astrocytes in the white matter, H&E, x 400

Once the cells which have increased in number have been recognized as astroglial on the basis of their nucleo-cytoplasmic appearance, they must be identified as tumor or reactive astrocytes. Reactive astrocytes are identifiable for their clear and large nuclei and the eosinophilic cytoplasm and for its long and thick GFAP-positive processes (Figure 4B). When the gliosis is not recent and it has not been renewed for some time, all the astrocytes show the same aspect and are found at regular intercellular distances. However, they are frequently generated at different times, as happens in infiltrating tumors for example, so that not all of them are at the same stage of reaction and the recently formed ones could not be recognized as reactive. The pre-existing GFAP-positive glial network may still be visible, and it may create some difficulties in the diagnosis when small reactive astrocytes are contemporarily present, simulating a low cell density astrocytoma (Figure 5A, B). Positive staining for MIB-1 may help, taking into account the fact that also reactive astrocytes may show positive nuclei, with a mean LI slightly lower than that of astrocytomas, but with more or less similar LI ranges (Wessels et al., 2001). Also positive staining for p53 may be of some help in favor of an astrocytoma when it is positive.

Even after recognizing reactive astrocytes as the major cell component in the section, the occurrence of a tumor cannot be discounted with certainty, because a massive peripheral gliosis my mask scattered neoplastic astrocytes. The examination of MRI is mandatory and it can maybe orientate the pathologist according to the type of lesion, with or without contrast enhancement or simply given by a small hypo-intense spot. The progression over time of the lesion or the appearance of contrast enhancement will confirm the neoplastic nature.

Studies with *in situ* hybridization showing chromosomal aberrations in all interphase cells have been instructive. Alterations of chromosome 1, 7 and 10, characteristic of low-grade astrocytomas (Rosso et al., 1997; Perry et al., 1997; Hopman and Ramaekers, 1998) have been found in 80% of astrocytomatous areas, but never in reactive astrocytes (Wessels et al., 2001). In the last series, three out of four cases with inconclusive diagnosis between glial reaction and low-grade astrocytoma, showed aneusomies and underwent rapid progression indicating that they were unrecognized high-grade tumors. It is important to note that no single genetic aberration has been shown to be responsible for tumor progression. Coupling microdissection techniques with PCR reactions for individual polymorphic microsatellites situated at 7 genomic regions (1p, 3p, 5q, 9p, 1oq, 17p, 19q), no allelic loss occurred in reactive gliosis against at least one loss in gliomas of every grade. In 73% of cases of uncertain attribution, the occurrence or not of LOH allowed a correct prediction (Finkelstein et al., 2004). This observation is very important, because it demonstrates that molecular genetics can be applied to minute formalin-fixed and paraffin-embedded specimens for a molecular analysis in everyday pathology practice.

It is of paramount importance to consider that, even when quite certain, the diagnosis of gliosis does not exclude the possibility of an adjacent tumor. In this case, the careful examination of clinical and imaging data is mandatory and in some cases, if the latter are in contrast with the pathology, another biopsy can be prompted.

It must be recalled that sometimes gliosis corresponds to a demyelinating disease or to an infarction; and the differential diagnosis must take into account the fact that the occurrence of macrophages, even if constant in these lesions, is not their exclusive characteristic, being frequent also in malignant tumors (Figures 6A, B; 7A, B).

A

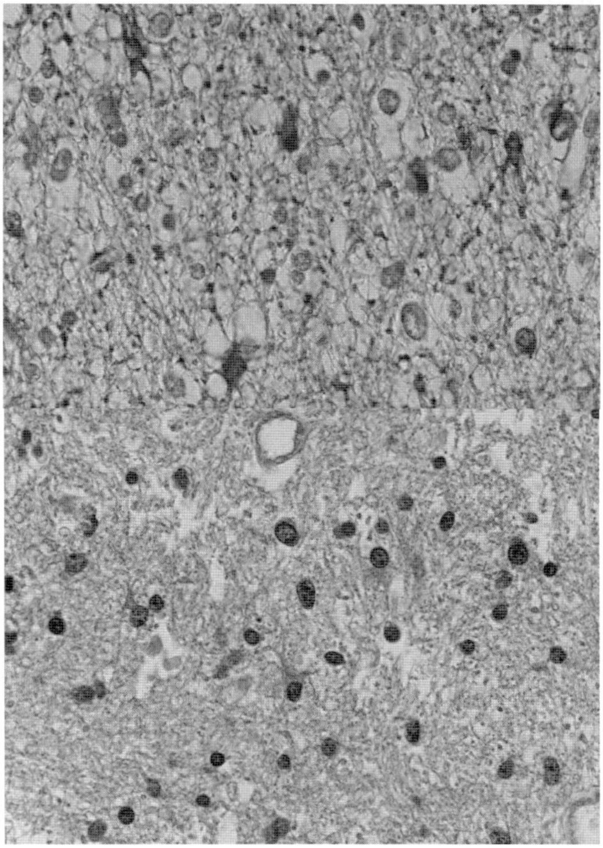

B

Figure 5. Glioma periphery. A. Pre-existing GFAP-positive reticulum and small reactive astrocytes, DAB x 400; B. Early reactive astrocytes, H&E, x 400

A

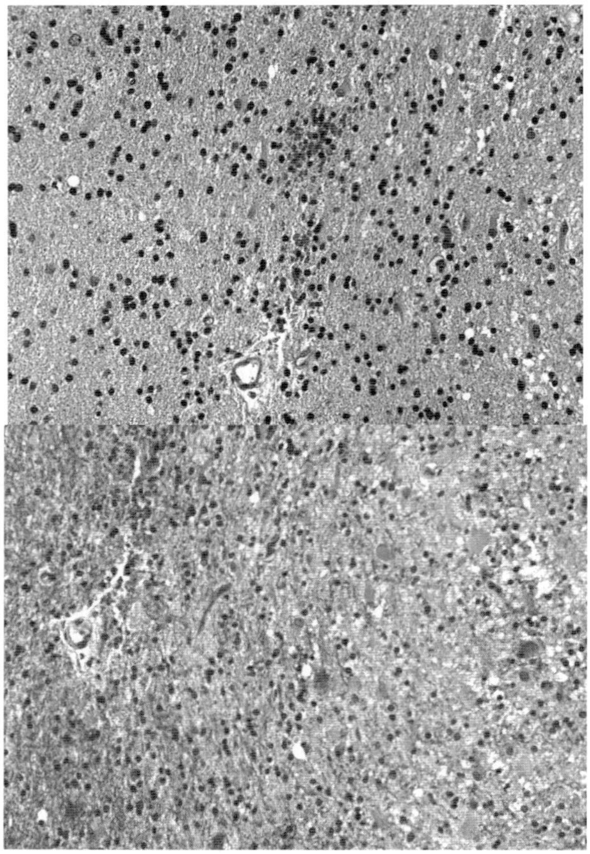

B

Figure 6. Demyelinating area. A. Increased number of nuclei with reactive astrocytes on the right, H&E, x 400; B. Myelin loss on the right, Luxol Fast B, x 400

A

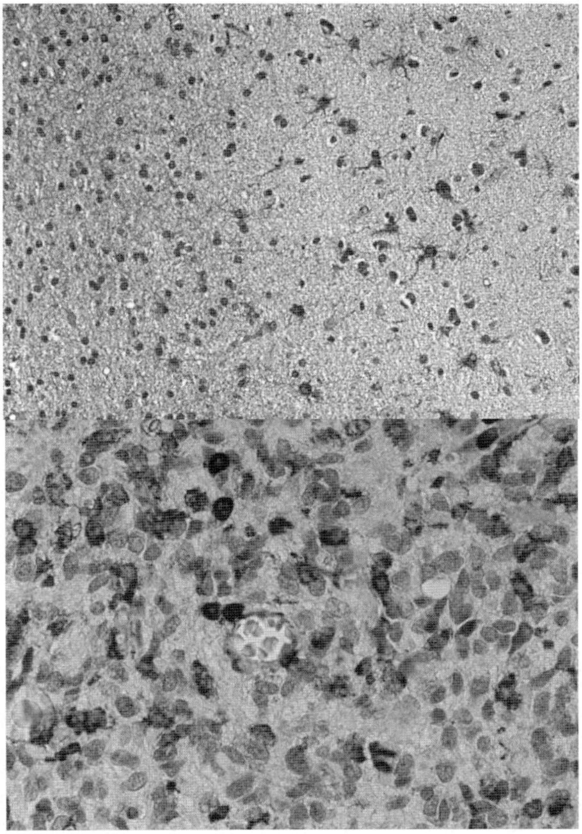

B

Figure 7. A. Reactive astrocytes in the demyelinating area, GFAP, DAB, x 400; B. Macrophages in glioblastoma, CD68, DAB, x 400

2. DIFFUSE ASTROCYTOMA VS OLIGODENDROGLIOMA

Frequently, the differential diagnosis using a small fragment from the periphery of a tumor, that at MRI appears as a circumscribed hypo-intense lesion (Figure 8), must be carried out between a diffuse astrocytoma and the diffuse growth of an oligodendroglioma.

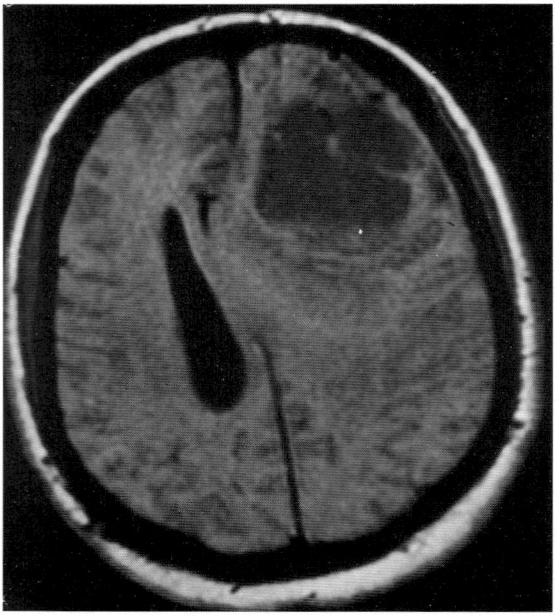

Figure 8. Hypodense frontal lesion, MRI. From the Neuroradiology Unit, Dpt Neuroscience, University of Turin

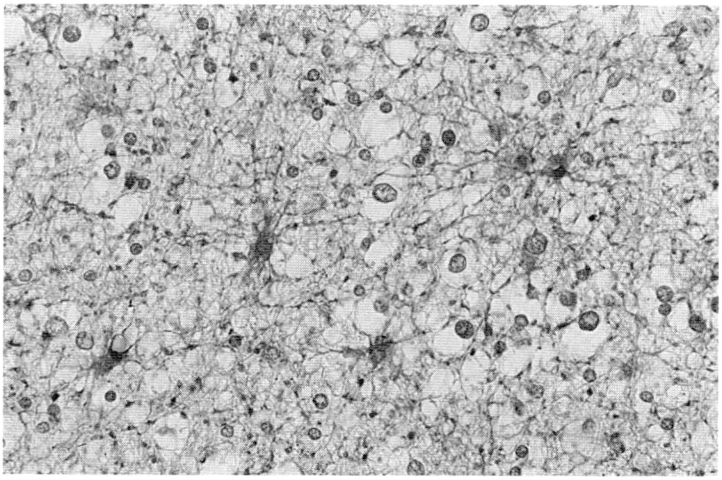

Figure 9. Reactive astrocytes with oligodendrocytes not increased in number, but with cytoplasmic halos. GFAP, DAB – Hematox, x 400

Classically, the distinction should be based on the aspect of the nuclei, the "chicken wire" distribution of small vessels (Figure 10), the occurrence of "honeycomb" appearance indicating oligodendroglioma (Figure 9) and on larger and more vesiculous nuclei and GFAP-positive staining indicating astrocytoma, but very rarely are these aspects so clear-cut (Figures 11, 12). In particular, nuclei are small and round, with a thick membrane and a central small nucleolus in oligodendrogliomas (Figure 15) and larger and vesicular with a very small nucleolus in astrocytomas. However, confusion can originate from the possibility that tumor cells are GFAP negative in astrocytoma and positive in oligodendroglioma, like minigemistocytes and GFOC (gliofibrillary oligodendrocytes) (Figures 6-13, 14) (Herpers and Budka, 1984; Schiffer, 1997).

A

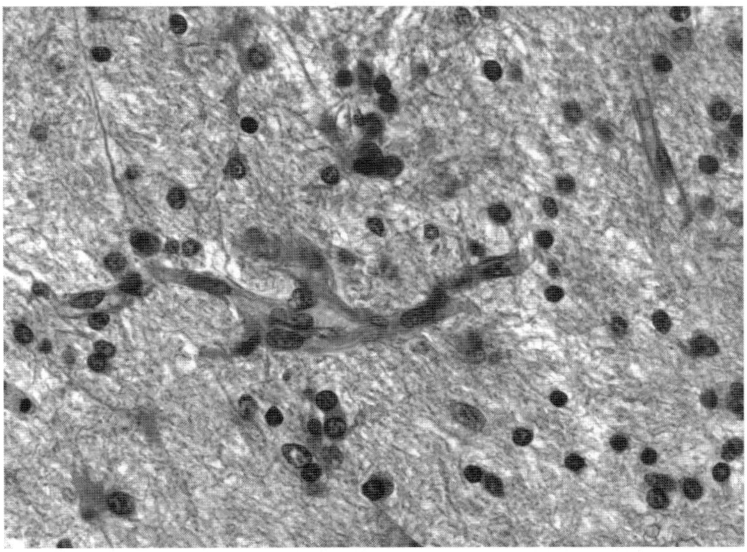

Figure 10. Small vessels of oligodendrogliomatous type, H&E, x 400

When tumor proliferation is in the white matter and small reactive astrocytes are present, the pre-existent GFAP-positive glial network being still visible, normal oligodendroglial nuclei could be confused with tumor oligodendrocytes and this could lead to the erroneous diagnosis of oligodendroglioma or of oligoastrocytoma (Figures 11-13). The opposite is also possible when small reactive astrocytes are overestimated, especially when associated with a GFAP-positive glial network, and oligodendroglial nuclei are considered as belonging to normal oligodendroglia of the white matter (Figures 11, 14). Sometimes it is difficult to establish whether GFAP-positive, astrocyte-like cells are minigemistocytes or tumor or reactive astrocytes (Figure 16).

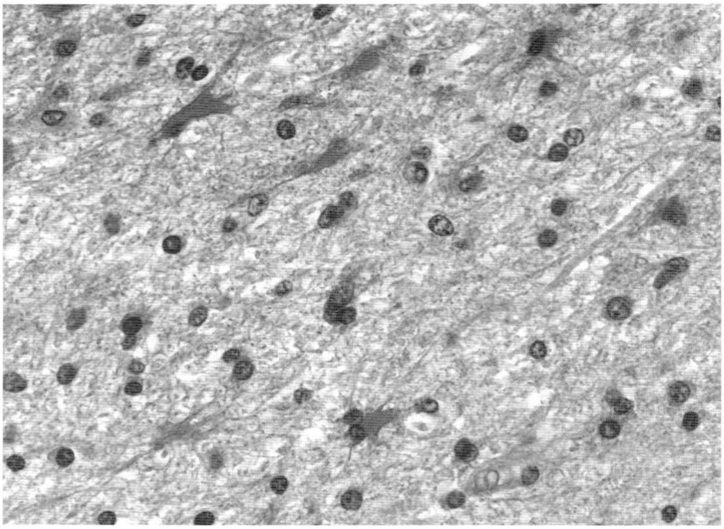

Figure 11. Reactive astrocytes and normal or tumor oligodendroglial nuclei, H&E, x 400

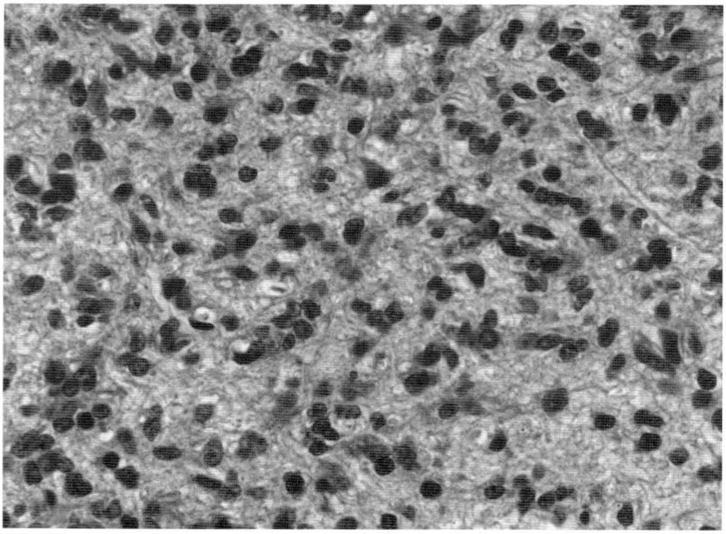

Figure 12. Field uncertain between diffuse astrocytoma and oligodendroglioma, H&E, x 400

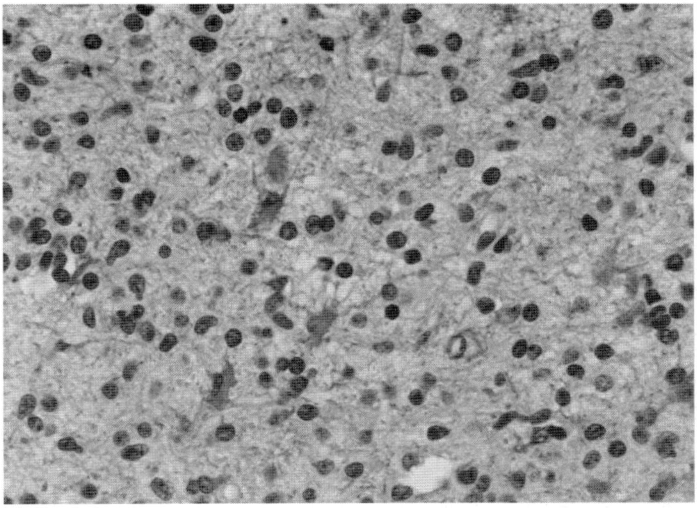

Figure 13. Field close to that of Figure 9: the tumor appears to be an oligodendroglioma with GFAP reactive astrocytes, DAB, x 400

Sophisticated cytometric procedures on Feulgen-stained sections could help in distinguishing astrocytic from oligodendrocytic nuclei (Deckaestecker et al., 1997), but they are not easily applicable, especially when the diagnosis must be given without delay. The old procedures with acetic carmin could be of help and are more practicable (Schiffer and Fabiani, 1971), but they are today obsolete. Since there is no marker for the recognition of tumor oligodendrocytes in paraffin sections, the distinction between normal and tumor oligodendrocytes may be important and this can be achieved by the immunohistochemical demonstration of Cyclin D1 which is positive in normal oligodendrocytes and not in most tumor oligodendrocytes, unless these are cycling cells (Bosone et al., 2001; Fiano et al., 2003) (See chapter VI).

In some cases, not only for the reduced dimensions of the specimen, but also because of the diffuse hypo-intensity at MRI that shows nothing characteristic for one tumor or the other, the distinction is really difficult and a diagnostic compromise is chosen with the diagnosis of oligoastrocytoma. From the point of view of the therapeutic decisions to be taken by the clinician, this uncertainty has no treatment-affecting consequence. It becomes a real problem when possible anaplastic features occur and an interpretation must be given, for example, to the MIB-1 LI, for establishing the malignancy grade. MIB-1 LI is $<4\%$ in astrocytomas, whereas in oligodendrogliomas it can reach 15% in grade II tumors (Schiffer et al., 1997). With a MIB-1 LI of 10%, for example, the grade will be the IIIrd if the

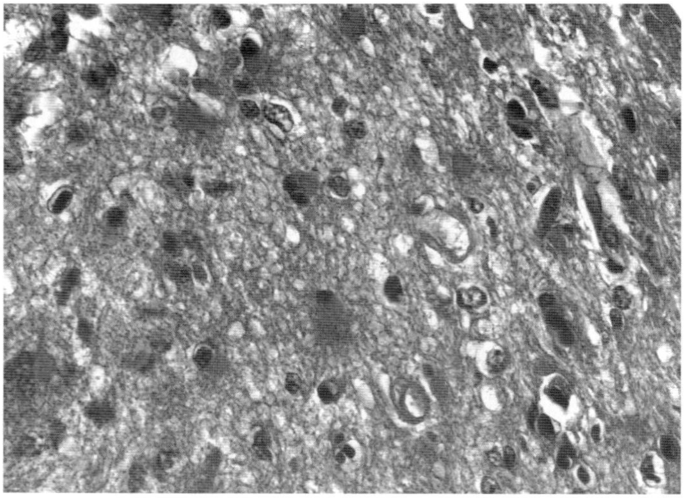

Figure 14. Reactive astrocytes and oligodendreoglial nuclei, H&E, x 400

diagnosis is that of astrocytoma and the IInd if it is that of oligodendroglioma. The confusion may have grave therapeutic consequences with irradiation and chemotherapy of an oligodendroglioma grade II as if it were an anaplastic astrocytoma and vice versa. Another serious consequence could be that of expecting an outcome of anaplastic astrocytoma from a tumor that is a grade II oligodendroglioma. This can be a source of mistakes when survival is used as a criterion for evaluating the efficacy of therapies. This matter will be further discussed with consideration of the oligodendrogliomas.

The possibility to have some help from non histological procedures is time-consuming and costly. In research directed to the identification of molecular markers of astrocytomas and oligodendrogliomas it has been shown that the frequency of galactosyltransferase (CGT) transcripts were twice as frequent in oligodendrogliomas than in astrocytomas and that GA1 (asialo GM1) was most frequent in oligodendrogliomas associated with the absence of paragloboside (Popko et al., 2002). The feasibility of this test however, must still be checked and its usefulness in individual cases seems low. More important can be the procedures for the identification of 1p and 19q losses, provided that the time required is reasonable. Of these more will be said later on.

3. MALIGNANCY GRADE IN RELATION TO PROGNOSIS AND THERAPIES

Basically, the starting point for every neuropathologist facing the problem of histological diagnosis of astrocytic tumors in small fragments is that most diffuse

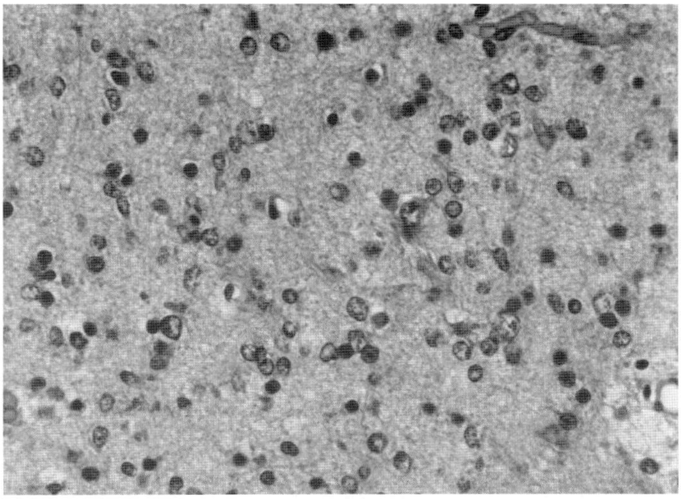

Figure 15. Increased number of frankly oligodendroglial nuclei, H&E, x 400

astrocytomas transform in time into glioblastomas, through the occurrence of anaplasia, and that this transformation may already be in progress at the moment of the diagnosis, even when either phenotypic or neuro-imaging alterations are not yet present (Figure 17). The occurrence of non-detectable transformed foci in a diffuse astrocytoma grade II has been considered among the rationales for radiotherapy of these tumors, with the goal of sterilizing them from malignant cells. Many studies have been dedicated to the detection, by every means and recently especially by molecular genetics, of prognostic factors when the phenotype is not or not yet informative.

It is common knowledge that >60% diffuse astrocytomas show TP53 mutations which do not increase with tumor progression. Not every cell contains mutations and with the increase of malignancy grade there can be a clonal expansion (Sidransky et al., 1992). Astrocytoma grade II cannot be further characterized from the molecular genetics point of view. On the contrary, in anaplastic astrocytoma, alterations of pRb pathway are present and can be demonstrated also in histological sections, even though none of them is characteristic of the tumor. They are interchangeable. The cell cycle can be deregulated by the alteration of one or more of the oncogenes/proteins involved.

None of the four criteria for the recognition of malignancy, i.e. nuclear pleomorphism, mitoses, circumscribed necroses and microvascular proliferations, is present in diffuse astrocytomas. In these cases, MRI shows circumscribed areas

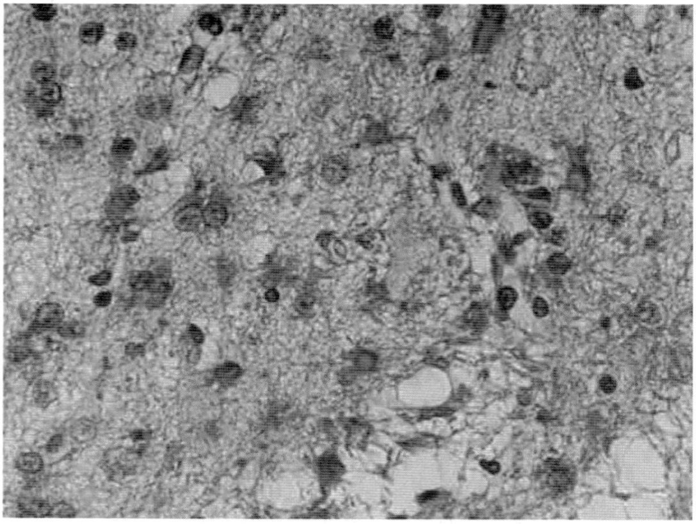

Figure 16. GFAP- positive cells, uncertain between GFOC and tumor or reactive astrocytes, DAB, x 400

hypo-intense in T1 and hyper-intense in T2 weighted images; but when it shows non homogeneous aspects and, especially, contrast enhancement, most probably the suspicion of a sampling error is legitimate, because the histological picture should be of greater malignancy. This confirms incidentally the necessity for the neuropathologist to receive at least some education in neurological clinics and neuro-radiology in order to be able to manage the neuro-imaging information. The possibility of being disavowed by the follow-up of the patient in this way can be reduced, but not completely abolished, because in some cases, in spite of the homogeneous iso- or hypo-intense aspect at MRI and a quiescent histological aspect, the outcome will be that of a malignant tumor, including contrast enhancement appearance in the course of time. The opposite situation is also possible i.e. that a quiescent MRI aspect disguises the occurrence of malignancy signs, maybe limited to nuclear pleomorphism and mitoses. Therefore, at the moment of diagnosis, anaplasia could already have been in progress or impending, but not yet phenotypically expressed, or it may develop later on (Figure 17).

In the differential diagnosis between astrocytoma grades II and III, the greatest importance is attributed to nuclear pleomorphism and mitoses (Figure 19A, B and Figure 20), because circumscribed necroses and microvascular proliferations indicate glioblastoma or grade IV (Figure 18A, B).

In the binary system (Daumas-Duport et al., 1988) applied to astrocytic tumors, nuclear pleomorphism or mitoses alone indicate grade II, whereas together they

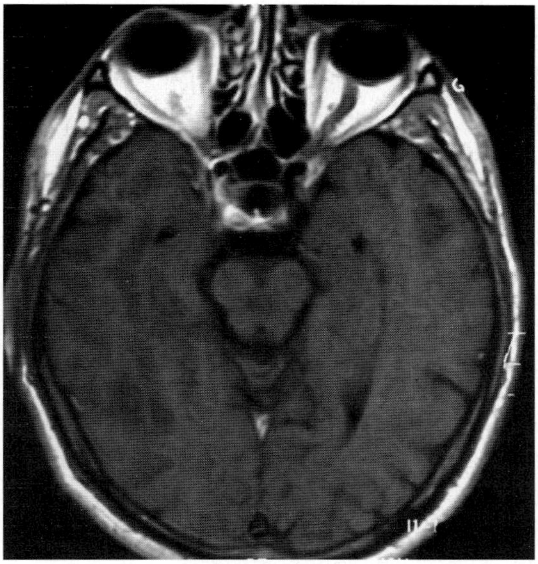

Figure 17. Hypodense temporal lesion, with histological characteristics of Grade III astrocytoma, MRI, T1. From the Neuroradiology Unit, Dpt Neuroscience, University of Turin

indicate grade III. In this regard, it has been discussed whether one single mitosis could be enough for recognizing grade III or whether a cut-off of the number of mitoses tolerated in grade II should be used. Almost everybody agrees that one single mitosis cannot indicate grade III, but it is also known that in a sample of astrocytoma grade III it may happen that not more than one single mitosis can be found. The latter observation is of paramount importance, but limited by the questionable representativeness of the sample. In our experience, the cut-off is < 5 mitoses per 10 HPF, but for others it is lower. The significance of a single mitosis found in a sample varies not only according to the extension of the sample, but also and mainly on how many fields must be examined before finding it. It has been observed, for example, that the number of fields at 400 x to be examined for finding one mitosis in astrocytoma grade III is 50, whereas in astrocytoma grade II and IV is 20 (Coons and Pearl, 1998).

The use of mitoses for the evaluation of the malignancy grade is a widespread, very simple and quick system, but many remarks have been addressed as to its reliability (Prayson, 2002): the number of mitoses can decrease for delayed and inadequate fixation and be influenced by the type of staining and the section thickness; different interpretations can be used in defining HPF; mitotic index (MI) can be calculated from the mean value of mitoses counted in all the extension of the section or in fields selected for containing the highest number of mitoses. As has been said before, it is generally acknowledged that it is possible that in an anaplastic astrocytoma not even a single mitosis is found, whereas some may be present in a differentiated astrocytoma.

A

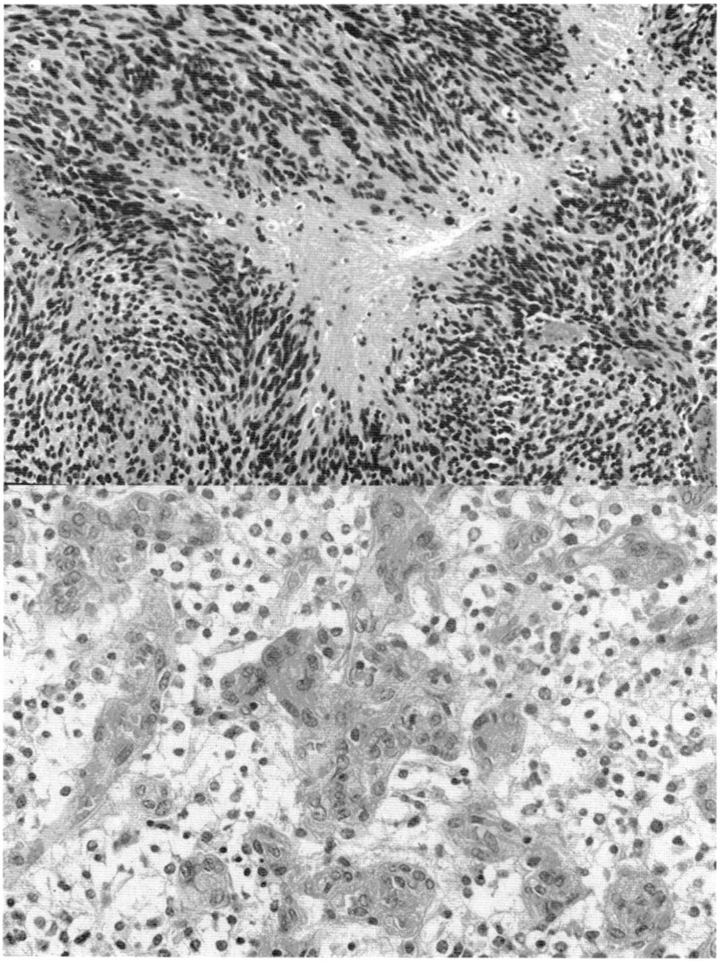

B

Figure 18. Glioblastoma. Circumscribed necroses, H&E x 200; B. Microvascular proliferations, H&E, x 200

A

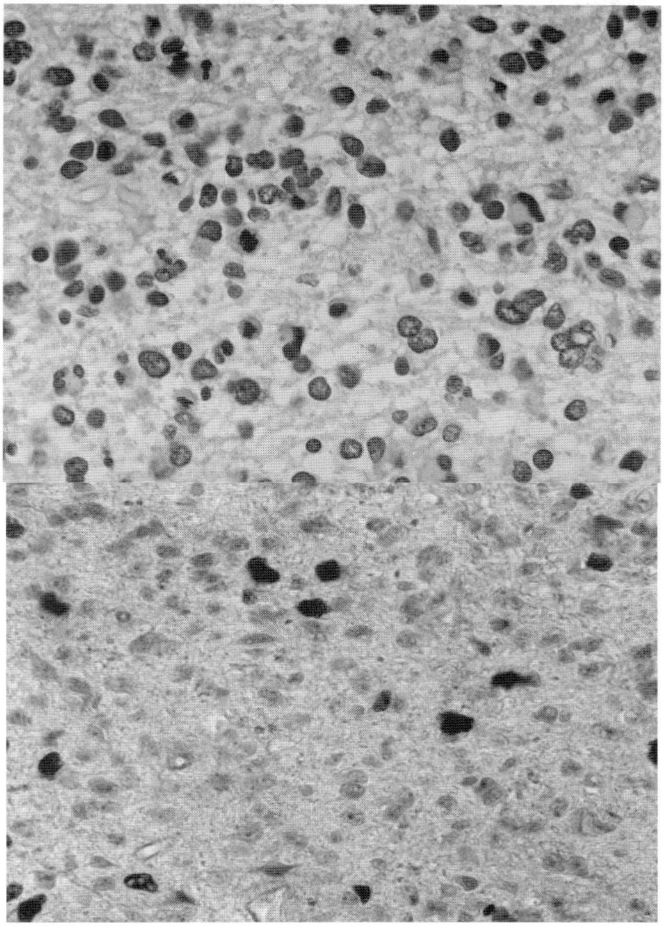

B

Figure 19. Astrocytoma Grade III: nuclear pleomorphism and mitoses, H&E, x 400; B. Astrocytoma grade III, MIB-1 LI > 10%, DAB x 400

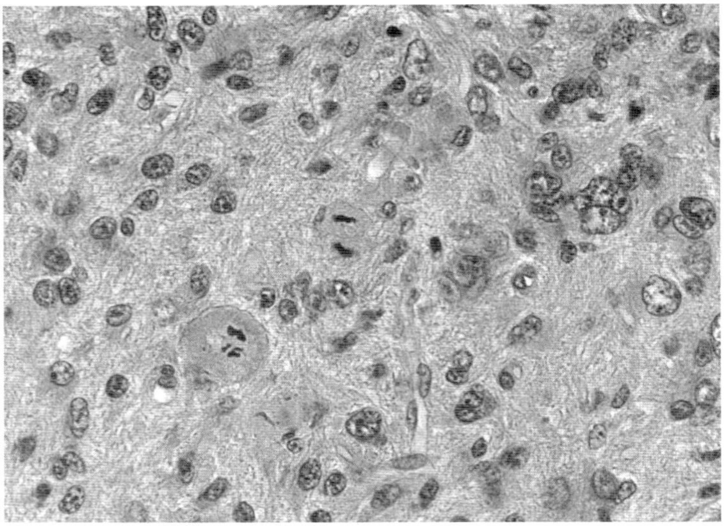

Figure 20. Anaplastic astrocytoma. Two mitoses and pleomorphic nuclei. H&E, x 400

The observation that statistically high-grade astrocytomas show contrast enhancement and more mitoses than low-grade gliomas does not help to exclude in single cases an ominous evolution when the mentioned features are not present. The evaluation of the malignancy grade cannot be based on the occurrence of factors identified as prognostic after multivariate analysis. For example, following the St. Anne system one single mitosis in a section can increase the grade of one (Daumas-Duport et al., 1988), as already said, but it has been also demonstrated that grade III astrocytomas with one single mitosis do not show different survival from those of grade II astrocytomas (Giannini et al., 1999). In conclusion, in small fragments of a case with MRI without contrast enhancing and heterogeneity, the occurrence of one single mitosis does not indicate anaplasia.

Evaluation of mitoses is the current method for establishing the proliferation capacity of tumors. Setting aside other more precise, but time-consuming, expensive or now obsolete methods such as labeled Thymidine, flow cytometry, BrDU, AgNor, DNA polymerase-α or PCNA, which put in evidence other phases, longer than mitosis, of the cell cycle, most currently used is Ki-67, expressed in phase G1-S- G_2 (Gerdes et al., 1983), in its clone MIB.1 working also in formalin fixed material (Cattoretti et al., 1982) which has become widespread and routine (Prayson, 2002). The main problem with this method is the calculation of the LI, based on counting positive nuclei in every field of the section, as said before for mitoses, for a general mean or counting only selected fields with the highest number of positive nuclei, recognized after a visual analysis. Obviously, vessel cell or non-tumor cell nuclei must not be counted and one can choose imaging analysis or a manual system. The inter-observer reliability in calculating the LI has been considered either moderate (Grzybicki et al., 2001) or good (Prayson et al., 2002).

High-grade gliomas have Ki-67/MIB.1 LIs higher than low-grade ones and the mean LI increases from grade II to grades III and IV (Coons and Johnson, 1993). For astrocytoma II grade MIB-1 LI has been found to be < 2% (Burger and Scheithauer, 1994), < 3% (Montine et al., 1994) or <4% (Watanabe et al., 1997). For II, III and IV grades the mean LI was found to be 3.8%, 18.4% and 31.6% respectively (Wakimoto et al., 1996). The ranges are very wide with a great amount of overlapping, especially between grades III and IV. There is no doubt that Ki-67/MIB.1 LI is an independent prognostic factor after multivariate analysis, but in the single case the distinction of grades III and IV and, especially, of grades II and III, which is the most important one, can be obtained only by establishing a cutoff-point. A reliable recognition of grade III, for example, can be achieved only if the LI is higher than its highest value found for grade II. Given the wide LI ranges and the regional variability, in small samples certainty can only be reached for grades III and IV; alternatively, the risk is that of undergrading high grade astrocytomas (Tihan et al., 2000). For the cutoff-point between grades II and III every laboratory should elaborate its own value. In our experience it was 8% (Schiffer et al., 1997). Between III and IV grades the cutoff-point was 15% (Tihan et al., 2000). It is also very important not to consider the limits as absolute, borderline values being less reliable than the others.

MIB-1 LI can be used for prognostic purposes only after the histological diagnosis has been made. In other words, if the diagnosis on limited tumor samples is that of astrocytic glioma, the prognosis on the basis of MIB-1 LI will be more reliable when it is in the range of grade III-IV tumors. When diagnosis is of a tumor grade II, with an MIB-1 LI in the range of grade II, there are the same reserves expressed for histology. The distinction of grade III from grade IV using MIB-1 LI is really difficult, because there is a much overlapping of the LI ranges, even though within grade III high LI values, for example >15%, are associated with the outcome of grade IV tumors (Tihan et al., 2000).

Also p53 LI has been used for the identification of the malignancy grade (Figure 21A), but not with great success, because of the great variability of its LI values (Iuzzolino et al., 1994; Ellison et al., 1995; Tihan et al., 2000) and also because malignant progression can involve pathways other than p53 mutations, for example amplification of Mdm2 (Figure 21B) (Biernat et al., 1998). Theoretically, the immunohistochemical demonstration of p16 inactivation by the absence of the protein in the sections, amplification of CDK4 for its over-expression, inactivation of pRb for its absence, or an increased LI for cyclins, especially for cyclin D1 could help in the recognition of anaplasia (Figure 22). The demonstration of a deregulation of the cell cycle could be a confirmation of anaplasia. However, till now these procedures have not been demonstrated to have reliability as indicators of anaplasia, confirmed by specific investigations including multivariate analyses for survival.

A

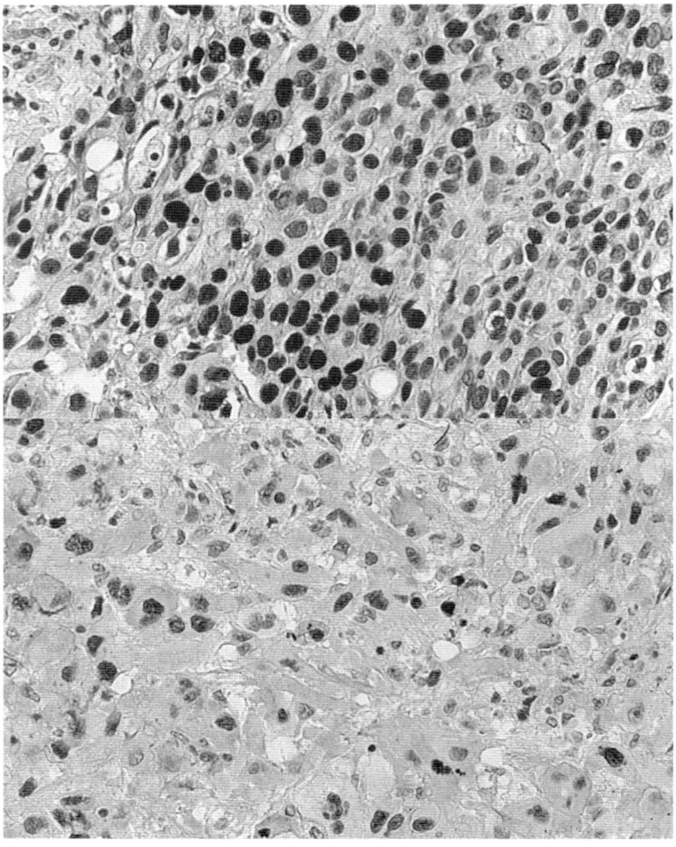

B

Figure 21. Glioblastoma. A. p53 DO-1; B. MDM2, DAB, x 400

An intriguing problem is the finding of proliferated small vessels in a tumor with nuclear pleomorphism and mitoses, but without necroses, with MRI not showing contrast enhancement. On the one hand, the diagnosis of glioblastoma could not be made, because of the absence of necroses, but on the other hand, also that of anaplastic astrocytoma should not be made, because of proliferated vessels. Once excluded the diagnosis of pilocytic astrocytoma, where microvascular proliferations coexist with a benign nature, the problem, however, can jeopardize the tumor categorization and the expected survival more than the therapeutic strategies.

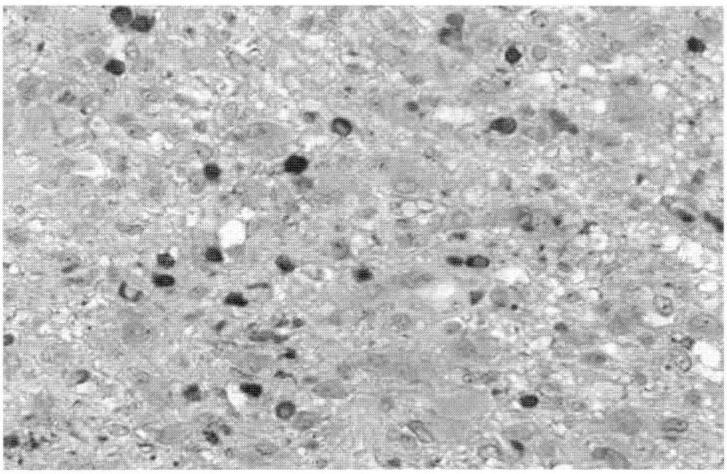

Figure 22. Glioblastoma. Cyclin D1, CDC6, DAB, x 400

This problem is typically encountered in stereotactic biopsies and it has been discussed in regard to the comparison of their histology with that of the corresponding surgically resected specimens. An agreement occurs in 92% of cases, the undergrading of glioblastomas being the major responsible factor for failures (McGirt et al., 2003). Since the presence of necrosis is mandatory for the diagnosis of glioblastoma, if it is lacking in the sample, the diagnosis should be that of grade III astrocytoma, even though it has been proposed, for practical purposes, to consider as glioblastomas all astrocytomas grade III on biopsy, just to avoid the undergrading of glioblastomas (Chandrasoma and Apuzzo, 1989).

The problems discussed till now reverberate on the management of patients. Following strict criteria for the indication of radiotherapy, based on a correct diagnosis of tumor type and grade, its application to grade II astrocytomas is still debated and mostly not performed, because of the lacking of a biological rationale. On the other hand, chemotherapy is today of poor efficacy: for astrocytomas grade II, a few investigators noted significantly improved survivals (Shaw et al., 1989; Shibamoto et al., 1993), but most of them did not (Piepmeyer, 1987) or found benefit concerning the duration of time to progression (TTP), but not of overall survival (Karim et al., 2002). Review analyses demonstrated that at 5 years there is no difference in survival between irradiated and not irradiated tumors, whereas differences can exist for the duration of TTP, as well as between low and high radiation doses (Shaw and Wisoff, 2003) or when radiotherapy is applied early post-operatively or in subtotally resected cases (Hanzély et al., 2003). The decision to irradiate an astrocytoma grade II must also take into consideration the occurrence of long-term side-effects of the therapy, such as neurocognitive sequelae, post-radiotherapy somnolence and even a deterioration syndrome etc. (Taphoorn, 2003; Whittle, 2004).

Some effects of chemotherapy with CCNU, procarbazine and vincristine have been shown to take place on anaplastic astrocytomas (Kortmann et al., 2003). Low-risk patients, i.e. young patients with good performance and with tumors < 5 cm, non-contrast enhancing, especially of oligodendroglial or oligoastrocytic subtype (Stupp et al., 2003), should not be treated by the radio- and chemotherapy which are indicated for high-risk tumors. Generally, for low-grade gliomas there are pros (Kortmann, 2003) and cons (Mirimanoff and Stupp, 2003) of chemotherapy, but till now, even with the advent of temozolomide, no substantial benefit has been shown (Whittle, 2004). At the moment, the possibility that chemotherapy replaces radiotherapy, when obviously this is indicated, remains still an open question (Van Den Bent, 2003).

4. MOLECULAR GENETICS IN THE RECOGNITION OF ASTROCYTOMA GRADE III

When the diagnosis on the surgical sample is not needed for immediate therapeutic measures, there is the theoretical possibility to demonstrate either in fresh surgical samples or in paraffin sections one of the several alterations affecting the regulators of the cell cycle: RB, CDKN2A, CDK4, etc. No specific study has been expressly dedicated to this matter.

5. PROGNOSTIC FACTORS OF GLIOBLASTOMA AND ITS VARIANTS

Age, location and extension of surgical removal are generally recognized as prognostic factors of glioblastoma. MIB-1 LI is significant in the differentiation among the three grades of astrocytic tumors as an independent prognostic factor (Sallinen et al., 1994; Wakimoto et al., 1996), as are cellular DNA content parameters (El-Rayes et al., 2005). MIB.1 LI is usually high in glioblastomas, but with a wide range of values starting from 0. Therefore, in single cases the LI can be used for identifying glioblastoma only when its value is higher than the highest value recognized to anaplastic astrocytoma and, needless to say, of astrocytoma. A problem of great practical importance is the usefulness of MIB.1 LI as a prognostic factor within the category of glioblastoma (Reavey-Cantwell et al., 2001), because it is not significant in the single case, due to the dispersion of the values within the category. Many studies have been dedicated to verify whether MIB.1 LI is predictive of survival and of sensitivity to radiotherapy and this would be very important for establishing post-surgical therapies. The studies are divided with respect to this point (Litofsky et al., 1998; Vaquero et al., 2000; Schroder et al., 2002; Bredel et al., 2002; Ho et al., 2003; Feveash and Spencer, 2003). Our experience is consistent with those who do not consider MIB.1 LI as predictive, regardless of the modalities of counting positive nuclei in histological sections, in random areas or in areas with the highest LIs.

Of some interest is the observation that the morphology of tumor cell nuclei can be related to survival time of patients with glioblastoma (Nafe et al., 2005). This observation may be important in the evaluation of early anaplasia in tumor tissue.

Several contributions indicate today the prognostic importance of various genetic alterations, considering glioblastoma in comparison with the other two malignancy grades of astrocytic tumor, even though they are of much less help in individual cases, as already said. Before discussing the importance of the existence and the recognition of categories of glioblastoma with different survival, it must be stressed that in every series there is a certain number of glioblastomas with longer duration. These cases may influence the results of studies based on survival for evaluating new therapies or for identifying new prognostic genetic factors. Recently, a number of glioblastomas with a TTP > 12 months have been reviewed and 25% have been re-classified as anaplastic oligodendrogliomas or astrocytomas or anaplastic pilocytic astrocytomas (Kraus et al., 2000). These results not only indicate that the mean survival of glioblastomas has been distorted in the considered series, but also that oligodendroglial tumors escaped an efficacious therapy. Another prognostic factor has been recognized in the pattern of vascularization: the classic glioblastoma pattern with prevailing small vessels is associated with longer survival than glioblastoma pattern with glomeruli and bizarre vascular formations (Birner et al., 2003).

In glioblastoma, many genetic alterations contribute to its nosography and many of them are discussed in relation to prognosis. Between the two glioblastoma types, the primary one arising as such in older patients and showing shorter pre-operative durations and the secondary one arising in younger patients from a previous astrocytoma and showing longer preoperative durations, important differences exist (Kleihues and Ohgaki, 1999). In the first type, EGFR amplifications and in the second type TP53 mutations prevail (Von Deimling et al., 1993), beside many other minor alterations. In our experience, based on the study of 100 glioblastomas using the "whole brain mounting technique", the two glioblastoma types, identified from their phenotypic aspect, compatible or not with an astrocytic derivation, showed different biological characteristics: infiltration modalities, diffusion to the contralateral or homolateral hemisphere, delimitation, etc. However, their distinction in the single case at the biopsy level is almost impossible, unless the tumor has been preceded by an ascertained astrocytoma.

A recent investigation using the microarray technique was addressed to verifying whether gene profiling expression associated with computational methodology of class prediction could identify molecular markers capable of refining methods used for classification of malignant gliomas: it was found that the diagnostic classification obtained correlated with clinical outcome better than pathology (Nutt et al., 2003).

A very important question is the existence of genetic prognostic factors in glioblastoma. Positive and negative data are equally available. There has been a general trend towards the correlation of molecular genetic alterations with phenotypical aspects, prognosis and also therapeutic strategies. In the attempt to find a phenotypical substrate to single molecular genetic alterations, it has been found, for example, that the demonstration by FISH of EGFR amplification, a typical feature of primary glioblastomas occurring in 40-50% of cases (Wong et al., 1987; Herbst et al., 2002), that in small cell tumors it helps to attribute these cells to the primary tumor type (Burger et al., 2001). However, over-expression of EGFR, as well as TP53 mutations, as assessed by immunohistochemistry, did not emerge as prognostic; they showed only a trend towards worse prognosis in young people when over-expression of the former is associated with wt of the latter, and better prognosis of EGFR over-expression in old people (Simmons et al., 2001; Smith et al., 2001). A meta-analysis of EGFR amplification did not succeed in establishing whether it is prognostic or not (Huncharek and Kupenick, 2000), but recently it has been shown that it has no prognostic significance (Quan et al., 2005). In 40% of GBM with EGFR amplification the receptor is truncated with elimination by splicing of exons 2–7 in its mRNA (Schwechheimer et al., 1995). Truncated receptor is specific for GBM and, when it is located at the cell surface and called ΔEGFR or EGFRvIII, it is autophosphorylating, not internalized and degraded after binding the ligand so that it gives rise to a continuous stimulation on the *ras* pathway. It promotes cell proliferation, reduces apoptosis by down-regulating p27 through activation of the PI3-Akt pathway (Narita et al., 2002) and gives the tumor resistance to chemotherapy through Bcl2-XL (Nagane et al., 1998).

Tumors from malignant glioma cells with retrovirally introduced EGFRvIII showed a higher labeling index for proliferation markers and a higher invasive capacity (Lai et al., 2002). GBM expressing EGFRvIII also have shorter survival expectancy in comparison with those without it (Diedrich et al., 1995; Feldkamp et al., 1999). There is a series of demonstrations that specific EGFRvIII antibodies, single-chain Fv, also radiolabeled or with *Pseudomonas* toxin, bind the receptor *in vitro* and *in vivo* and increase the survival of animals implanted with malignant glioma cells expressing the variant receptor (Kuan et al., 1999; Sampson et al., 2000; Mishima et al., 2001). Amplified EGFR is demonstrable by immunohistochemistry in most tumor cells of GBM, but EGFRvIII is found only in a small number of tumors and in less extensive areas (Biernat et al., 2004). This indicates that, cells with EGFRvIII may not be predominant in GBM and that, from a therapeutic point of view, both receptors should be targeted for silencing. In a study comparing RT-PCR with immunohistochemistry it was found that 41.3% glioblastomas showed positive immunohistochemistry for vIII and that all the positive tumors expressed vIII and amplified EGFR. In anaplastic astrocytomas, only 21.4% showed this alteration. Interestingly in glioblastomas, vIII was not associated with a reduced survival, whereas in anaplastic astrocytomas the association was very high. At the same time, however, vIII was associated with age (Aldape et al., 2004). Neither EGFR over-expression nor EGFRvIII emerged as prognostic in patients with extensive tumor resection (Heimberger et al., 2005).

PTEN mutations and EGFR amplification, on the contrary, turned out to be important prognostic factors in anaplastic astrocytomas. Moreover, the latter also indicated longer survival in glioblastomas older than 60 years (Smith et al., 2001). MDM2 amplification was found to be a negative factor (Schiebe et al., 2000), whereas TP53 mutations, nuclear masses of p53, Waf/p21 and CD95 (APO-1/Fas) turned out not to be prognostic factors in primary glioblastoma (Kraus et al., 2001). In a large series of tumors, TP53 mutations emerged as favorable and PTEN LOH as poor prognostic factors, independently of the tumor type, whereas LOH of 1p and 19q were not associated with better survival, unless they were in combination, even with no oligodendroglial morphology. Interestingly, the combination of TP53 mutations and EGFR amplification was not found to be as rare as usually believed and EGFR amplification did not prove to be prognostic. In the same series of 97 cases, no morphological parameter was found to be of prognostic importance, whereas TP53 mutations and LOH of 10q came out as favorable or poor prognostic factors, respectively. LOH of 1p was not prognostic, unless associated with LOH of 19q for better survival (Schmidt et al., 2002). In a recent series of 129 cases, CDKN2A, CDKN2B, RB1 and CDK4 proved to be associated with shorter survivals, especially if PTEN was also altered at the same time (Bäcklund et al., 2003).

PTEN mutations were found in 20–40% of glioblastomas, especially in primary ones (Tohma et al., 1998), but PTEN inactivation can be achieved by other mechanisms than mutation, such as methylation and LOH at the gene locus. It has been demonstrated that there is an inconsistent correlation between PTEN mutations, methylation, LOH and protein expression and that a PTEN pseudogene present on chromosome 9p21 may co-react with PTEN antibody and be partially responsible for the inconsistency (Baeza et al., 2003). Many genetic alterations have been described in relation to PI3K-Akt pathway (Knobbe and Reifenberger, 2003) and a correlation was found between Akt and proliferation and apoptosis through its gene effectors, mTOR, FKH etc. (Choe et al., 2003). One important observation has been that whereas the PTEN gene is mutated in 20–40% of glioblastomas, the Akt pathway is activated in over 80% of tumors (Holland et al., 2000). However, it has been shown that the PI3K-Akt pathway can be activated independently of PTEN mutation, for example by genetic alterations, mutations and deletions of class IA PI3K subunit genes-AKT (Mizoguchi et al., 2004).

The CDKN2A locus on 9p21, which is frequently altered in glioblastoma, encodes for two proteins translated from alternative spliced mRNAs: $p16^{INK4a}$ and $p14^{ARF}$. The former is the product of exon 1α, 2 and 3 and the latter of 1β, 2 and 3. $p14^{ARF}$ binds to mdm2 and stabilizes p53; its inactivation produces p53 inactivation. Homozygous deletion and promoter hypermethylation of $p14^{ARF}$ do occur in glioblastoma, but $p14^{ARF}$ methylation can be detected already in low-grade astrocytomas, being an early event in the progression to secondary glioblastoma (Nakamura et al., 2001).

Glioblastomas with differentiated oligodendroglial areas, GBMO (Figure 23), besides typical genetic alterations of GBM show a high rate of 1p and 19q LOH (He et al., 2000) and in comparison with classic glioblastomas show a significantly higher CDKN2A/p16 inactivation, in line with the observation that this genetic alteration is the most frequent one in anaplastic oligodendrogliomas (Figure 24) (Ghimenti et al., 2003).

A

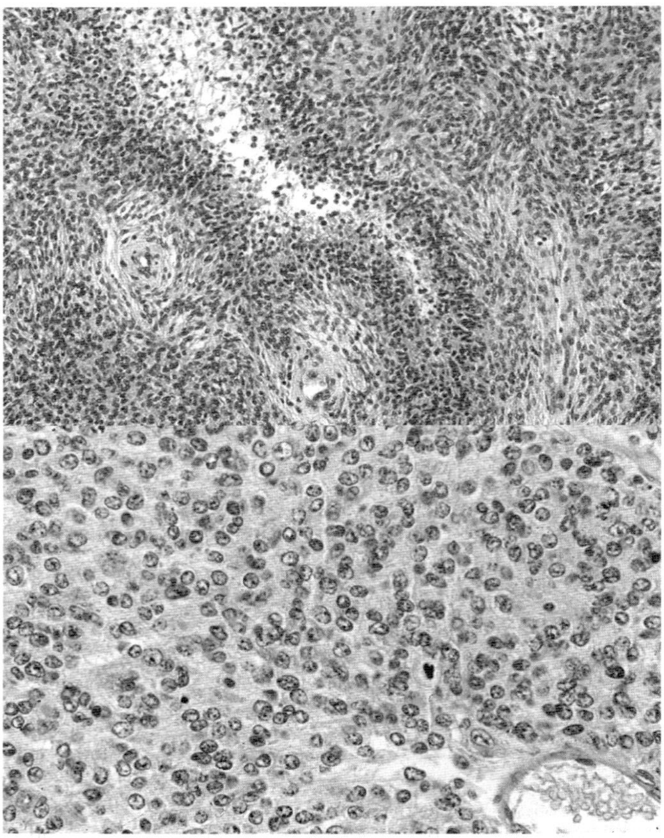

B

Figure 23. A. A typical glioblastoma area, H&E, x 200; B. Oligodendroglial area in the same glioblastoma, H&E, x 400

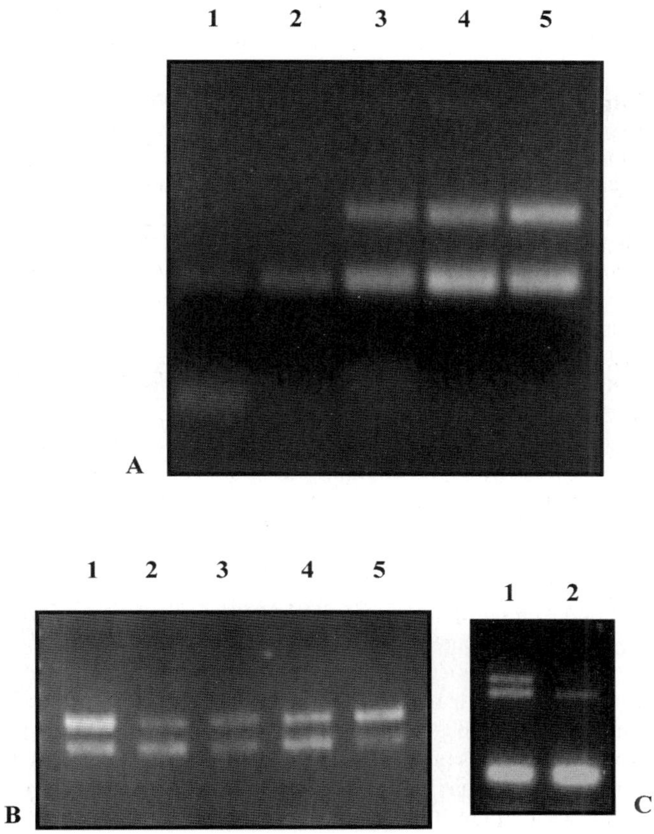

Figure 24. Glioblastoma with oligodendroglial areas. A. Lanes 1, 2: CDKN2A/p14 homozygous deletion; B. Lane 1, 5: MDM2 amplification; C. Lane 2: CDKN2A/p16 homozygous deletion. Multiplex PCR

All the alterations listed above represent good opportunities to discriminate prognostic subtypes of glioblastoma, but still today there is no codified procedure to define subgroups of malignant gliomas in relation to therapies in a more precise way than to pathology, let alone using them in the single case. However, it is of great interest that using a subset of glioblastomas with classical pathology as a pattern of class prediction after gene profiling for classifying glioblastomas with nonclassical pathology, the prognosis was better defined with gene profiling (Nutt et al., 2003).

Radiotherapy for glioblastomas is mandatory. Chemotherapy is also extensively used, even though with limited or contrasting results. A recent meta-analysis demonstrated that chemotherapy produces a moderate, but certain increase of survival that corresponds to 6% per year (Glioma meta-analysis trialists (GMT) group, 2002); another meta-analysis did not show any efficacy for tamoxifen and carboplatinum added to radiotherapy in comparison with radiotherapy alone (Puchner et al., 2000). Other reviews do not recognize chemotherapy with PCV, paclitaxel, temozolomide or gemcytabin as having any effect on glioblastomas (Kortmann et al., 2003). In particular cases, for example in unresectable glioblastomas, CCNU associated with radiotherapy increased survival in comparison with radiotherapy alone (Fazeny-Dorner et al., 2001). In general, the results of adjuvant chemotherapy in high-grade gliomas are disappointing (Grossman, 2003). In a very large multicenter study it has been shown that radiotherapy plus temozolomide gives a median survival of 14.6 months against 12.1 of radiotherapy alone with a two-year survival of 26.5% against 10.4% (Stupp et al., 2005).

Glioblastomas with strong expression of MGMT are more resistant to chemotherapy with ACNU and show shorter TTPs (Anda et al., 2003). MGMT has been demonstrated to be absent in the normal nervous tissue (Silber et al., 1996) and it is the most important DNA repair enzyme, catalyzing the transfer of the methyl group from O^6-methylguanine and O^4-methylthymine adducts of double-stranded DNA induced by alkylating agents to its cystein residue and preventing G:C to A:T transition (Pegg and Byers, 1992). The MGMT wt in combination with the allelic variant V1 is particularly expressed in primary glioblastomas (Inoue et al., 1993).

Recently, a targeted molecular therapy of GBM is developing based on the well known molecular pathways such as EGFR/ΔEGFR, PDGFR, PTEN, RAS/MAPK-Akt/PKB, TP53, pRb and on peculiar aspects of the tumor such as migration, invasion, angiogenesis. “Unarmed” or “armed” antibodies with immunotoxins, radionuclides and immunoliposomes, inhibitors of the different pathways, oncolytic viruses, anti-angiogenetic agents etc. are the main tools under study (Mischel and Cloughesy, 2003). The success of these therapies will depend on a deeper knowledge of the molecular machinery of the tumor and also on the possibility of identifying molecular prognostic factors.

In spite of the enormous quantity of contributions on the problem of prognosis of glioblastomas, no factor has been identified with certainty which could help in single cases, once the tumor has been histologically recognized. However, if a series of tumors is used for testing the efficacy of a therapy, based on the difference between the expected and the verified survival, it is advisable to check for the principal genetic alterations in order to take precautions against the fortuitous concentrations of cases with factors suspected to be prognostic.

6. PROBLEMS IN DIFFERENTIAL DIAGNOSIS

Glioblastomas must be differentiated from various other neoplastic lesions and their variants must be recognized, being characterized by different prognoses. One of the most important problems is that concerning “small cell” glioblastomas (Burger et al.,

2001) which contain EGFR alterations and must be kept separate from other tumors, for example anaplastic oligodendroglioma. A series of these tumors were studied: they do not show enhancement, necrosis and microvascular proliferations; but have the aspect of anaplastic astrocytomas and undergo rapid progress. These tumors express more frequently than other glioblastomas EGFR and EGFRvIII (Perry et al., 2004).

Giant cell glioblastoma (Figure 23), pleomorphic xanthoastrocytoma, metastatic tumors are easily recognizable, unless the surgical sample is so small as not to contain characteristic features. Beside xanthoastrocytoma, one particular tumor falls into this category and is hardly recognized, especially in very small samples: glioblastoma with epithelial differentiation (Kepes et al., 1982). Focal areas of epithelial differentiation with adenoid formations or a papillary aspect, positive for cytokeratins and devoid of GFAP staining, may occur requiring a differential diagnosis with carcinoma metastases. The dimensions of the surgical sample are very important in this case. Molecular genetic analysis of TP53 proved that the lesion is a product of clonality and is not a collision tumor (Müller et al., 2001). In a recent case identical losses on 17p13 and identical base pair deletion in TP53 gene, codon 209, exon 6 were found demonstrating the same genetic lineage for the two components (du Plessis et al., 2004).

Giant cell glioblastomas are recognized because of the extensive occurrence of giant cells with large cytoplasms and monstrous nuclei (Figure 25). Mitoses are not so frequent and often show pathologic features, for example multipolarity.

Glioblastomas with oligodendroglial areas will be considered with oligodendrogliomas.

A

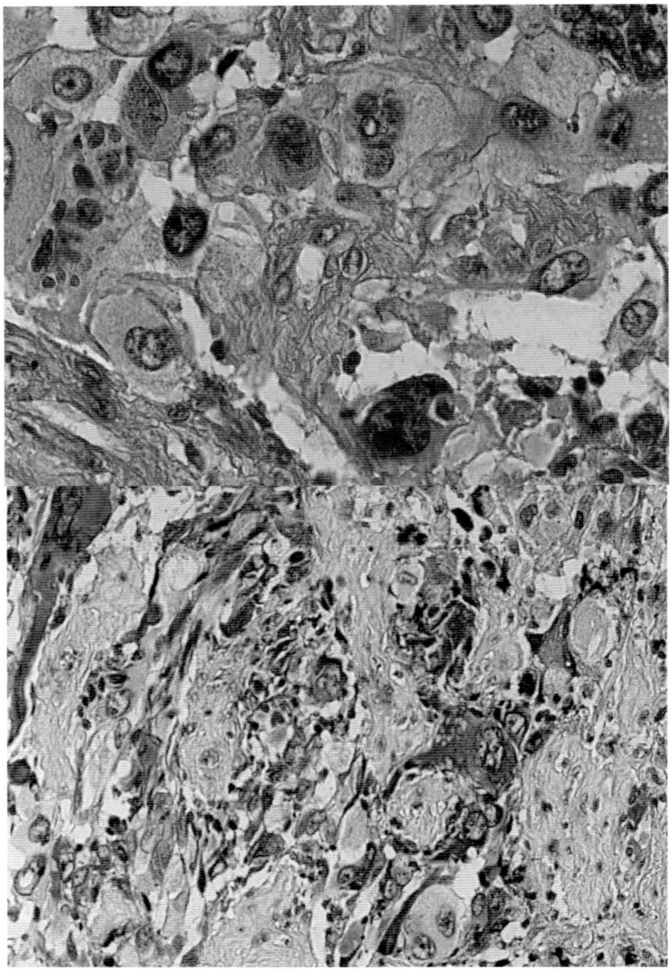

B

Figure 25. Giant cell glioblastoma, H&E, x 400; B. Id for GFAP, DAB, x 400

Chapter 5

ASTROCYTIC TUMORS II

1. PLEOMORPHIC XANTHOASTROCYTOMA (PXA)

This affects mainly young people and is superficially located. At MRI and CT scan it appears as a solid mass with contrast enhancement and is associated with a cyst (Figure 1).

It can involve the dura and arrive at the inner surface of the bone. The histological diagnosis is not difficult, when all the features are present: beside a general GFAP-positivity,(Figures 6A, 7B), multinucleated giant cells, xanthomatous cells, granular bodies (Figures 2A, 3, 5), and reticulin fibres around cell nests or giant cells (Figure 4) (Kepes et al., 1979). Mitoses are uncommon (Fouladi et al., 2001). Problems arise when the diagnosis has to be done on small fragments.

If giant and xanthomatous cells and the reticulin component are lacking, the diagnosis of tumor type becomes difficult and the differential diagnosis must be performed towards desmoplastic glioblastoma, fibrohistiocytoma (Cerda-Nicolas and Kepes, 1993), meningeal sarcomas (Schiffer, 1997) and also pilocytic astrocytoma. Usually, if no malignancy sign is present, the nosographic uncertainty has no therapeutic consequence, because the tumor is recognized as astrocytic and the diagnosis will not be treatment-affecting. Problems arise when malignancy signs are present and the malignancy grade has to be declared.

PXA may recur and undergo anaplastic transformation in 15-20% of cases (Pahapill et al., 1996; Kepes et al., 1997), but its assessment is not precisely known. In the WHO classification book a grade III is accepted, but no indication is given how to recognize it, even though the occurrence of necroses (Figure 6B) and the increased number of mitoses (Figure 7A) are usually indicative of anaplasia (Pahapill et al., 1996) and recurrence. After multivariate analysis the number of mitoses > 5 x 10 HPF is a negative prognostic factor both for survival and recurrence (Giannini et al., 1999); however, in the single case the cut-off points of the mitotic number and of the MIB-1 LI values are not known. As a matter of fact, it is difficult to make decisions when it is considered that MIB-1 LI has been found to be 1.9% for the entire group, only in 6 cases exceeding 2%. In a series of 5 cases (Sugita et al., 2000) MIB-1 LI was <1% in those with longer and 3-4% in those with shorter survival, with values obtained from tumor areas with the highest number of positive nuclei.

It would be really important to know whether the outcomes of classic PXA and of the anaplastic variant were different receiving completely different post-surgical

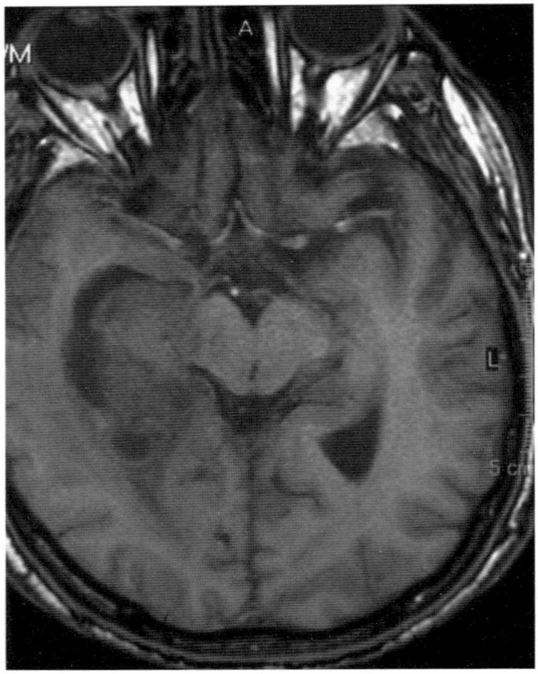

Figure 1. PXA. Temporal location of the tumor, MRI. From the Neuroradiology Unit, Dpt Neuroscience, University of Turin

treatments. The effects of radiotherapy are not yet completely known, because of the reduced number of cases studied. Overall survival at 5 and 10 years was 81% and 70%, respectively, and the only prognostic factor, after multivariate analysis, till now known was the extension of surgical removal (Giannini et al., 1999; Fouladi et al., 2001). Radiotherapy was observed to reduce the frequency of recurrences, but not the mortality rate (Macaulay et al., 1993). It can be proposed for grade II tumors after subtotal resection, if one believes that radiotherapy of grade II astrocytomas has a rationale, and of course for tumors with anaplastic features (Giannini et al., 1999). The hotspot of this tumor when it is histologically diagnosed is therefore its recognition as PXA "with anaplastic features", as has been proposed instead of "anaplastic PXA", because of a less aggressive connotation (Giannini et al., 1999). The size of the surgical specimen plays a significant role in this uncertainty.

The occurrence of cells with neuronal differentiation or expressing neuronal antigens can be either disturbing or, on the contrary, useful for a more detailed diagnosis (Hirato et al., 1994; Powell et al., 1996; Giannini et al., 2002). The antigens are class III β-tubulin, synaptophysin, neurofilaments, MAP2 and chromogranin A (Giannini et al., 2002; Im et al., 2003). Under the electron microscope, microtubules,

A

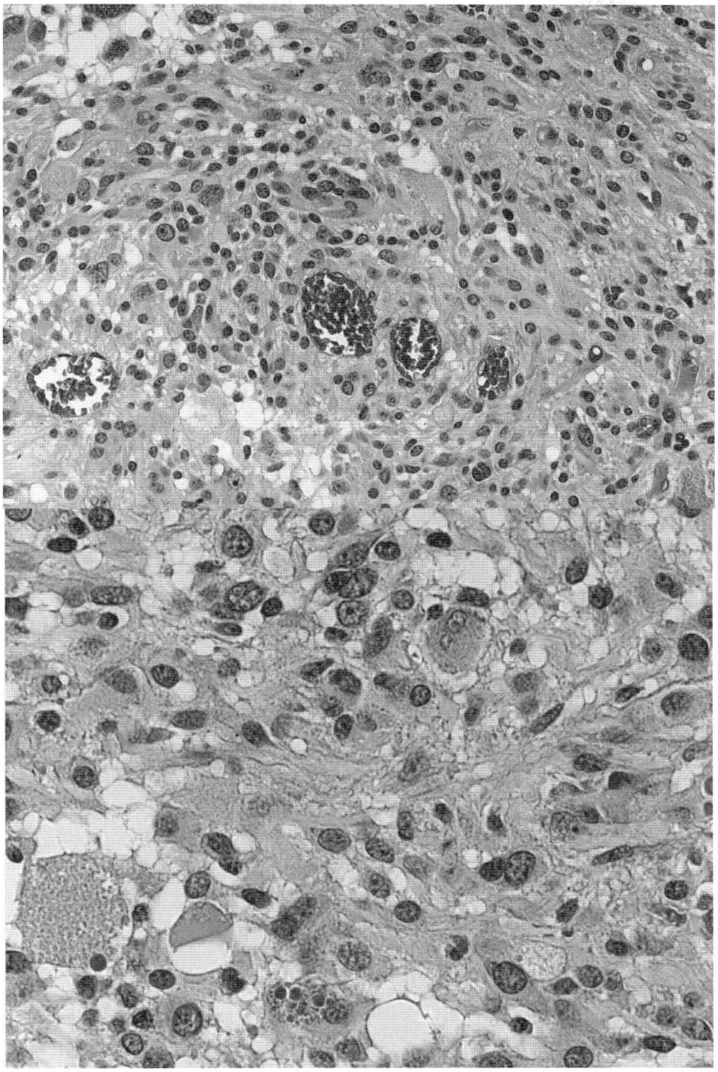

B

Figure 2. PXA. A. General astrocytic aspect, H&E, x 200; B. Large xanthomatous astrocytes, H&E, x 400

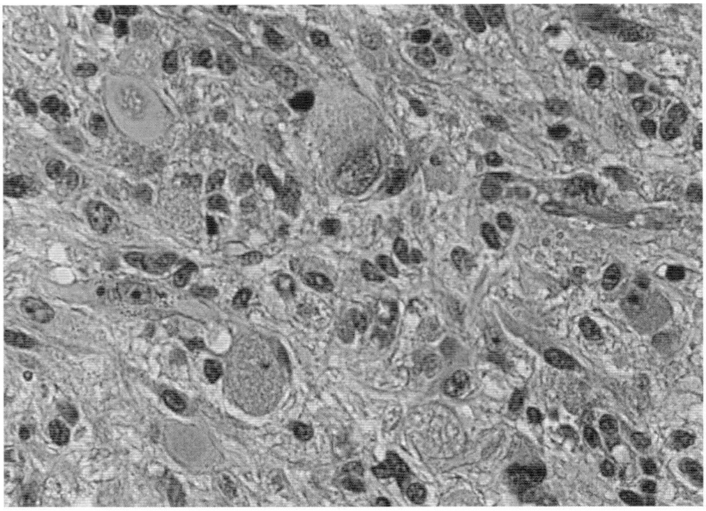

Figure 3. PXA. Large xanthomatous astrocytes, H&E, x 400

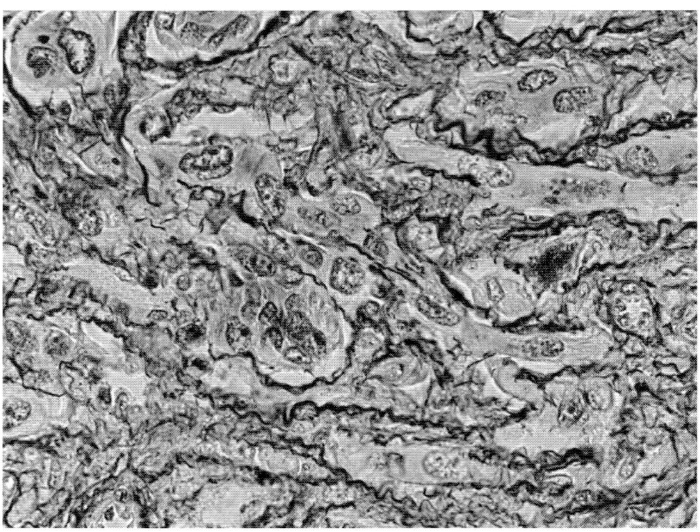

Figure 4. PXA. Reticulin fibres around astrocytes, Gomori for reticulin, x 400

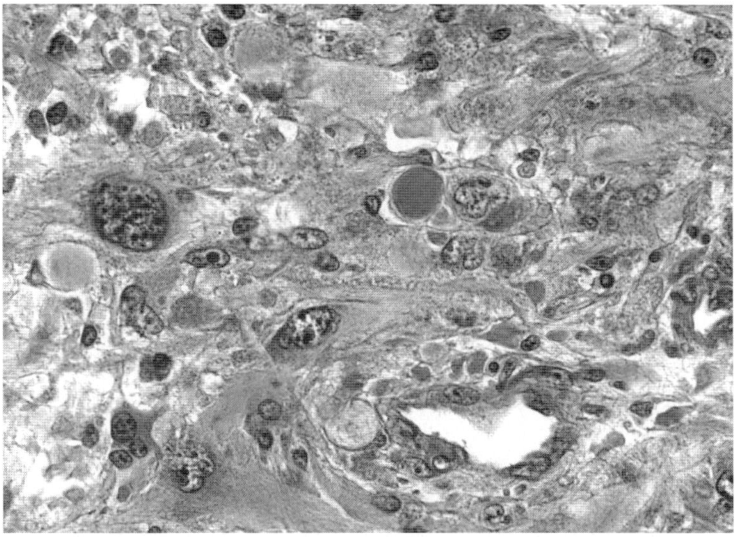

Figure 5. PXA. Eosinophilic or granular bodies, H&E, x 400

dense core vesicles and clear vesicles have been observed. Class III β-tubulin has been found in 73% of cases, but without any relation to malignancy (Giannini et al., 2002). The expression of the latter antigens can facilitate the differential diagnosis with glioblastoma, where a limited expression of class III β-tubulin can occur, always associated with TP53 mutations and p53 expression which, on the contrary, are rarely found in PXA (Giannini et al., 2001; Martinez-Diaz et al., 2003).

As in DNTs and gangliogliomas, cortical dysplasia may be associated with the tumor more frequently than believed, because it goes undetected in surgical samples which do not contain cortical tissue. This finding adds something to the hypothesis that the tumor is a lesion borderline between malformation and neoplasia (Lach et al., 1996). It can derive from the proliferation of multipotent precursor neuroectodermal cells associated with developmental cortical abnormalities (Im et al., 2003).

Recently, the expression of CD34 in vascular endothelial cells, in tumor cells, in septae and also in dysplastic neural cells in the cortex adjacent to the tumor has been reported and high levels of CD34 mRNA, observed for both full-length molecule or the truncated form, have been found (Reifenberger et al., 2003). This finding could indicate a relationship of PXA with gangliogliomas or their origin from a previous hamartomatous lesion, besides being useful in the differential diagnosis towards malignant gliomas. Pleomorphic xanthoastrocytomas may have a gangliogliomatous component that evolves from bipotent PXA cells along astrocytic and neuronal lineages (Vajtai et al., 1997).

The possible occurrence of neuronal cells in the tumor has made the problem of its cell composition more complicated. The distinction of "infantile desmoplastic astrocytoma" (Taratuto et al., 1984) and "infantile desmoplastic ganglioglioma" (VandenBerg et al., 1987), where neurons are not evident for some reasons, has long

A

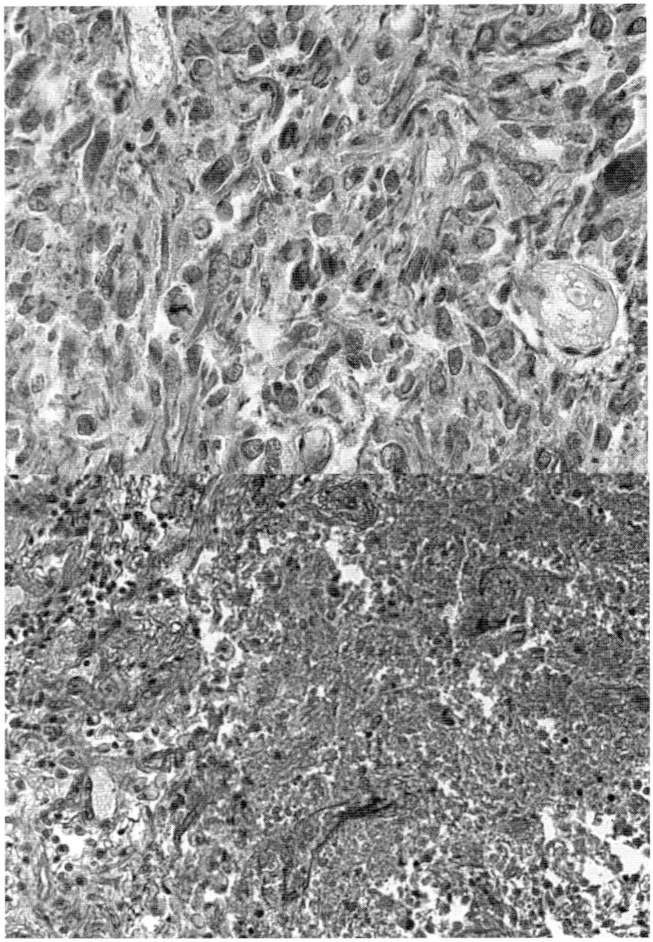

B

Figure 6. A. GFAP-positive staining, DAB, x 400; B. Necrotic focus, H&E, x 200

A

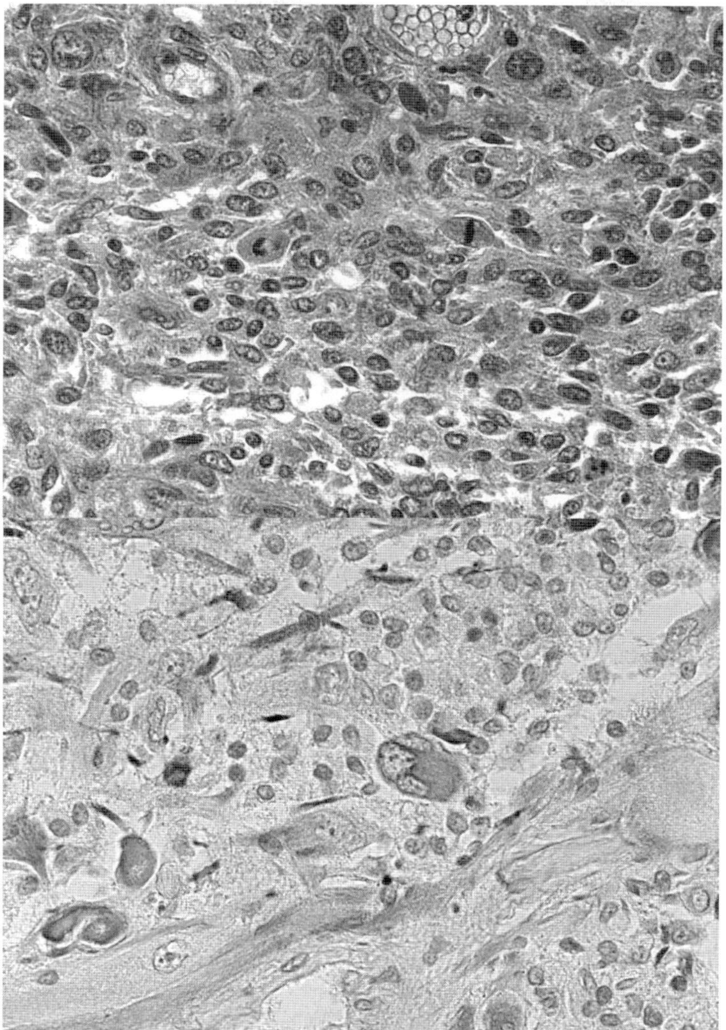

B

Figure 7. PXA. A. Nuclear pleomorphism and mitoses, H&E, x 400; B. GFAP-positive astrocytes, DAB, x 400 been discussed. A category has been proposed of "desmoplastic, supratentorial, neuroepithelial tumors" including tumors containing neurons and also malignant and desmoplastic PXA (Sperner et al., 1994). The differential diagnosis of PXA in this way widens, but the most important problem remains the recognition of the anaplastic variant.

There are no data from molecular genetics which could help in the diagnostics, because till now they are really scarce (Paulus et al., 1996; Munoz et al., 1996). In particular, mutations of TP53 are infrequent, 2/8 cases (Paulus et al., 1996; Giannini et al., 2001) and in general genetic alterations characteristic of astrocytomas and oligodendrogliomas, typically infiltrating tumors, are not present in PXA (Kaulich et al., 2002).

2. PILOCYTIC ASTROCYTOMA

This is a tumor typical of children with preferential locations: optic nerve and chiasm, hypothalamus, thalamus and basal ganglia, cerebral hemispheres, cerebellum, brainstem and spinal cord with some differences between one series and the other. Histologically, two aspects are prominent: one is composed by compact bipolar cells with long, GFAP-positive, processes and with Rosenthal fibres (Figures 8, 10B) and eosinophilic granular bodies (Burger et al., 2000). The other shows a loose texture, cells with scarce cytoplasm and short processes, weakly GFAP-positive and with microcysts (Figure 9). There is also a pilomyxoid variant with bipolar cells immersed in a myxoid magma, more frequent in brainstem (Tihan et al., 2000) and a variant with mucoid degeneration and an oligodendroglia-like aspect (Figure 10A). The latter can give serious problems of differential diagnosis towards ependymomas and oligodendrogliomas.

The tumor belongs to grade I of the WHO classification and very rarely undergoes malignant transformation. If the diagnosis is carried out on small fragments, its recognition may be difficult as well as the tumor grade. Sometimes a clear pleomorphism and hyperchromaticity and even some mitoses and apoptosis (Figure 14) may be present, especially in the cerebellum, which do not necessarily indicate malignancy or can be confused with nuclear regressive changes, such as nuclear pseudoinclusions.

Cerebellar tumors frequently recur, especially after subtotal resection, but this does not necessarily indicate malignization. Tumors with malignant transformation occurring after decades are rare as well as primary malignant tumors. They are characterized by nuclear pleomorphism (Figure 11), increase of mitoses and of LI for proliferation markers (Figure 13), microvascular proliferations (Figures 11A, 12A) and circumscribed necroses with pseudo-palisadings (Figure 12B). Microvascular proliferations alone, however, have no ominous meaning, because they can be found also in classic tumors. It is difficult to establish whether one is dealing with glioblastomas or anaplastic astrocytomas or even with benign tumors, because there is no certainty that these malignant aspects are associated with bad prognosis. Often they are associated with survival similar to that of tumors with a completely benign aspect (Schiffer, 1997). It has been proposed to put them into one category or to call them anaplastic astrocytomas (Burger et al., 2000). Sometimes tumors with these

A

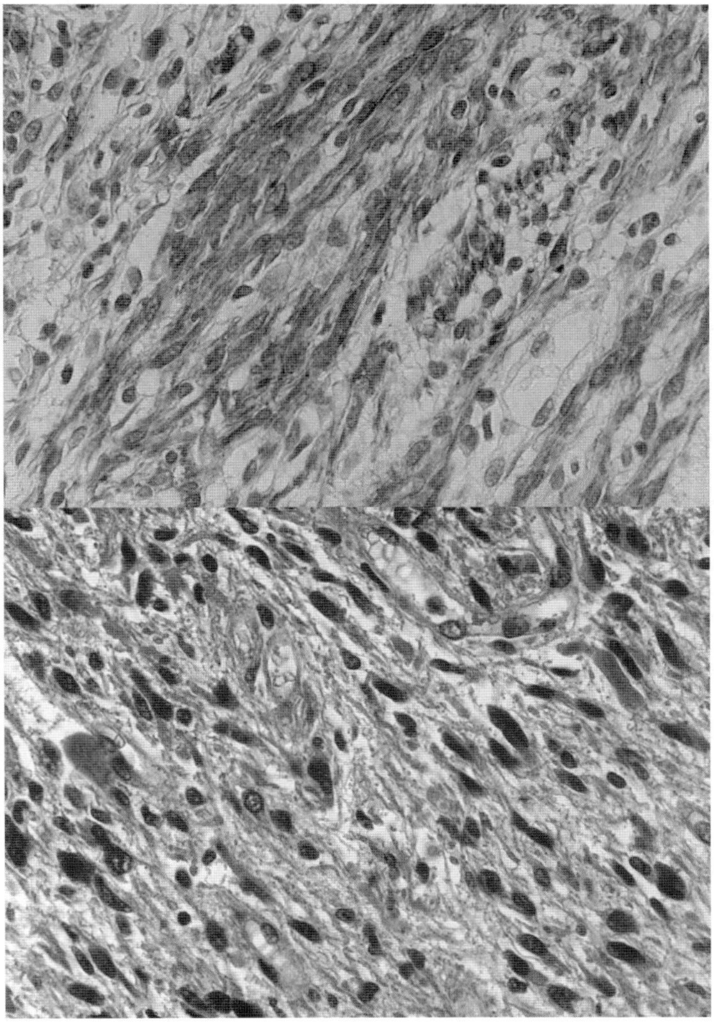

B

Figure 8. Pilocytic astrocytoma; A. Bipolar, compacted, GFAP-positive cells, DAB, x 400; B. Compact cells and Rosenthal fibres, H&E, x 400

A

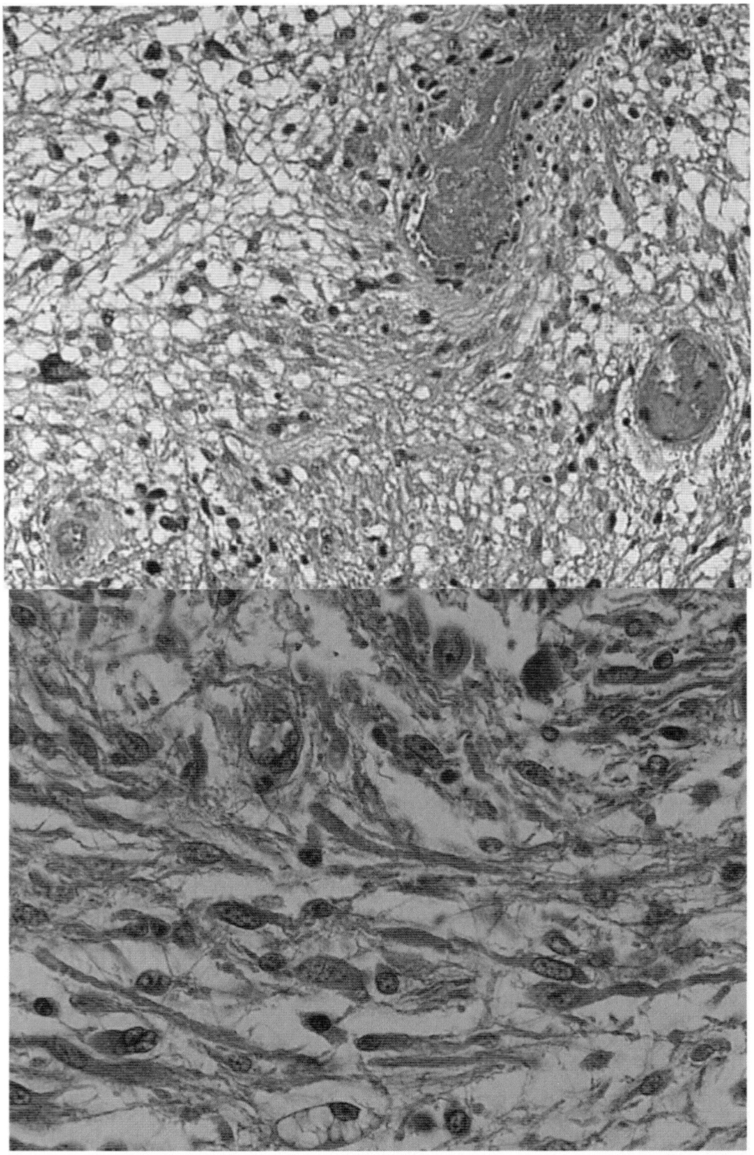

B

Figure 9. Pilocytic astrocytoma. A. Loose texture, H&E, x 400; B. Bipolar dissociated cells, H&E, x 400

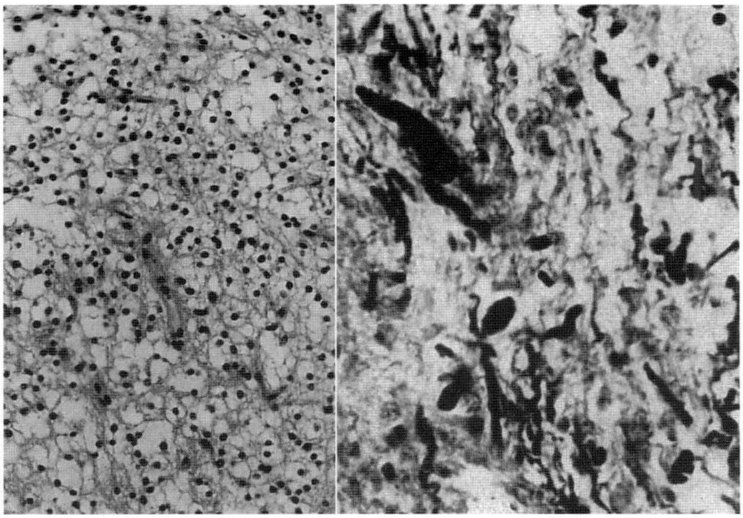

A B

Figure 10. Pilocytic astrocytoma. A.Mucous degeneration, H&E, x 400; B. Rosenthal fibres, Ferric hematox , x 400. From Schiffer, 1997

A B

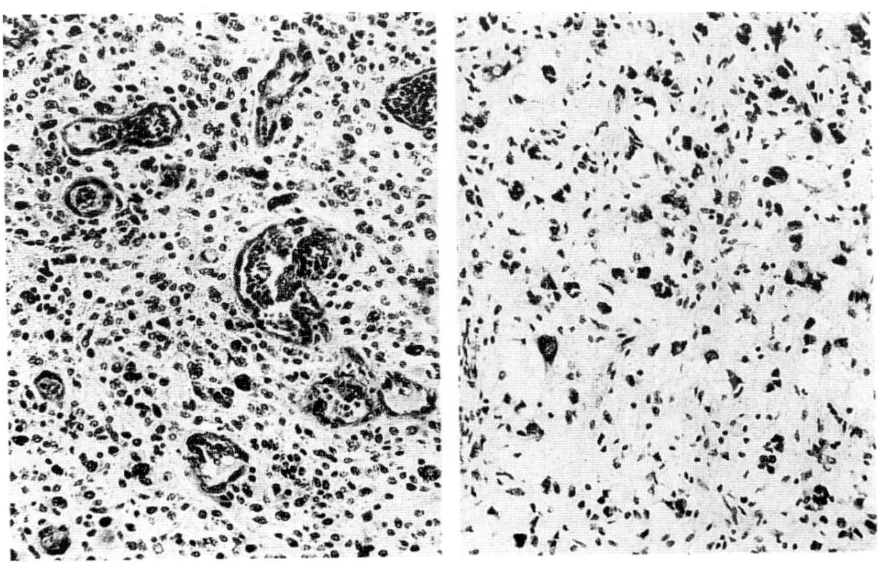

Figure 11. Pilocytic astrocytoma. A. Nuclear pleomorphism and vessel proliferations, H&E, x 200; B. Nuclear pleomorphism, H&E, x 400. From Schiffer, 1997

A

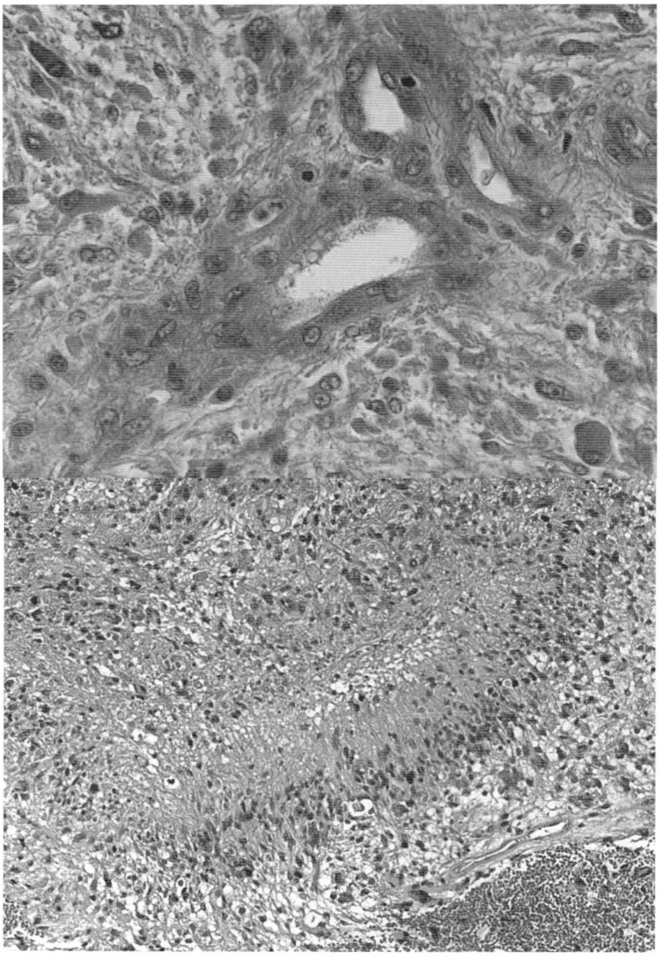

B

Figure 12. Pilocytic astrocytoma. A. Microvascular proliferations, H&E, x 200; B. Circumscribed necrosis, H&E, x 200

aspects have been previously irradiated (Dirks et al., 1994; Tomlinson et al., 1994) and it is wise in every case to gather information about treatments previously performed.

In a wide series (Giannini et al., 1999), 32% of the tumors presented mitoses: single in 55% and more than one in 45% with a maximum of 4 x 10 HPF. Endothelial proliferations and glomeruloid formations were present in 18% of cases

A

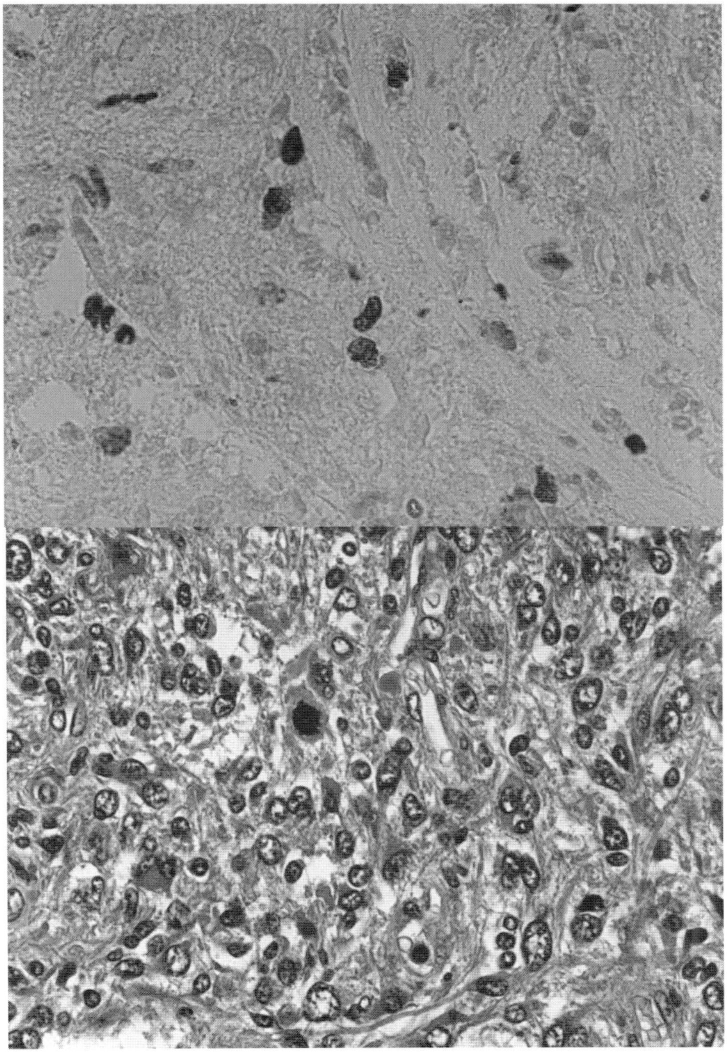

B

Figure 13. Pilocytic astrocytoma. A. MIB-1 positive nuclei, DAB, x 400; B. Mitoses, H&E, x 400

and necroses with pseudo-palisadings in 8% of cases. MIB-1 LI was low and not in conformity with histological aspect or survival. In the same series there was no anaplastic tumor; and survival at 5 and 10 years was 87% and 83%, respectively. These data correspond to those of another series (Fernandez et al., 2003) in which the pilomyxoid variant is considered, occurring mostly in the opto-chiasmatic location and in the brainstem. Setting aside location and total resection that are associated with longer TTP, in this series no more important prognostic factors exist. Still in another series, including 4 atypical cases (?) a relationship between MIB-1 LI and TTP was not found, even though in one case the LI was high. The mean was 4.4% with a range of 0.6% –12% (Roessler et al., 2002). Apoptosis is more frequent than in diffuse astrocytoma (Figure 14).

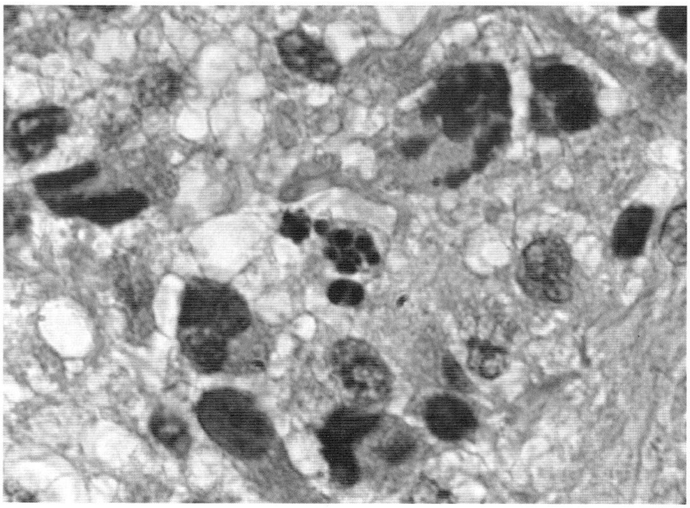

Figure 14. Pilocytic astrocytoma. Apoptotic figures, H&E, x 400

Less than 100 cases of malignant transformation have been reported in the literature (Schiffer, 1997; Endo et al., 2003). Between 1975 and 1994 the mean survival of the 78 reported cases was 18 months (Djalilian and Hall, 1998). Five cases with leptomeningeal diffusion have been reported recently. Treated by radiotherapy, gamma-knife and BCNU their survival was 22 months (Endo et al., 2003). A recent infantile series of 250 cases of posterior fossa, included 11 anaplastic astrocytomas and 2 glioblastomas (Bertolone et al., 2003).

Practically, a prognosis does not come out with certainty from the histological diagnosis in single cases, even when it concerns a high grade. No help till now comes from cytogenetics and CGH (Bigner et al., 1997; Sanoudou et al., 2000) or from molecular genetics: the studies have been too few and have shown allelic losses on both 17p and 17q including TP53 and NF1 loci applied to sections. Rare TP53 mutations have been reported (Patt et al., 1996) and inconsistent data have been given as for methylation (Uhlmann et al., 2003). A recently described cerebellar glioblastoma showed p53 positivity and a series of changes of PDGFR, RB, PTEN and LOH of 19q and 10q, but no EGFR amplification, as secondary glioblastomas do (Kawarabuchi et al., 2005).

How to recognize astrocytomas of a grade higher than II is very problematic, since increased cell density, appearance of nuclear atypias, mitoses, unless pronounced, and even occasional necrosis (Burger et al., 1997), may be destitute of definite prognostic meaning (Tomlinson et al., 1994). As already said, anaplastic tumors are rare, but they do exist. Only it is difficult to recognize them, because the criteria used for diffuse astrocytomas do not work with them. The missed recognition of anaplastic tumors or the attribution of malignancy to tumors that are not malignant represent normal diagnostic pitfalls (Burger et al., 1997) which can lead either to omitted therapies or to unnecessary treatments. MIB-1 LI may be useful, but its interpretation requires a previous correct diagnosis and this may represent a vicious circle. If, on the one hand, MIB-1 LI and the number of mitoses, even high, may not be associated with malignancy and microvascular proliferations and necroses have very little prognostic indication, on the other hand, Rosenthal fibres may be lacking or they may occur, but with no relevance to pilocytic astrocytoma, because simple expression of degeneration of reactive sub-ependymal glia. Necroses and high MIB-1 LI can be found in the pilomyxoid variant of optic nerve and chiasm, III ventricle and brainstem (Fernandez et al., 2003).

Neuro-imaging may help in the diagnosis, because the tumors are often cystic, but it may also be misleading, because the tumors may assume contrast enhancement. Also the location to the brainstem may be misleading, because tumors, especially in the pons, are often diffuse astrocytoma and not pilocytic astrocytomas. The therapeutic mistake that can follow is to treat a tumor grade I or not to treat a tumor grade III. The latter event, however, is less serious than believed, because of the uncertainty, persisting still today, on the effects of radio- and chemotherapy on posterior fossa gliomas (Bertolone et al., 2003). It must be mentioned that these tumors not only rarely recur locally and need a local irradiation (Kopelson, 1982), but also spread in the subarachnoidal space requiring a cranio-spinal irradiation (Salazar, 1981).

The smaller is the surgical sample the more difficult is the tumor recognition, because it may get confused with a pleomorphic infiltrating glioma. The evaluation of Ki-67 MIB-1 LI and p53 LI have been used for its recognition. MIB-1 LI is lower than that of diffuse astrocytomas, but with a similar range, whereas p53 has always been found negative (Tihan et al., 2000).

The last source of mistakes can be the differentiation between a pilocytic astrocytoma and a diffuse astrocytoma in the brainstem. This can be true also in hemispheric tumors in young people, but in the first location surgical samples are usually very small and sometimes the distinction between the two tumors is almost impossible. Leaving aside the occurrence of Rosenthal fibres, which can even be misleading, as already said, it has been shown that maybe the expression and positive immunohistochemistry of Galectin-3 can be of help, because it is higher in pilocytic than in diffuse astrocytomas and oligodendrogliomas (Neder et al., 2004).

3. GLIOSARCOMA

This is a variant of glioblastoma with which it shares clinics, neuro-imaging and prognosis. By definition it is a biphasic tumor with a glial and a mesenchymal component (Figure 16) and it has been arranged in a glio-mesenchymal spectrum of lesions which includes: the meningeal reaction to an infiltrating glioma; the desmoplastic response of the meninges; the scars of glioblastoma; the confluence of vessels thickened by protein exudates (Cerda-Nicolas and Kepes, 1993). Described by Stroebe in 1895, it was defined as a glioblastoma with the vessels proliferating in a sarcomatous sense (Feigin et al., 1955, 1958). It is not easy to calculate the incidence of the tumor, because of the uncertainty in defining the mesodermic component. The tumor varies from 1.2% of malignant gliomas (Feigin et al., 1958; Kochi and Budka, 1987) to 8% (Morantz et al., 1976). By neuro-imaging (Figure 15) and macroscopically it cannot be differentiated from glioblastoma, if not for a harder consistency.

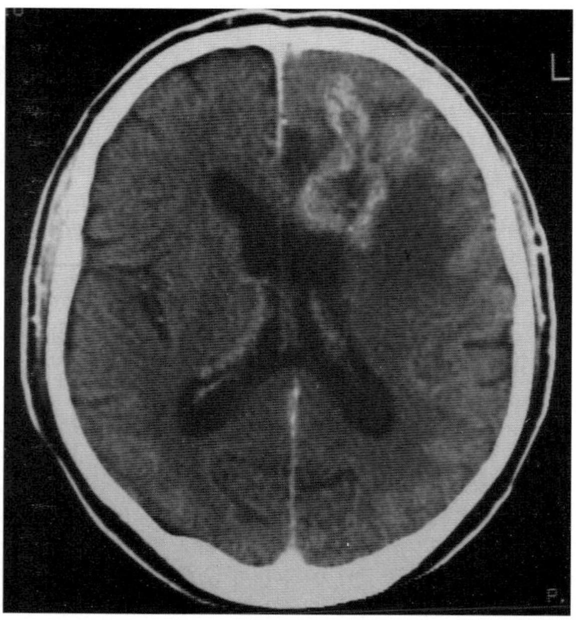

Figure 15. Gliosarcoma. CAT scan with contrast enhancement. From the Neuroradiology Unit, Dpt Neuroscience, University of Turin

A

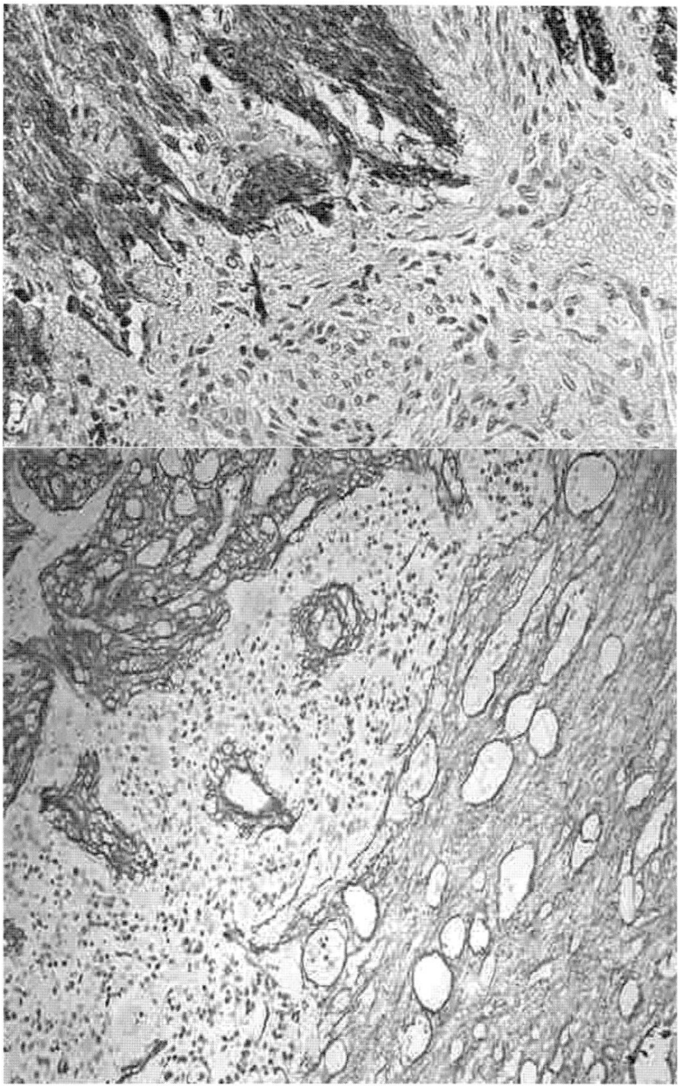

B

Figure 16. Gliosarcoma. Glial and mesodermic component. A. GFAP, DAB, x 200; B. Gomori for reticulin, x 200

The glial component is usually a glioblastoma, but oligodendroglioma (Feigin et al., 1976; Pasquier et al., 1978) and ependymoma (Louis et al., 1990) have been described. The mesenchymal component must necessarily be neoplastic and produce reticulin (Figure 18B). Usually it appears as a fibrosarcoma, or as a malignant histiofibrocytoma. Necroses can be present also in the mesodermic component, with typical clear-cut borders, as well as mitoses. The delimitation of the two components is sharp, but frequently one is larger than the other. For example, in very small surgical specimens the glial component may be limited to a few nests of GFAP-positive cells immersed in a fibrosarcomatous proliferation (Figures 17, 18A). On the other hand, a fibrosarcomatous small area, hardly distinguishable from a particularly active microvascular proliferation of a glioblastoma, appears to be inserted in a full glioblastomatous tumor. As a matter of fact, the mesenchymal component must be differentiated from a mesodermic scar of glioblastoma and from a large glomeruloid proliferation of the same tumor. In the first case, the distinction is easy when, besides the lack of any sign of malignancy in the scar, this also shows a loose texture with the presence of histiocytes, lymphocytes, macrophages etc. (Figure 19).

In the second case, the distinction can be difficult, because in large glomeruloid proliferations of glioblastoma, besides many vascular canals and endothelial cells, other cell types are present, including α-sm-actin-positive cells, macrophages, fibroblasts etc.

The occurrence of many small vessel lumina may help in the differential diagnosis, but the distinction in this case raises the entire problem of the origin of gliosarcoma.

The mesenchymal component was once regarded as arising from the endothelial hyperplasia of glioblastoma vessels (Feigin et al., 1958; Schiffer et al., 1984; Slowik et al., 1985), but this interpretation was not confirmed by the observation that FVIII/RAg was positive only in endothelial cells abutting on the lumina, even though the factor could have got lost in the other cells with proliferation (Guarda et al., 1982). The cells of the mesenchymal component could derive also from adventitia cells (Grant et al., 1986) or from histiocytes, because these cells, positive for α_1-antitrypsin and α -antichymotrypsin, are diffusely present in the mesenchymal component (Kochi and Budka, 1987). Many cells are also positive for α-sm-actin and are considered as smooth muscle cells (Haddad et al., 1991), but it was also suggested that they are astrocytes that, while keeping to express GFAP, could assume an elongated form and express α-sm-actin (Jones et al., 1991). This opinion was not shared by others (Paulus and Jellinger, 1992). Finally, the hypothesis was put forward that the two components derive from a common precursor cell (Perry et al., 1995).

In glomeruloid proliferations of glioblastoma, the many cells expressing α_1-chymotrypsin, α_1-trypsin and mainly α-sm-actin were regarded in favour of the vessel origin of the mesenchymal component of gliosarcoma (Schiffer, 1997). The hypothesis of the origin of both components of the tumor from a common precursor, i.e. the monoclonal hypothesis, has been supported by the observation that the two components share the same genetic changes in the same tumor (Biernat et al., 1995; Boerman et al., 1996; Reis et al., 2000). A complete study of the problem

A

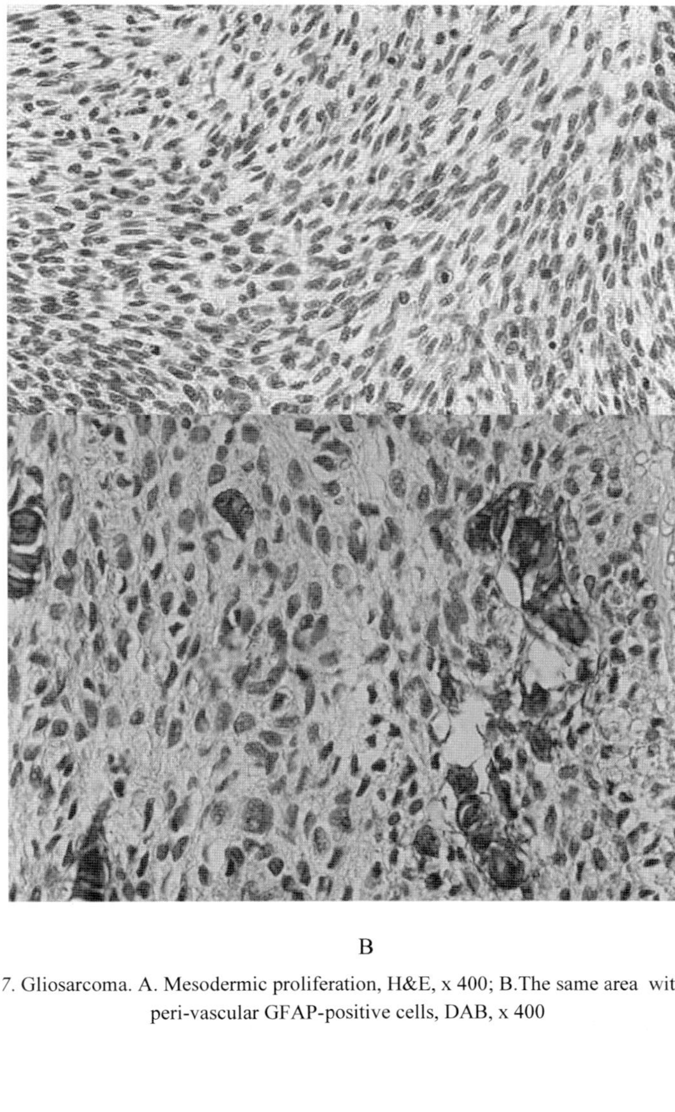

B

Figure 17. Gliosarcoma. A. Mesodermic proliferation, H&E, x 400; B.The same area with some peri-vascular GFAP-positive cells, DAB, x 400

A

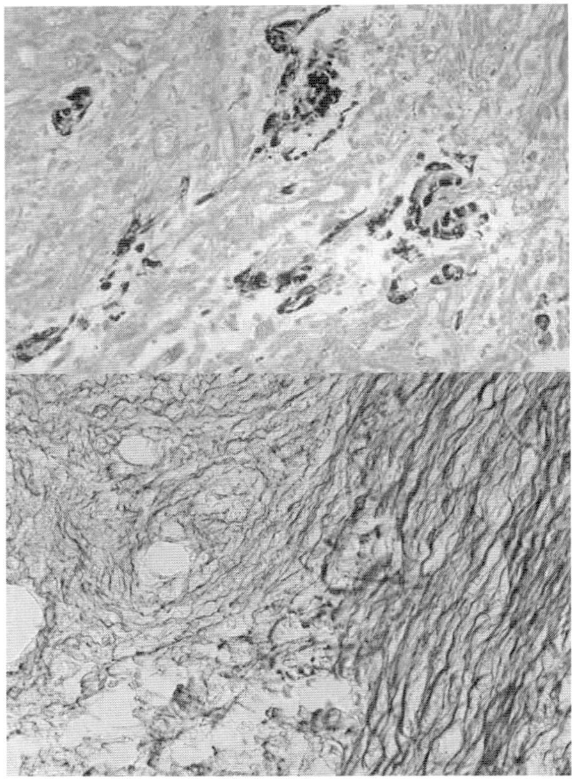

B

Figure 18. Gliosarcoma. A. GFAP-positive nests of few glioma cells. DAB, x 400; D. Positive reticulin staining, Gomori, x 400

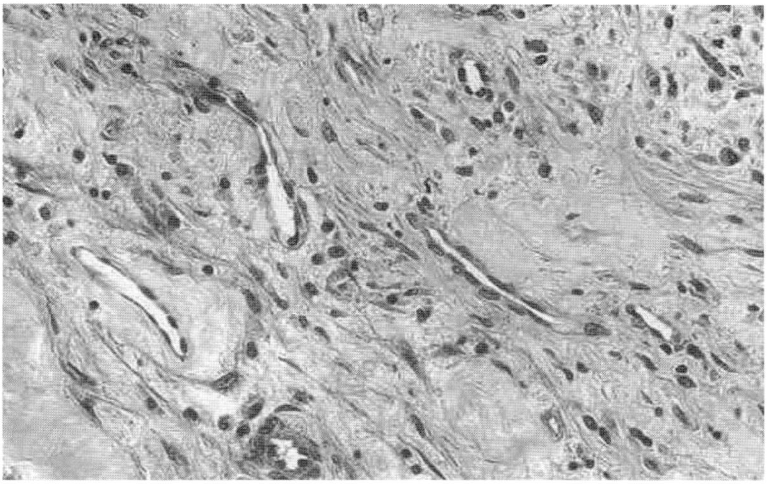

Figure 19. Gliosarcoma. Sarcomatous cell population originating from proliferated vessels of irradiated glioblastoma, H&E, x 400

demonstrated that chromosomal imbalances identified by CGH in gliosarcoma, i.e. gains on 7, X, 9q and 20q and losses on 10, 9p and 13q, were not different from those of glioblastoma, that there was no genomic alterations which distinguish both tumors and that a greater instability was present in glioblastoma than in gliosarcoma. TP53 mutations and PTEN mutations are equally frequent in the two tumors, whereas CDK4 and MDM2 were more and EGFR less amplified in gliosarcomas (Actor et al., 2002). Genomic alterations develop, therefore, in the two components after the origin from a common precursor. This could be a multipotent neural stem cell or the glial precursor that kept the capacity of differentiation into a muscular phenotype (Maher et al., 2001). There is a long discussion on the "mesenchymal paradox of glioma cells" (Nobel and Mayer-Pröschel, 1997), but here the capacity of embryonal cortical neural stem cells to differentiate in GFAP-producing cells and muscle cells (Tsai and McKay et al., 2000) must be emphasized and this corresponds to the observations on the mesenchymal transformation of glioma cells in culture (Bigner et al., 1981; Shapiro and Shapiro, 1984; Westphal et al., 1988). Till now there is no molecular interpretation of this event. It must be remarked that p53 is positive in the two components (Figure 20).

Cases have been described of gliosarcomas developed from glioblastomas after irradiation, but their number is not so high as it would be expected from the great amount of irradiated glioblastomas. It must be considered that fibrosarcomas or malignant fibrohistiocytomas not rarely arise after irradiation of other intracranial

A

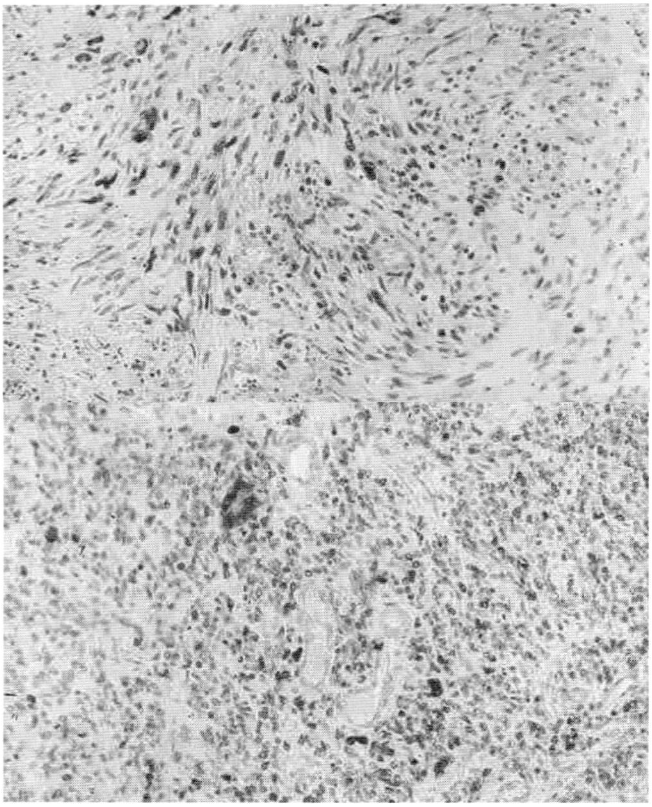

B

Figure 20. Gliosarcoma. Positive staining for p53. A. Glial component; B. Mesodermic component, DAB, x 400

tumors. In this regard, the time interval between the irradiation of a glioblastoma and death is too short for the organization of a mesodermic tumor. After irradiation of glioblastomas, vessel responses take place going from degenerative changes of the walls and endothelium alterations to complicated cell proliferations not easily distinguishable from fibrosarcomas with positivity for α-sm-actin, as myofibroblasts (Schiffer et al., 1990). This finding corresponds to the observation that 60% of tumors express α-sm-actin and on this basis it has been hypothesized that one origin of gliosarcomas could be from smooth muscle cells of glioblastomas (Perry et al., 1995).

Very important are gliosarcomas which arose after irradiation of glioblastomas. As a matter of fact, they are rather rare: 6 out of 31 (Ang et al., 1996) and 7 out of 32 gliosarcomas (Perry et al., 1996). Out of 129 radiation induced intracranial neoplasms of the literature, 4 were gliosarcomas (Kashten et al., 1995). The rarity of the tumors can be explained by the duration of the latency period in comparison with the survival of glioblastomas. A latency period of 9.6 months has been calculated with a radiation dose of 40.5 Gray.

The hotspot of this tumor at the moment of histological diagnosis does not reside in its differentiation from glioblastoma, because there is no consequent different therapeutic strategy, but just in its origin. As already said, the origin of both components from a common glial precursor capable of mesenchymal differentiation cannot be considered as definitely settled. It needs further contributions and demonstrations. On the other hand, the origin of the mesenchymal component from vessel cells cannot be disregarded as out-dated. In irradiated recurrent glioblastomas, the sarcomatous cell population seems to originate from preexistent glomeruloid formations (Schiffer et al., 1990). Finally, in very small tumor samples, it is possible that one of the two components is not represented or it goes unrecognized and, when this happens with the mesodermic components, the gliosarcomatous transformation of an irradiated glioblastoma could simply represent a sampling error. Less frequent is the opposite, i.e. the missed recognition of the glial component, as it is too small or limited to only a few cells in the sample.

Chapter 6

OLIGODENDROGLIAL TUMORS

1. THE DIAGNOSIS OF OLIGODENDROGLIOMA

Oligodendroglioma is today the object of great interest because it is the only glioma for which an effective chemotherapy has been found. Therefore, the major diagnostic problem is that of not missing one. It has always been a controversial tumor type for its uncertain nosographic delimitations, but today LOH of 1p and 19q are of great help in recognizing it and its responsiveness to therapy.

The frequency of the tumor varied enormously in the past in different series, from 5% to 20%, testifying to its indefinite borders (Schiffer, 1997). It is a hemispheric tumor, with a characteristic macroscopic aspect, even though not patognomonic. Its nosographic position oscillated according to the restricted or permissive criteria used for its recognition and according to the philosophy adopted by neuropathologists in the distinction between astrocytic and oligodendrocytic cells, since reliable markers of the oligodendroglial nature of cells in histological sections are still lacking. The real diagnostic conflict has been masterly depicted (Burger, 2002) and opened the door to the use of genetic analysis in the diagnosis of uncertain infiltrating tumors.

With restricted criteria, diagnostic signs are the "honeycomb" appearance of the cells and the "chicken-wire" distribution of small vessels (Figures 1, 3B). Maybe it is worth saying that the "honeycomb" aspect is a regressive sign, even though characteristic, that is not always present. Its absence does not exclude the diagnosis of oligodendroglioma, whereas its presence is definitely indicative, once the possibility that a vacuolar degeneration from another source giving the cell a clear perinuclear halo is ruled out, as may happen in mucoid degeneration of pilocytic astrocytoma. It has been observed that the occurrence of the halo in >50% of the cells of a glioma is definitely associated with LOH of 1p and 19q, whereas in <50% the association is with TP53 mutations (Watanabe et al., 2001).

Nuclear morphology is very important for the recognition of the tumor. Normal oligodendroglial nuclei can be pale or dark, the latter representing fully mature cells (Pannese, 1994) and tumor nuclei are very similar, with a characteristic thickness of the membrane, a central nucleolus with adherent chromatin. These aspects have a

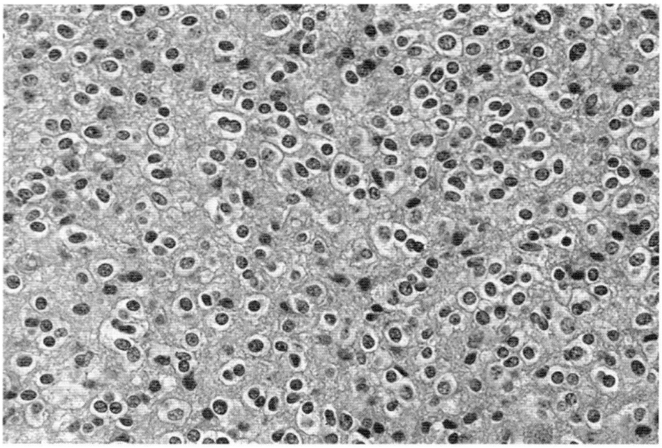

Figure 1. Oligodendroglioma. Typical "honeycomb" appearance, H&E, x 400

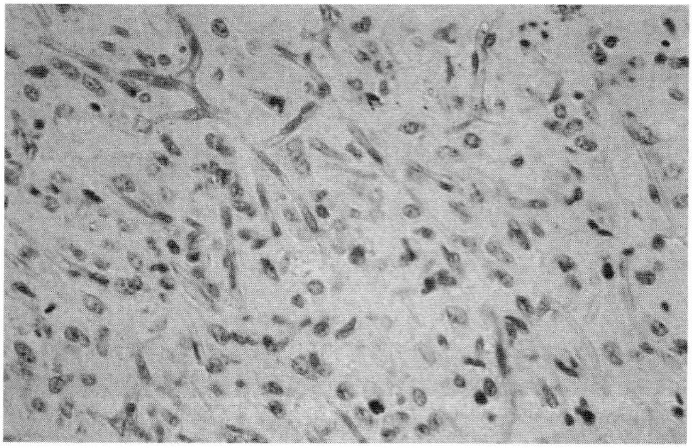

Figure 2. Oligodendroglioma. Small vessels with endothelial hyperplasia, H&E, x 400

diagnostic importance and they have been the object of quantitative cytophotometric studies on Feulgen stained sections for their differentiation from astrocytic nuclei (Decaestecker et al., 1998). This methodology, however, is too complicated for a practical application. The distinction can be made, on the other hand, with simple histological observation (Daumas-Duport et al., 1997) and it is worth saying that in earlier studies and in a coarser way the typical and characteristic nuclear structure had already been quickly demonstrated by the acetic carmin procedure (Schiffer, 1971). If nuclear morphology is useful and reliable for the recognition of the tumor in classic oligodendrogliomas, it is much less so in the anaplastic variant (Figure 4A). With the appearance of anaplasia, the variations of the nuclear aspect increase to the point where the nucleus is no more recognizable as oligodendroglial, and the nuclear criterion does not help any more (Schiffer et al., 1971).

With more permissive criteria, more indulgent on the nuclear characteristics and including the absence of processes and the presence of a thin cytoplasm rim (Coons et al., 1997; Daumas-Duport et al., 1997; Fortin et al., 1999), the frequency of oligodendrogliomas obviously increases together with the decrease of diffuse astrocytomas (Coons et al., 1997), because astrocytic gliomas are misclassified as oligodendrogliomas (Burger, 2002).

The recent molecular genetics observations that LOH of 1p and 19q are associated with an oligodendroglial nature, with benign lesions and with good response to therapy with PCV – which, on the contrary, has no efficacy on astrocytic tumors – increased the importance of the differential diagnosis between oligodendroglioma and diffuse astrocytoma. The greater sensitivity of oligodendroglioma has been interpreted as in line with the observation that precursor oligodendrocytes in the rat would be more sensitive to nitrosourea, because their MGMT is down-regulated in comparison with astrocytic precursors (Nutt et al., 2000).

The distinction of oligodendroglioma from diffuse astrocytoma is important for the longer survival of the former and for the different prognostic interpretation that the number of mitoses and of nuclei positive for proliferation markers may receive in the two tumors. In diffuse astrocytoma, mitoses are absent or very low in number. Its increase indicates anaplasia, whereas in oligodendroglioma the number of mitoses tolerated by the classic variant is definitely higher. The same MI, therefore, may indicate anaplasia or not according to the diagnosis of astrocytoma or oligodendroglioma (Schiffer, 1997; Kleihues and Cavenee, 2000). The explanation why oligodendrogliomas show higher MI and MIB-1 LI than astrocytomas and at the same time longer survivals could be that their AI (apoptotic index) is much higher than in astrocytomas and apoptosis can be elicited by different pathways (Schiffer et al., 1997).

In oligodendrogliomas, two types of tumor growth have been identified (Daumas-Duport et al., 1997): a compact one with isolated tumor cells at the periphery

A

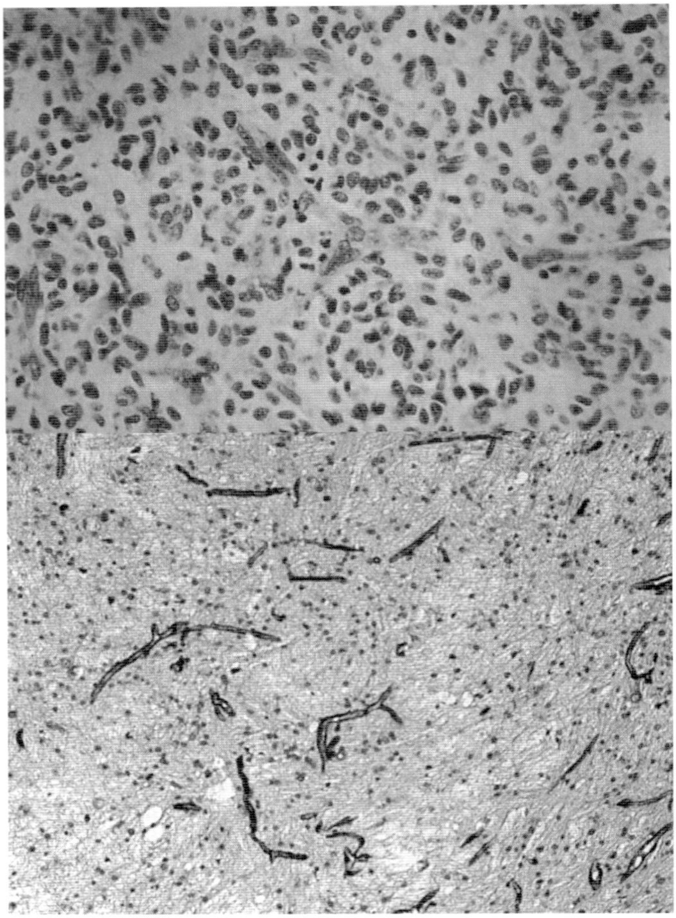

B

Figure 3. Oligodendroglioma. A. Endothelial hyperplasia, H&E, x 400; B. "Chicken wire" aspect of small vessels, CD31-DAB, x 200

A

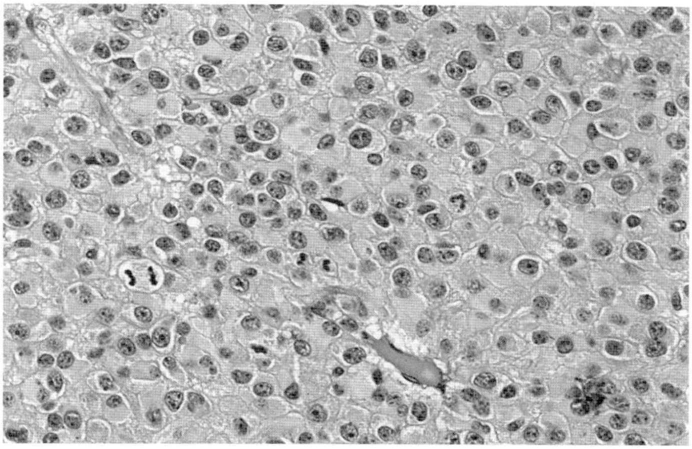

Figure 4. A. Typical oligodendroglial aspect of nuclei is preserved in spite of the elevated number of mitoses, H&E, x 400

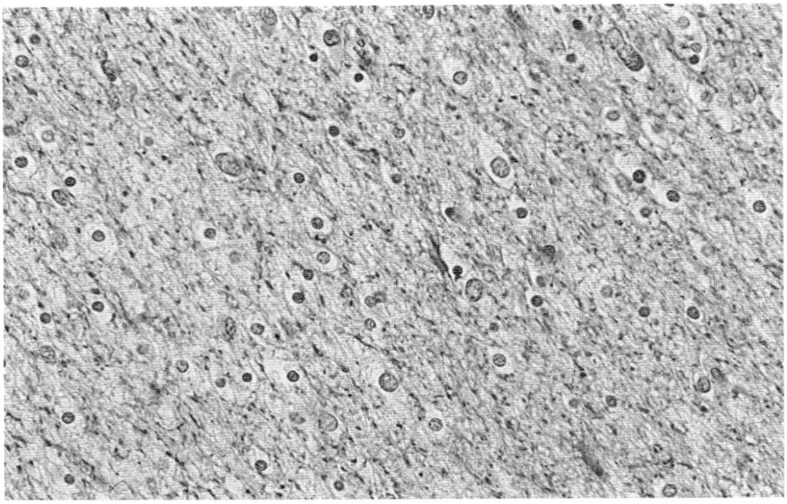

Figure 5. Oligodendroglioma. Isolated tumor cells with the glial network visible. GFAP, DAB – Hematox, x 400

and a diffuse type with only isolated tumor cells (Figure 5) which correspond to nodular and diffuse growths (Schiffer et al., 1997). When the cell proliferation in the diffuse growth is very loose or when tumor cells in small fragments are isolated and scattered, they appear immersed in a fibrillary GFAP-positive network which corresponds to the pre-existing glial net; and, especially if small GFAP-positive, reactive astrocytes are present, the diagnosis may be that of diffuse astrocytoma (Figure 5). The missed recognition of tumor oligodendrocytes could be responsible for the mistake. In a similar situation the missed recognition of normal oligodendrocytes could be responsible for the less frequent misdiagnosis of oligodendroglioma. Often a diffuse oligodendroglioma growth in the white matter and its recognition is based only on a higher density of oligodendroglial nuclei which are not distinguishable as normal or belonging to the tumor. In the cortex, a diffuse growth can be more easily detected because of the typical satellitosis.

Specific markers for oligodendroglia in histological sections, as GFAP for astrocytes, do not exist. MBP, MAG and galactocerebroside as well as other properties of oligodendroglial precursors are not expressed in tumors. Two recent observations contributed to solve this problem. Normal oligodendrocytes express proteins involved in the regulation of the cell cycle. One of these is p27/Kip.1 acting as inhibitor at the passage G1-S. In the development of precursor oligodendrocytes, p27/Kip.1 accumulates at the moment of differentiation and arrests cell proliferation (Durand et al., 1997). Generally, p27/Kip.1 is expressed in tissues and tumors which do not proliferate and it is degraded into the proteasome in proliferating tumors. It is positive in normal oligodendrocytes and less positive in oligodendrogliomas where the staining decreases with anaplasia (Cavalla et al., 1998), because of the activation of the ubiquitin-proteasome system (Piva et al., 1999). The opposite happens with cyclin D1. In many non-neural tumors it is over-expressed and contributes to the cell cycle deregulation, but this is very rare in gliomas (Buschges et al., 1999). In the latter, and therefore also in oligodendrogliomas, cyclin D1 expression is very low and increases with anaplasia, i.e. with the increase of cycling cells (Cavalla et al., 1999) (Figure 6A, B). In normal oligodendrocytes of the optic nerve of the rat during development, with the beginning of differentiation, cyclin D1 disappears (Durand et al., 1997), whereas it is strongly expressed in normal human oligodendrocytes, not associated with cyclin A and B1, beside microglia cells, both in the cortex and white matter (Bosone et al., 2000) (Figure 7A, B). Combined together, the two observations may be useful to distinguish normal from tumor oligodedrocytes and also, in some conditions, oligodendrogliomas from diffuse astrocytomas.

By in-situ hybridization, normal and tumor oligodendrocytes strongly express OLIG1 (Marie et al., 2001), a protein codified by a gene of the oligodendroglial line for elix-loop-elix factors (Lu et al., 2000; Zhou et al., 2000). It is expressed both in oligodendroglial precursors and in mature cells. OLIG1 mRNA was found to be strongly expressed in oligodendrogliomas, but not in other gliomas (Lu et al., 2001), and present also in oligoastrocytomas, becoming thus a good marker for oligodendroglial gliomas (Marie et al., 2001). However, recently it was shown to be positive also in astrocytomas, especially in the pilocytic variant. It would be, therefore, a marker for gliomas (Bouvier et al., 2003). In a further systematic study, OLIG1 and OLIG2 mRNAs turned out to be present in oligodendrogliomas, astrocytomas, oligoastrocytomas and anaplastic oligodendrogliomas and astrocytomas, much less in glioblastomas. By immunohistochemistry, OLIG2 was positive in 80% oligodendrogliomas and in 34% astrocytomas and more uniformly and intensely expressed in anaplastic oligodendrogliomas than in other glial tumors. Therefore, even though not specific for oligodendroglial tumors, it can contribute to the differential diagnosis (Ohnishi et al., 2003), but in practice the markers cannot be used as reliable for the recognition of oligodendroglioma in the single case (Riemenschneider et al., 2004). The discussion about the possibility of using these antibodies for the recognition of the oligodendroglial lineage in tumors is not yet completed. OLIG1, for example, has recently been found by immunohistochemistry to be positive in oligodendroglial cells of oligodendrogliomas, oligoastrocytomas and DNT, appearing again as a good marker. In particular, it was positive in GFOC, demonstrating their oligodendroglial nature, but was negative in minigemistocytes demonstrating their astrocytic origin (Figure 8A, B) (Azzarelli et al., 2003).

The problem of the potential role of the marker as a diagnostic tool has been studied by antibodies specifically recognizing OLIG2 protein in mouse and human tissue by immunohistochemistry and micro array analysis. It has been found that OLIG2 protein is limited to normal oligodendroglia and its progenitors in the human brain, but it is highly expressed in all diffuse gliomas. Its expression is also higher in anaplastic oligodendrogliomas than in gliobastomas which are heterogeneous with respect to OLIG2 (Ligon et al., 2004). In particular, of great interest is the finding that there is no obvious difference in OLIG2 expression between "pure" astrocytomas and mixed oligoastrocytoma. High expression of OLIG1 and OLIG2 was also found in 90% oligodendroglial tumors of another series, in astrocytomas

A

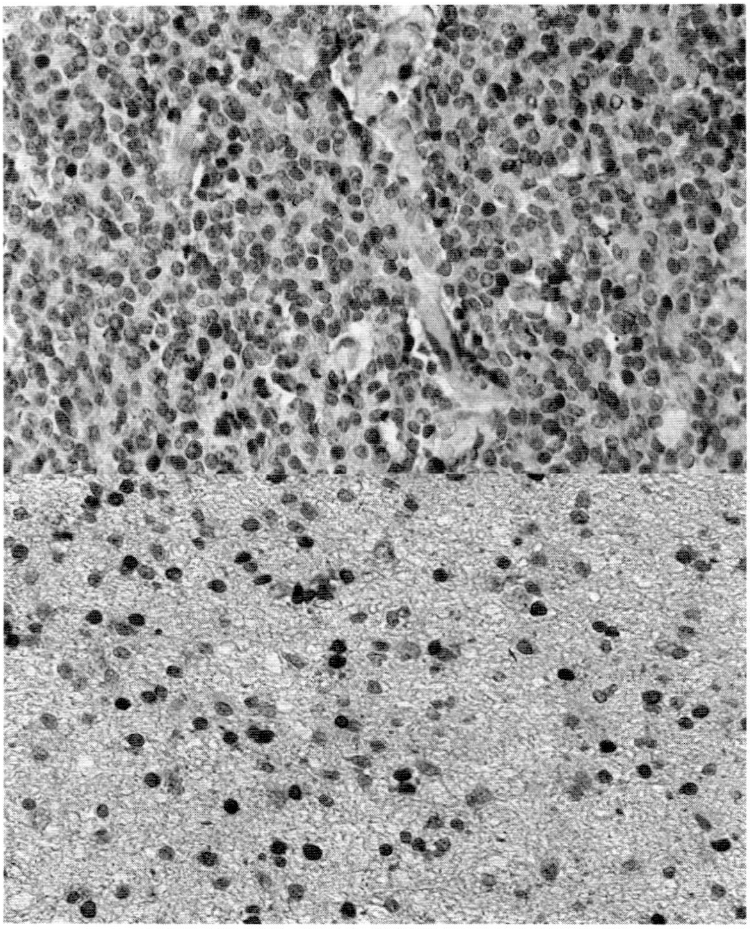

B

Figure 6. Cyclin D1. A. High LI in anaplastic oligodendroglioima; B. Positive normal nuclei and negative tumor nuclei in the with matter, DAB, x 400

A

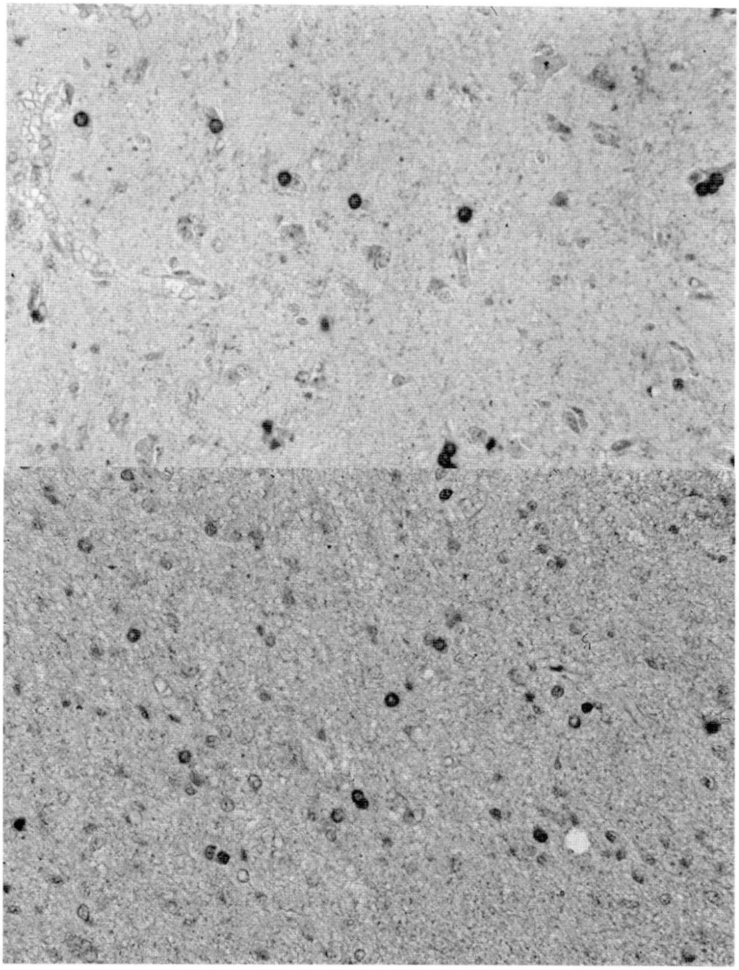

B

Figure 7. Cyclin D1. A. Positive normal oligodendrocytes in the cortex and (B) in the white matter. DAB, x 400

A B

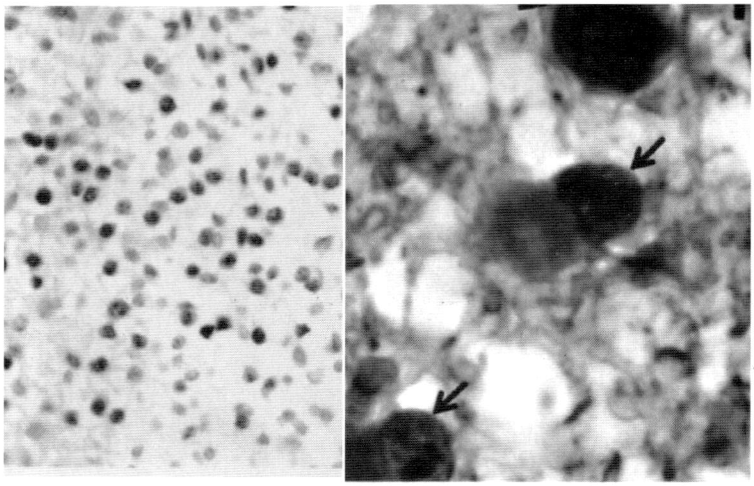

Figure 8. A. Moderate activity of OLIG1 in many nuclei, x 730; B. Minigemistocytes positive for GFAP and negative for OLIG 1, x 2400. From Azzarelli et al., 2003, reproduced by permission from the *Journal of Neuropathology and Experimental Neurology*

and anaplastic astrocytomas, but only in 9% glioblastomas. It could distinguish glioblastomas with an oligodendroglial component (Aguirre-Cruz et al., 2004).

In the single case, when the diagnostics has to be made between an infiltrating diffuse astrocytoma and a diffuse oligodendroglioma in a small fragment, the doubt that isolated OLIG2-positive cells represent normal oligodendrocytes of the white matter cannot be ruled out.

Another protein positive in oligodendroglioma is MAP-2. During development it characterizes precursor cells in sub-ventricular germinal zones, whereas in the adult it is limited to cortical neurons. By immunohistochemistry and in-situ hybridization it can be put in evidence in cytoplasms of tumor oligodendrocytes, in the oligodendroglial component of oligoastrocytomas, including the anaplastic variant, but also in astroctyomas (Blümke et al., 2001). Also Class III β-tubulin is expressed in oligodendrogliomas, especially in the anaplastic ones with no indication of neuronal differentiation, whereas it is negative in normal oligodendrocytes (Katsetos et al., 2002).

Positivity for neuronal markers has been repeatedly described in oligodendrogliomas (Ng et al., 1994; Patt et al., 1996; Wharton et al., 1998; Perry et al., 2002) indicating the existence of a common oligodendroglial/neuronal precursor line. Recently, Mytl1 usually related to neuronal development and expressed in neurons of the adult brain, was found in oligodendrogliomas with 1p loss that would thus have both glial and neuronal differentiation (Mukasa et al., 2004). Some oligodendrocytes would share the same progenitor cells with neurons (Lu et al., 2002; Zhou et al., 2002).

The identification of oligodendroglioma is of paramount therapeutic importance and the possibility of detecting its molecular characteristics in paraffin sections seems today feasible (Figure 9). Multiplex PCR amplification of microsatellite loci followed by high-resolution PCR product sizing by capillary electrophoresis have been used to determine 1p and 19q LOH in formalin-fixed and paraffin-embedded histologic slides. LOH can be detected even without normal tissue for comparison, provided that tumor cells represent 80–90% in the slides (Hatanpaa et al., 2003). This method represents a real improvement in the diagnosis of oligodendroglioma. Also FISH for 1p and 19q has been applied (Smith et al., 1999; Gelpi et al., 2003) as well as real-time quantitative PCR (Ginzinger et al., 2000; Nigro et al., 2001) which shows a good concordance with FISH. Micro arrays techniques are for the future and should be reserved to more sophisticated laboratories (Jeuken et al., 2004). A detailed protocol for the recognition of loss of heterozygosity has recently been put forward (Hartmann et al., 2005).

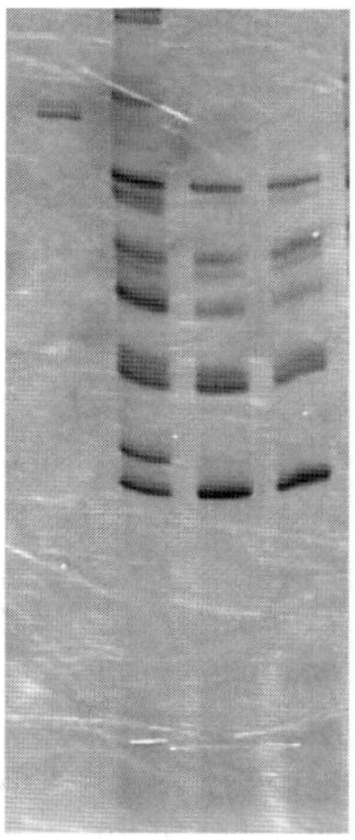

a b c

Figure 9. Microsatellite marker DIS1612. Loss of heterozygosity, Lane a = normal blood lymphocytes; lane b, c = oligodendroglioma

The last problem is the differential diagnosis of small cell glioblastoma versus glioblastomas of oligodendroglial origin or with an oligodendroglial component (GBMO) or versus anaplastic oligodendrogliomas. Glioblastomas with an oligodendroglial component show prolonged survival in comparison with conventional GBMs (Kraus et al., 2000; He et al., 2001). GBMO showed all the genetic characteristics of GBM, such as EGFR amplification, p16 deletion, PTEN mutations etc., but with a higher rate of 1p-19q loss, representing thus a subgroup of tumors of oligodendroglial origin, distinct from classic GBM (He et al., 2001). GBMO, moreover, show a higher frequency of CDKN2A homozygous deletion in comparison with other GBM, consistently with the finding that this genetic alteration is the main change in anaplastic oligodendroglioma (Ghimenti et al., 2003).

2. MIXED OLIGOASTROCYTOMA (MOA)

This is a mixed tumor with two cell components, astrocytic and oligodendrocytic (Figure 11), which can be found in distinct areas with a "biphasic" aspect or the two cell types intermingled (Figure 10A, B). The first aspect is rather rare. Fundamental for the distinction is GFAP, but it is important to consider that tumor GFAP-positive oligodendrocytes exist, like minigemistocytes and GFOC (Figures 13, 14). Minigemistocytes have been regarded as transitional forms between oligodendrocytes and astrocytes, corresponding to a bipotential precursor (Raff, 1989), or as GFAP expressing oligodendrocytes (Herpers and Budka, 1984), remnants of myelin forming glia of the developmental period (Choi e Kim, 1984).

Interestingly, typical large gemistocytes or transitional forms can also be found in oligodendrogliomas, even though criteria have been suggested for the distinction between the two gemistocytic types, based on the number of nuclei (Bigner et al., 1999), even though this is not of easy application (Figure 12). Another source of confusion is that GFAP-negative astrocytes may exist (Schiffer, 1997). The diagnosis of oligoastrocytoma is difficult, because beside the recognition of the two cell components, also their neoplastic nature must be demonstrated: normal white matter oligodendrocytes and small reactive astrocytes may be the source of error. Since no immunohistochemical marker is available to reliably distinguish the two cell types, alternative means are required.

The first approach to the problem is to exclude reactive astrocytes from the astrocytic component. These cells are frequently found at the infiltrating periphery of oligodendroglioma, distributed at regular intervals and showing large GFAP-positive cytoplasms and thick processes. Sometimes they are more frequent, smaller, and, as already mentioned, may not be in the same stage of reaction, so that their recognition is not at all easy. However, once ascertained that, besides reactive astrocytes, neoplastic astrocytes exist in the tumor, it remains to be established what is the lowest percentage of tumor astrocytes or alternatively of tumor oligodendrocytes required to make the diagnosis of oligoastrocytoma. In some findings it is 10% of oligodendrocytes (Coons et al., 1999), whereas in others (Krouwer et al., 1997) it is still 10%, but of astrocytes. These criteria leave unsolved the question of the recognition of normal from tumor oligodendrocytes, especially in infiltrating or diffuse growth where cell density is low. To what has already been said, it must be added that the "honeycomb" aspect, if present, represents good evidence in favor of the oligodendroglial nature of the cells. The influence of the extension of the surgical sample results from the observation that the increase of its dimensions parallels that of the frequency of the diagnosis of oligoastrocytoma.

A

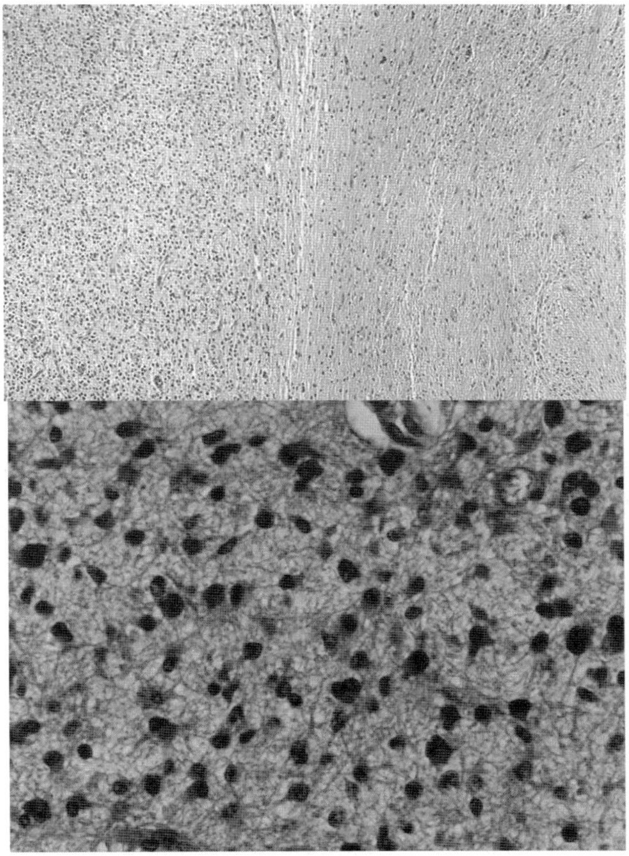

B

Figure 10. Oligoastrocytoma. A. The two components are separate. H&E, x 200; B. The two cell types are intermingled, H&E, x 400

A

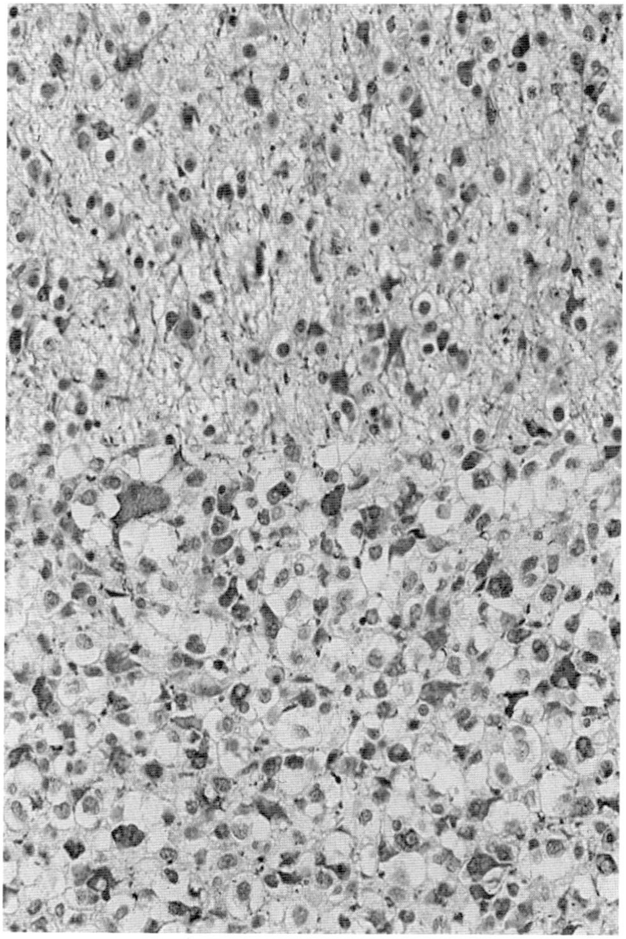

B

Figure 11. A. Admixture of astrocytes and oligodendrocytes n the cortex, H&E, x 400; B. Admixed tumor astrocytes and oligodendrocytes, GFAP-DAB, x 400

Nuclear morphology can be the object of personal bias. In summary, the recognition of MOA or of ambiguous gliomas in the single case is really very difficult and this is documented by the poor diagnostic concordance among five outstanding neuropathologists (Fuller et al., 2003).

The missed distinction between oligodendroglioma and oligoastrocytoma has no serious consequence at the prognostic level, because biologically the two tumors with the same grade do not differ greatly and their survivals overlap, but some doubts remain as for therapy.

Attempts are increasingly made to reach a clear distinction between the two cell components on the basis of molecular genetics or of known molecular markers. A subset of MOA harbors 1p and 19q deletions, like oligodendrogliomas, and another subset contain altered p16/RB/CDK4, typical of malignant astrocytic morphology (Ueki et al., 2002). The latter is characterized also by other molecular alterations, such as EGFR amplification, PTEN or DMBT1 inactivation etc. Using FISH in situ hybridization, 90 tumors with mixed appearance were studied by a series of oligodendroglial and astrocytic molecular markers; and individual genetic patterns could not be associated with specific morphologies, but with survivals. Tumors with 1p and 19q deletions were associated with prolonged survivals and those with EGFR amplification or PTEN/DMBT1 losses or p16 deletions with shorter survivals (Fuller et al., 2003).

3. THE ANAPLASTIC VARIANT

The diagnosis of the anaplastic variant of oligodendroglioma represents the crucial and most controversial point of this chapter for its prognostic and therapeutic consequences. The discussion about the many systems for classifying oligodendrogliomas has been repeatedly reported (Schiffer, 1997; Kleihues and Cavenee, 2000). In the major series, a number of clinical, biological, therapeutical and histological parameters have been compared with survival in multivariate analysis. Prognostic factors emerged and they were more or less always the same, also considering that the extension of surgical removal has been regarded more recently as one of them (Puduvalli et al., 2003). Systems with three (Smith et al., 1983) or four malignancy grades have been used, but the correlation of the grades with survival has been poor (Shaw et al., 1992); systems with only two grades appeared to be more effective (Kros et al., 1988; Shaw et al., 1992; Daumas-Duport et al., 1997; Fortin et al., 2001). The criteria for identifying the anaplastic variant vary from author to author. The indications of the WHO book (Reifenberger et al., 2000) are rather vague when they have to be applied to individual cases; and in general the criteria suggested must be discussed separately in their prognostic significance. For example, endothelial hyperplasia and contrast enhancement at MRI have been regarded as indicating anaplasia (Daumas-Duport et al., 1997); on the contrary in our experience endothelial hyperplasia (Figures 2, 15A), occurring in >70% of tumors, necroses and even capillary density, quantitatively measured, do not come out as prognostic factors after multivariate analysis.

A

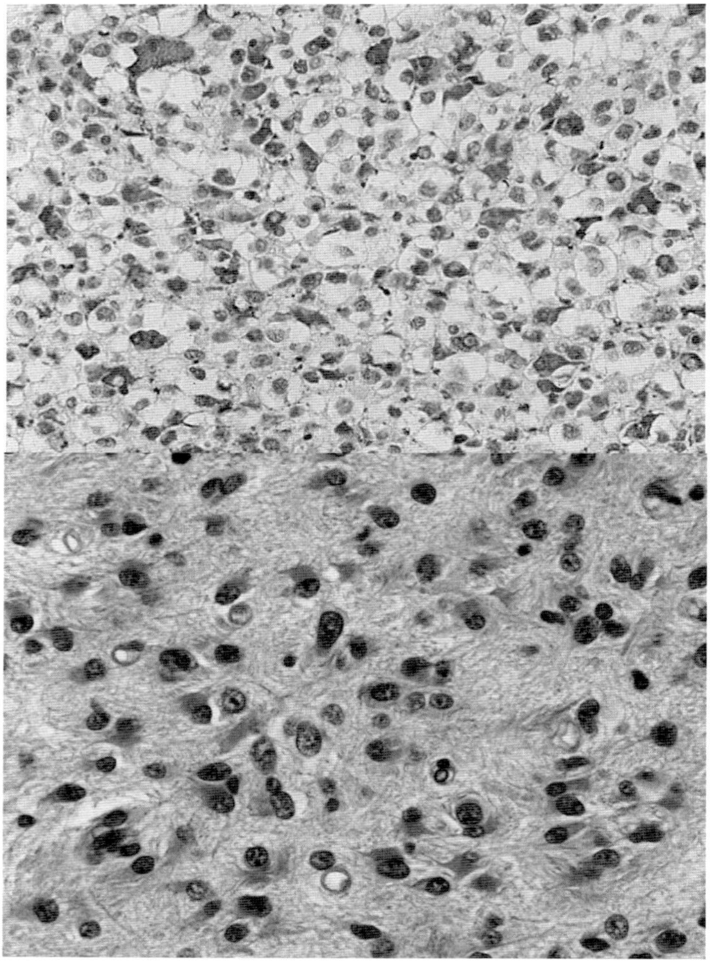

B

Figure 12. Oligoastrocytoma. A. Large GFAP-positive tumor astrocytes. DAB, x 400; B. Astrocyte-like cells with nuclei of oligodendroglial type: minigemistocytes or astrocytes? H&E, x 400

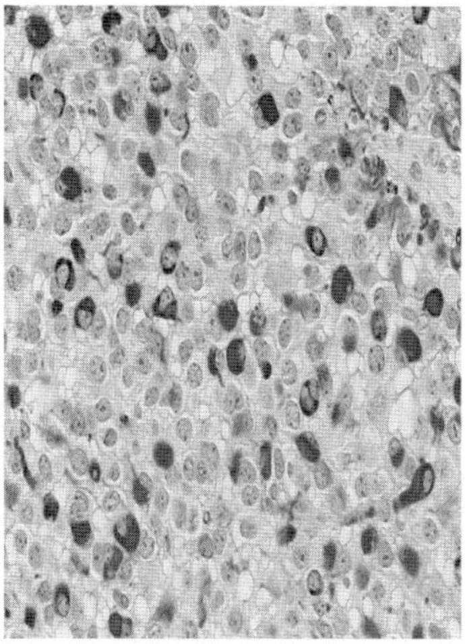

Figure 13. Classic minigemistocytes, GFAP- DAB, x 400

The only prognostic factors are MIB-1 LI (Figure 16A), the number of mitoses (Figure 15B) and, maybe, microvascular proliferations (Figure 17A) (Schiffer et al., 1997–99). These observations have been confirmed (Giannini et al., 2001). The most important parameters, which can also be utilized in the single case, are MI and MIB1 LI and maybe AI (Schiffer et al., 1997-2000; Coons et al., 1997; Miettinen et al., 2001) (Figure 16B). In the single case, the criteria used for recognizing malignancy in astrocytic tumors cannot be employed, because, even though independent prognostic factors after multivariate analysis, nuclear pleomorphism (Figure 18B), mitoses, necroses (Figure 17B), vascular proliferations (Figure 17A) can be found also in cases with the same long survival of the classic variant .

In the anaplastic variant, nuclei can still be recognizable as oligodendroglial or they lose such characteristics (Figure 19A, B).

The assessment of TTP and OS (overall survival) for oligodendrogliomas grades II and III in relation to different parameters, including radiation- and chemotherapy, has not been an easy task in retrospective studies, because more or less all the tumors have been treated and according to different modalities. To what has been said above, it can be added that TTP and OS of grade II tumors are 4.8-5 and 14.2 (Felsberg et al., 2004), that 5 years survival is 85% (Felsberg et al., 2004) or 74% (Shaw et al., 2992). OS reported in a recent series (Felsberg et al., 2004) was longer both for grade II and grade III tumors (14.2 and 7.6 years) than that reported in older series (3.5–10 years and 4 years) (Shaw et al., 1992; Dehgani et al., 1998). Other prognostic factors have been age and frontal location (Schiffer et al., 1997).

LOH of 1p and 19q was found to be associated with longer OS and TTP in grade III tumors (Cairncross et al., 1998) and the prolongation was greater than in grade II tumors. The association of 1p and 19q LOH with longer OS and TTP in oligodendrogliomas and oligoastrocytomas came out from many other retrospective studies (Ino et al., 2001; Smith et al., 2000; Van den Bent et al., 2003; Hashimoto et al., 2003).

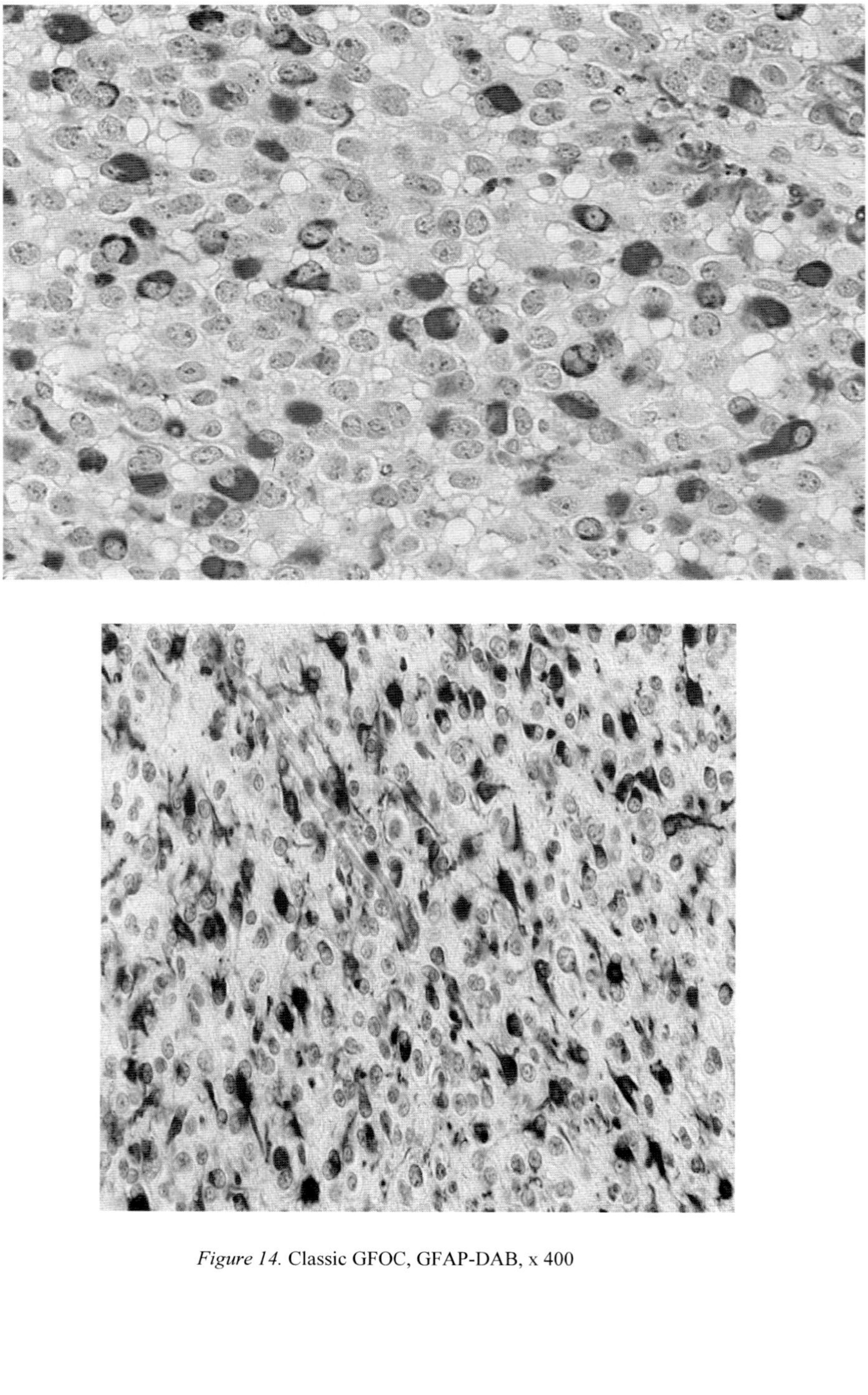

Figure 14. Classic GFOC, GFAP-DAB, x 400

A

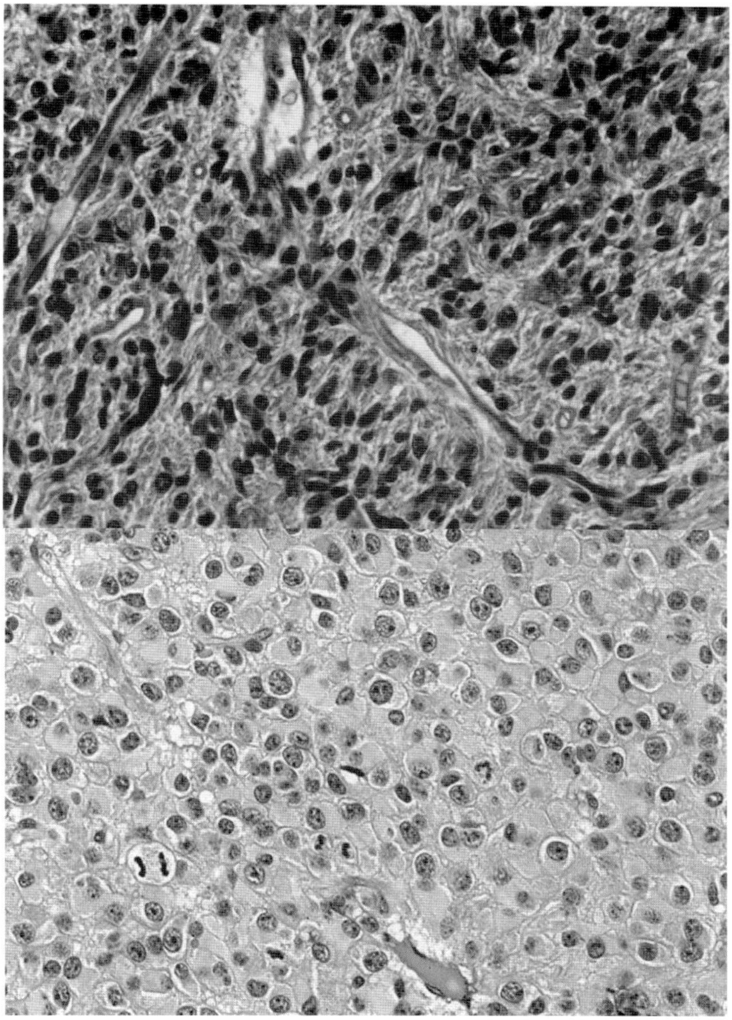

B

Figure 15. Anaplastic oligodendroglioma. A. Endothelial hyperplasia. H&E, x 400; B. High number of mitoses; x 400

A

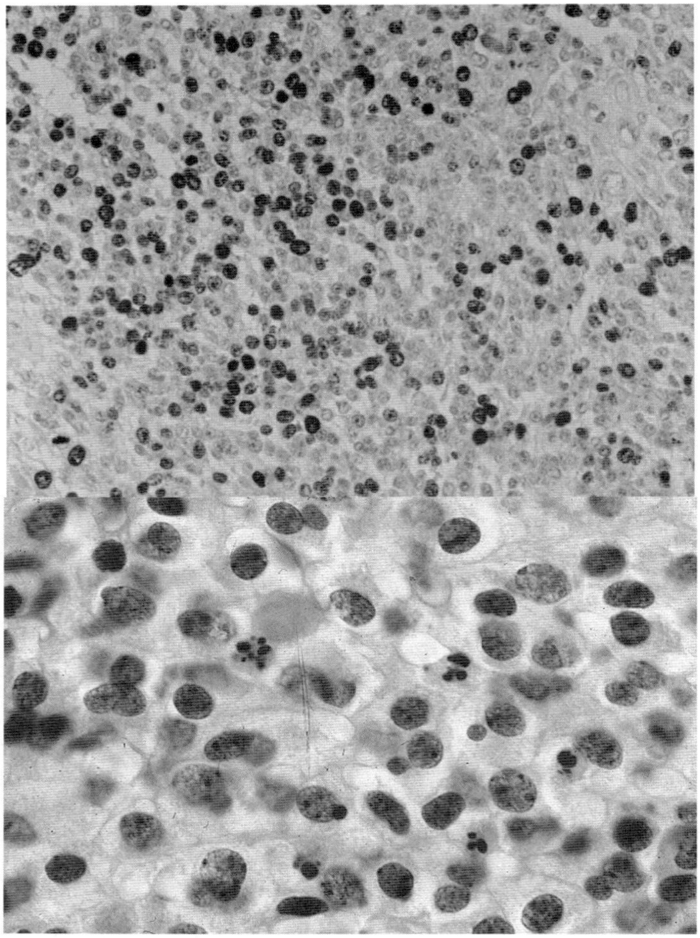

B

Figure 16. A. High MIB.1 LI. DAB, and B. Apoptosis, H&E, x 400

A

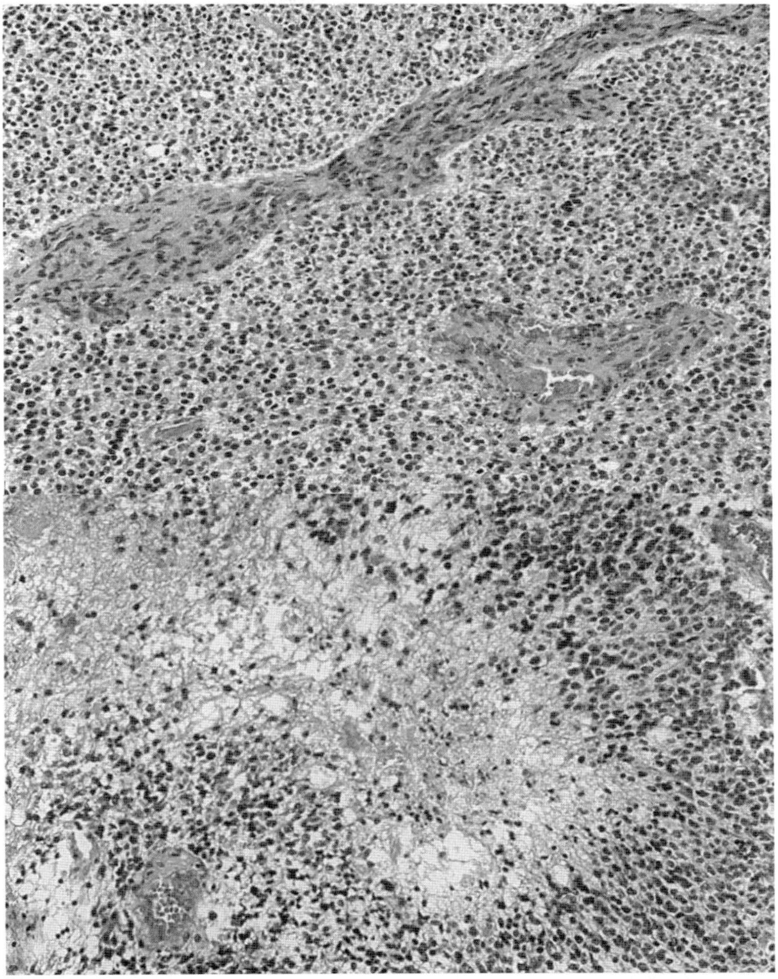

B

Figure 17. Anaplastic oligodendroglioma. A. Microvascular proliferations; B. Circumscribed necroses. H&E, x 400

A

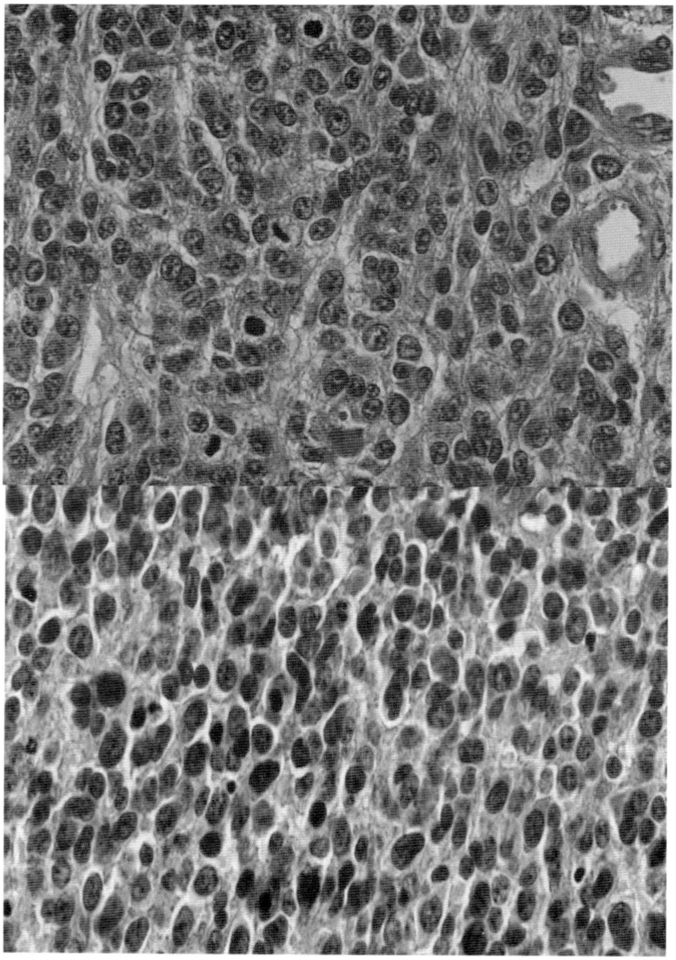

B

Figure 18. Anaplastic oligodendroglioma. A. Mitoses and nuclei still recognizable as oligodendroglial; B. Mitoses and nuclei no more recognizable as oligodendroglial. H&E, x 400

A

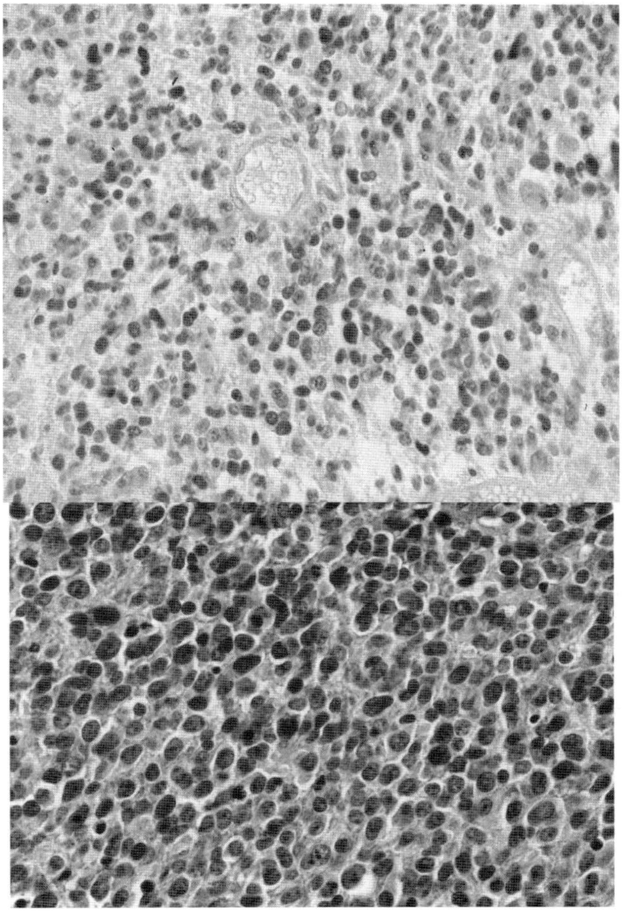

B

Figure 19. A. Intermediate aspect of nuclei, H&E, x 400; B. Highly pleomorphic nuclei, no more recognizable as oligodendroglial, H&E, x 400

In the single case, the recognition of the anaplastic variant has an enormous importance for prognosis, for the post-surgical therapy and for the expected survival or TTP, from the comparison of which with controls the evaluation of the efficacy of a given therapy will emerge. In the histological diagnosis of single cases, none of the parameters recognized by multivariate analysis as independent prognostic factors grant the diagnosis of anaplastic variant. Occurring, even though less frequently, also in the classic variant, they only increase the probability of such diagnosis. MI and MIB-1 LI show an overlapping between classic and anaplastic variant (Figure 20) and they themselves can be used in the single case only after establishing a cut-off point: they indicate the anaplastic variant only when higher than the highest value in the classic variant (Schiffer et al., 1998). The cut-off point that better statistically divides the two variants does not necessarily coincide with that of 100% specificity for malignancy. In our experience, for example, a cut-off point of 8% of MIB-1 indicates that the survival of tumors with a mean MIB-1 LI > 8% is significantly lower than that of tumors with a mean MIB-LI <8%. However, the value of MB-1 LI that indicates malignancy in the single case is >20%, because up to this value tumors can show survival of the classic variant. The same can be said for AI.

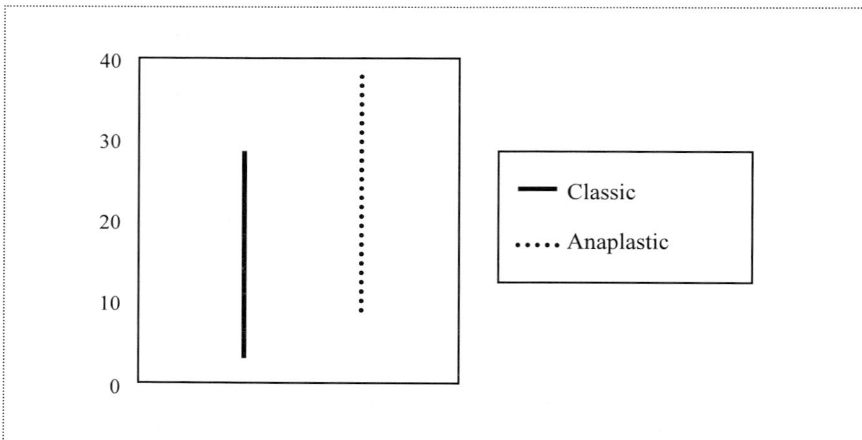

Figure 20. Oligodendroglioma. Ranges of MIB.1 LI in classical and anaplastic variant

One important question is whether analysis for 1p and 19q losses is useful in the single case for prognosis. In other words, are there oligodendrogliomas with no 1p and 19q LOH and long survival? Or what are they? They can be diffuse astrocytomas misdiagnosed as oligodendrogliomas, or oligodendrogliomas with more focal genetic changes or they can yet be oligodendrogliomas (Burger, 2002; Sasaki et al., 2002; Burger et al., 2003). As a matter of fact, oligodendrogliomas with no LOH of 1p and 19q would exist, but they show shorter survivals (Smith et al., 2000).

Other parameters have been studied for the identification of the anaplastic variant. For example, the expression of HIF-1α does not correlate with the histological grade, but with short survivals, because it would induce angiogenesis transactivating VEGF (Birner et al., 2001). This, in turn, would correlate with microangiogenesis and short survivals, reducing apoptotic death induced by hypoxia (Varlet et al., 2000). The correlation of both factors with neo-vascularization, as represented by capillary density, would not coincide with our demonstration that capillary density does not correlate with survival (Schiffer et al., 1997).

Differentiated oligodendroglial areas are a common finding in glioblastomas (Bigner et al., 1999) up to 15%–17% of cases (He et al., 2001). Whether they represent tumors showing both typical glioblastoma and oligodendroglioma features or the malignant evolution of oligodendroglial lineage is still a matter of discussion (Coons et al., 1997; Decaestecker et al., 1998). A distinction between glioblastoma and anaplastic oligodendroglioma has been proposed on the basis of a higher expression and positivity of Galactin-3 in the former than in the latter (Neder et al., 2004).

4. GENETIC ALTERATIONS IN OLIGODENDROGLIOMA AND THEIR DIAGNOSTIC USE

As has been mentioned before, oligodendroglioma grade II shows LOH of 1p and 19q in 80%–90% of cases, but tumor suppressor regions have not been identified (Jeuken et al., 2004). On the contrary, LOH on 17p and TP53 mutations are rare and practically mutually exclusive with 1p and 19q LOH (Watanabe et al., 2002). However, p53 pathway may be deregulated, because of inactivation of $p14^{ARF}$ by hypermethylation (Wolter et al., 2001). The question of 1p and 19q LOH has today become of paramount importance as for the recognition of the tumor, its prognosis and good response to chemotherapy. The initial observation was that allelic losses of 1p and 19q were associated with longer survival and a favorable response to chemotherapy with procarbazine, lomustine and vincristine (PCV) (Cairncross et al., 1998). A series of studies on retrospective material confirmed these findings (Smith et al., 2000; Hoang-Xuan et al., 2001; Ino et al., 2001; Hashimoto et al., 2003). Candidate regions on chromosome 19 have been mapped between markers D19S219 and D19S246 at 19q13.3 (Hartmann et al., 2002), whereas on chromosome 1 several candidate regions have been described (Husemann et al., 1999; Felsberg et al., 2004) and recently 3 of them have been defined after analyzing a great number of cases (Felsberg et al., 2004).

Malignant progression is generally characterized by accumulation of multiple genetic abnormalities. Molecular genetics of oligodendrogliomas shows today that the expression of a number of proteins, codified by oncogenes involved in neuro-oncogenesis, changes in the development of anaplasia. For example, it has been demonstrated that Bcl-2 is associated with progression in oligodendrogliomas and it is positive in anaplastic oligodendrogliomas (Deininger et al., 1999), that $p18^{INK4C}$, but not $p14^{ARF}$, correlates with histological malignancy and with survival (Korshunov and Golanov, 2001), and that protein mcm2 of the replicative complex MCM (Williams and Stoeber, 1999) correlates with Ki.67 LI (Wharton et al., 2001). It can be said that molecular genetics contributed to the identification of the anaplastic variant, because some alterations have been found associated with anaplasia.

These alterations concern genes/proteins involved in tumor progression and could not only have a prognostic significance, but can also indicate the anaplastic variant in the single case. Homozygous deletions of CDKN2A/p16 could satisfy these requirements (Figure 21), but it should be established whether they can be found, even though with a lower frequency, also in the classical variant (Figure 22). In a series, LOH of CDKN2A has been found, for example, in 45% of cases of anaplastic oligodendroglioma and never in the classic variant (Bigner et al., 1999), confirming previous data (Cairncross et al., 1988). Also in other series, CDKN2A inactivation, both by homozygous deletions or hypermethylation, has been found in 30% of anaplastic oligodendrogliomas and never in the classic variant (Watanabe et al., 2001), or only in grade III tumors where hypermethylation of CDKN2A was associated with that of p14 and CDKN2B (Wolter et al., 2001).

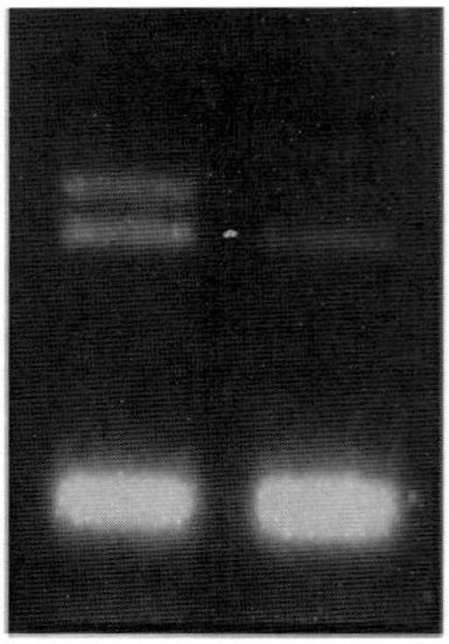

Figure 21. CDKN2A homozygous deletion in oligodendroglioma

By immunohistochemistry, negative staining for p16, on the contrary, was found also in classical oligodendrogliomas, even though less frequently than in the anaplastic variant: one third of tumors were negative for p16, whereas half the anaplastic cases expressed p16. Very important was the observation that in 6 patients with negative p16, survival has not been short (Miettinen et al., 1999); in still another series, the few II grade cases with negative p16 showed also LOH of CDKN2A (Bortolotto et al., 2000).

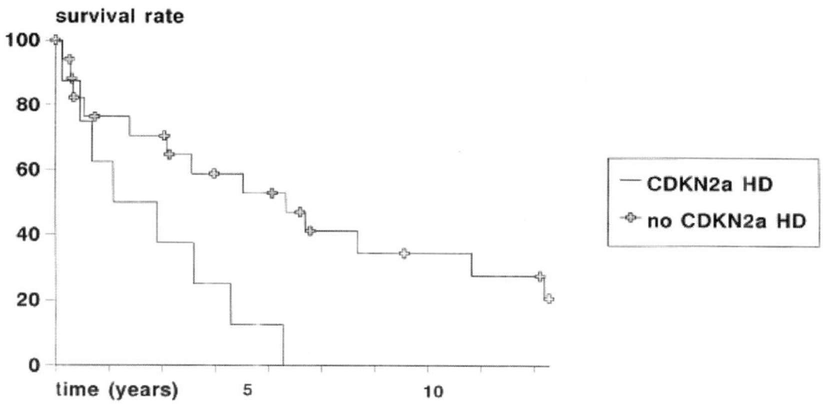

Figure 22. Survival of patients with oligodendrogliomas: with and without CDKN2A HD (Bortolotto et al., 2000)

As long as these alterations are found, even with low frequency, also in II grade tumors diagnosed according to classical criteria, a defect in the histological diagnosis can be invoked. If, on the contrary, they are found, even though with a very low frequency, also in cases with long survival, the genetic alteration cannot be used in the single case for prognostic purposes. Studied by microarray techniques, p53 and pRb did not appear as predictors of survival, whereas p21 was found in correlation with a greater proliferation capacity and a short survival; but its expression, again, is not exclusive of cases with short duration and, therefore, it is not useful in the single case (Miettinen et al., 2001). RB1/CDK4/p16^{INK4a}/p15^{INK4b} and TP53/p14^{ARF}/MDM2 were found alternatively altered in 20 anaplastic oligodendrogliomas (Watanabe et al., 2001).

PTEN mutations are found in <10% anaplastic oligodendrogliomas (Sasaki et al., 2001), as amplification of EGFR, PDGFRA, CDK4, even though EGFR proteins are over-expressed in >50% of tumors (Reifenberger and Louis, 2003).

At pediatric ages, in children from 0 to 9 years of age, none had any deletion of 1p-19q, whereas in tumors of children >9 years of age these alterations were present, but to a lesser extent than in adults (Raghavan et al., 2003).

The identification of the anaplastic variant in the single case is somewhat subjective and this affects the decision about therapy and, more importantly, may impair the evaluation of the efficacy of therapies. In the last few years, because of the long durations of oligodendrogliomas, in the practice TTP and recurrence at MRI with the criteria of McDonald et al. (1990) have been adopted. There are controversies about the reliability of the volume measurements and the meaning of contrast enhancement, but these do not change the problem which, on the one hand, has been clarified and, on the other hand, complicated by the observations of the association of 1p and 19q losses. The whole thing started with the observation that anaplastic oligodendrogliomas responded to PCV therapy (Levin et al., 1980; Cairncross and MacDonald, 1988; Cairncross et al., 1994; Peterson et al., 1996). Then it was found that benign oligodendrogliomas also responded to the same therapy

(MacDonald et al., 1990; Kyritsis et al., 1993; Mason et al., 1996), even if performed before radiotherapy (Kirby et al., 1996). Summarizing, 40–90% of oligodendrogliomas are characterized by 1p and 19q LOH (Reifenberger et al., 1994; Kraus et al., 1995; Maintz et al., 1997; Bigner et al., 1999; Smith et al., 1999; Von Deimling et al., 2000), whereas one third responded to PCV therapy; (Mason et al., 1996; Streffer et al., 2000). Practically, LOH of 1p and 19q is associated with sensitivity to PCV therapy, and longer survival in anaplastic tumors (Cairncross et al., 1998), but also in the classical ones (Smith et al., 2000). From all these contributions it is difficult to understand whether 1p and 19q LOH simply characterizes their oligodendroglial nature, or denounces sensitivity to chemotherapy or indicates benignity of the tumors. In all the experiments carried out, there never was a branch with untreated oligodendrogliomas.

It is very important to establish what has to be done at the time of diagnosis. After examining a series of genetic alterations, oligodendrogliomas have been divided into four categories. In the first one, characterized by 1p-19q LOH, PCV therapy is mandatory even without radiotherapy. In the last one, characterized by 10q loss, PTEN mutations, CDKN2A deletion, EGFR amplification, but without 1p loss and with an ominous prognosis, PCV therapy is not recommended (Ino et al., 2001). From the practical point of view it would be necessary to carry out diagnostic testing for all the genes/proteins mentioned, but it would be laborious and not as useful as for 1p-19q carried out either by LOH or by FISH (Reifenberger and Louis, 2003).

On the recognition of the anaplastic variant the subjective interpretation of the histological criteria plays a role, as already mentioned. As a matter of fact the disagreement of neuropathologists on this variant concerns 63% of the cases (Ueki et al., 2002). However, new observations are contributing to the problem. If 1p and 19q LOH characterize the classical variant, they are associated with 10q LOH in the anaplastic variant.

In the experience of the GMH, the oligodendroglial aspect, recognized by strict criteria, is associated with 1p and 19q losses; if enlarged criteria are used, LOH is found only in 57% of cases (Sasaki et al., 2002). In another series selected with strict oligodendroglial aspects, i.e. with "honeycomb" appearance in >50% of the cells and "chicken wire" pattern of vessel distribution, 1p and 19q LOH were found in 93% of the cases and TP53 mutations were present in 7% of the cases only (Watanabe et al., 2002). With enlarged criteria the percentage of oligodendrogliomas obviously increases (Coons et al., 1997) and TP53 mutations appear. As a matter of fact, oligodendrogliomas with no LOH of 1p and 19q would exist, but they show shorter survivals (Smith et al., 2000). What is surprising is that the LOH mentioned before would indicate a better survival also in tumors without any oligodendroglial characteristic (Ino et al., 2000). It is important that if LOH gets lost, chemosensitivity and better survivals disappear, as confirmed in 6 cases by CGH, FISH and microsatellites (Burger et al., 2001). More information can be added: in a series of 53 cases the efficacy of PCV therapy was apparent from the comparison of oligodendroglioma with fibrillary astrocytomas (Fortin et al., 2001); and in another finding PCV treated oligodendrogliomas and oligoastrocytomas showed the same TTP as nontreated tumors (Olson et al., 2001). Finally, some observations seem to demonstrate that 1p and 19q losses do not denounce chemosensitivity, but merely their oligodendroglial nature. The oligodendroglial precursors would have a down-regulated MGMT in comparison with the astrocytic ones, thus showing sensitivity to nitrosourea (Nutt et al., 2000). It has been demonstrated that MGMT promoter hypermethylation was more frequent in tumors with loss of heterozygosity on 1p and 19q and it contributes together with a low MGMT expression to chemosensitivity (Mollemann et al., 2005).

Independently of the recognition of the anaplastic variant or of prognostic factors that can ameliorate prevision of outcome and therapies, oligodendrogliomas are an example of how molecular genetics can contribute to the identification of subsets in a tumor type with different prognosis. The study of gene expression profiles demonstrated two distinct molecular groups corresponding to the two degrees of malignancy in a series (Watson et al., 2001); and in another series, that subgroups with and without 1p LOH show different expression profiles (Mukasa et al., 2002). Interestingly, most of the genes with distinct expression in oligodendrogliomas with 1p loss were expressed in normal nervous tissue in relation to neuronal functions. The most important one is MYT1L which is expressed in normal neurons and in tumor cells, confirming the presence of neuronal characteristics in oligodendrogliomas (Mukasa et al., 2004). There are hopes that MYT1L can become not only a tumor marker, but also a prognostic factor of the tumor type.

Chapter 7

EPENDYMAL TUMORS

1. EPENDYMOMAS

Ependymomas originate from the ventricular walls and affect children or young adults. Four main locations are recognized: supra- and infra-tentorial, spinal and conus-cauda-filum regions. At neuro-imaging the tumor appears as a circumscribed mass with irregular contours and assuming contrast enhancement. Histologically, ependymomas show five variants, beside the anaplastic one: cellular, epithelial, papillary, tanycytic and "clear cell". The characteristic features are rosettes, pseudo-rosettes, canaliculi (Figure 1). The major prognostic factors are age, extension of resection and location in order of survival: conus-cauda-filum, spinal, supra- and infra-tentorial (Schiffer et al., 1997).

Ependymoma is a grade II tumor and its anaplastic variant a grade III tumor. With the exception of completely resected tumors of the spinal cord and of conus-cauda-filum region, ependymomas require radiotherapy after surgery. Good recent reviews are available (Teo et al., 2003).

In biopsies, the differential diagnosis of ependymoma must be carried out with some other neoplasias of the CNS and the availability of specific markers has long been awaited. EMA, a glycosylated transmembrane protein marker of epithelial differentiation and expressed by ependymal cells (Uematsu et al., 1989), shows in-ependymomas a punctate intracytoplasmatic and luminal immunopositivity (Figarella-Branger et al., 1991). The number of tumors showing these "EMA-dots" seems to be higher than previously reported, but with no correlation with the tumor grade. They can be used for differential diagnosis (Hasselblatt and Paulus, 2003). The occurrence of step-ladder rhythms (Figure 2) requires a differential diagnosis toward pilocytic astrocytoma. The existence of a polar spongioblastoma is no longer recognized in the WHO classification (Kleihues and Cavenee, 2000).

Little is known about molecular genetics alterations in ependymomas. Structural and numerical abnormalities occur in many chromosomes (Hamilton and Pollack, 1997), with frequent losses or gains by CGH (Scheil et al., 2001). A wide range of genomic alterations have been found by this method; the most frequent were loss of 22, 17 and gain on 4q, but no correlation with histological grade or ploidy was found (Gilhuis et al., 2003). By microsatellite markers, deletions on 22q have been shown (Huang et al., 2002); and recently deletions on 6 and 9 have been described as the

A

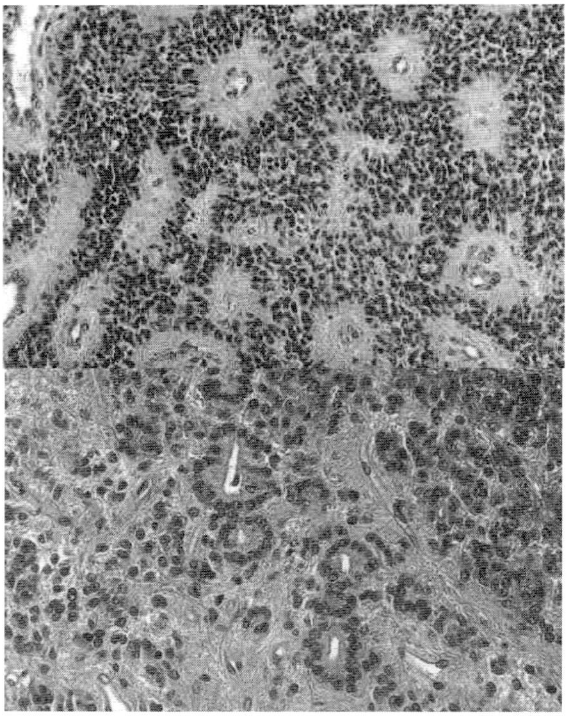

B

Figure 1. Ependymoma. A. Typical perivascular pseudo-rosettes; B. Typical canaliculi. H&E, x 400

most frequent alterations, but with no difference between the tumor grades (Huang et al., 2003). Interesting findings, even though not conclusive, have been obtained for CDKN2A, CDKN2B and $p14^{ARF}$. No point mutation has been found for them (Sato et al., 1996) and homozygous deletion of CDKN2A has been found only once in a series of tumors (Bortolotto et al., 2001). On the contrary, promoter hypermethylation has been described in 21% of tumors for CDKN2A, 32% for CDKN2B and 21% for $p14^{ARF}$, more frequent in low-grade tumors for CDKN2A and the reverse for CDKN2B and $p14^{ARF}$ (Rousseau et al., 2003).

Also α-synuclein has been found to be positive in nuclei and perinuclear cytoplasmic rings of ependymomas, especially in anaplastic ones (Fung et al., 2003).

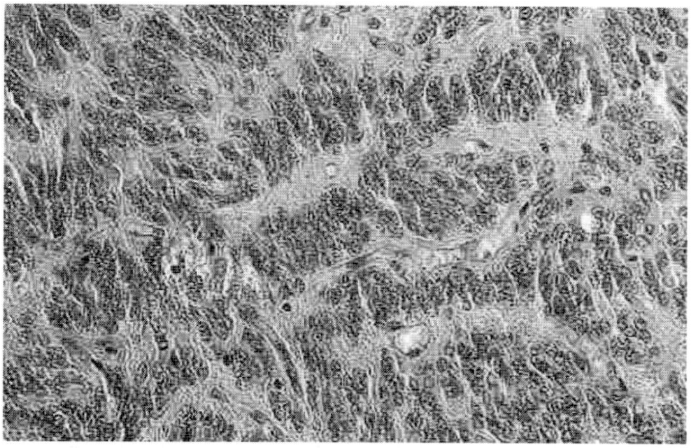

Figure 2. Ependymoma. Step-ladder rhythms. H&E, x 400

In this group of tumors the main problem is the recognition of the anaplastic variant which is important more for survival prediction than for therapeutic strategy. The usual treatment of ependymomas is surgery, followed by radiotherapy in case of partial removal. Cerebrospinal diffusion requires craniospinal irradiation. The role of chemotherapy is uncertain (Chamberlain, 2003).

2. RECOGNITION OF THE ANAPLASTIC VARIANT

The distinction between classic and anaplastic variant may be important for post-surgical treatment, especially in certain locations, and for prognosis. The application of the grading system to ependymoma, either with four or with three grades, has not been greatly successful, because the correlation between histology and survival turned out to be poor or barely useful (Mork and Loken, 1977; Chin et al., 1982). In 6 out of 9 series evaluated, the percentage of anaplastic tumors, diagnosed on the basis of nuclear pleomorphism, mitoses, necroses, vascular proliferations and cell density, varied from 40% to 90% (West et al., 1985). If only the largest series are considered, it can be said that definite data did not emerge (Ilgren et al., 1984; Rawlings et al., 1988; Ross and Rubinstein, 1989). In a series of 298 cases, using the same criteria for recognizing anaplasia in gliomas, a correlation between histology and survival was not found. After multivariate analysis for survival, age (<4 years), number of mitoses (>20 x 10 HPF) and cell density were prognostic factors for supra-tentorial ependymomas, whereas for infra-tentorial ones there was only the subependymoma variant (Schiffer et al., 1991a). In supratentorial tumors the anaplastic variant was characterized by high cell density, high number of mitoses and aspects of incomplete perivascular pseudo-rosettes (Schiffer et al., 1991b) (Figure 3). The use of proliferation markers showed high LI values, but without a clear cut-off point (Figure 4B). The PCNA LI >7 (Schiffer et al., 1993) and

MIB-1 LI > 4 (Prayson, 1999) were more frequent among high grade tumors. Ki-67 LI has been confirmed to be predictive of shorter TTP (Figarella-Branger et al., 2000).

The defective recognition of the malignant variant in supratentorial tumors has less consequences for radiotherapy, because this is carried out in incompletely resected tumors, regardless of their location, and more consequences for prognosis, especially when the expected survival on histological basis is compared with the observed survival for the evaluation of therapeutic tools. This uncertainty plays a greater role in IV ventricle tumors the irradiation of which is debated, whether they are completely or incompletely removed. Spinal cord tumors usually do not require TCT.

3. RECOGNITION OF CLEAR CELL EPENDYMOMA

It is a typical infantile tumor, already known as ependymoma of the foramen of Monro (Zülch, 1956), The tumor is supra-tentorial and localized in the lobes. At MRI it appears iso- hypointense with frequent contrast enhancement. Histologically it is characterized by a cytoplasmic halo resembling that of oligodendroglioma (Figure 4A).

The presence of rosettes and pseudo-rosettes is fundamental for the differential diagnosis toward not only oligodendroglioma, but also central neurocytoma, hemangioblastoma and metastases from clear cell kidney carcinoma. The distinction from oligodendroglioma is based on the circumscribed aspect the tumor shows at MRI and on the careful discrimination of the relevant cytologic features. The distinction from hemangioblastoma, especially when containing GFAP-positive cells, is based on its negative staining for EMA and scarcity of the reticulin net. As a matter of fact, when the reticulin net is abundant these tumors have been supposed to mimick hemangioblastoma, besides common clinical and biological features (Kawano et al., 1999). These very tumors, however, have been considered as true hemangioblastomas by others (Burger, 1999). The distinction from central neurocytoma is based on the negative staining for synaptophysin, whereas perivascular pseudo-rosettes are important for the distinction from renal carcinoma characterized by lobules of cells with round nucleus and a cytoplasmic halo which resembles oligodendroglioma.

The number of anaplastic tumors of the "clear cell" type is high and recurrences are frequent. In a recent series of 10 cases, true rosettes and canaliculi were lacking in the pathological picture and in 7 cases anaplastic features, such as high cell density, mitoses and microvascular proliferations were present (Fouladi et al., 2003). GFAP is positive around vessels and MIB-1 LI is high in anaplastic cases. TTP and survival rates at 5 years were 34% and 75% respectively. It was shown that patients with complete remission at 20 months had undergone total tumor resection and radiotherapy. The patients who died had not received radiotherapy. More or less the same observations have been made in other series (Kawano et al., 1989; Min and Scheithauer, 1997).

The cytogenetic and genetic information is very poor. In anaplastic tumors loss of CEP18 and DAL-1 has been observed (Fouladi et al., 2003).

Radiotherapy as in ependymoma of children is recommended (Merchant et al., 2002) and this makes the recognition of this variant very important, because some of the tumors with which it can be confused do not require radiotherapy.

A

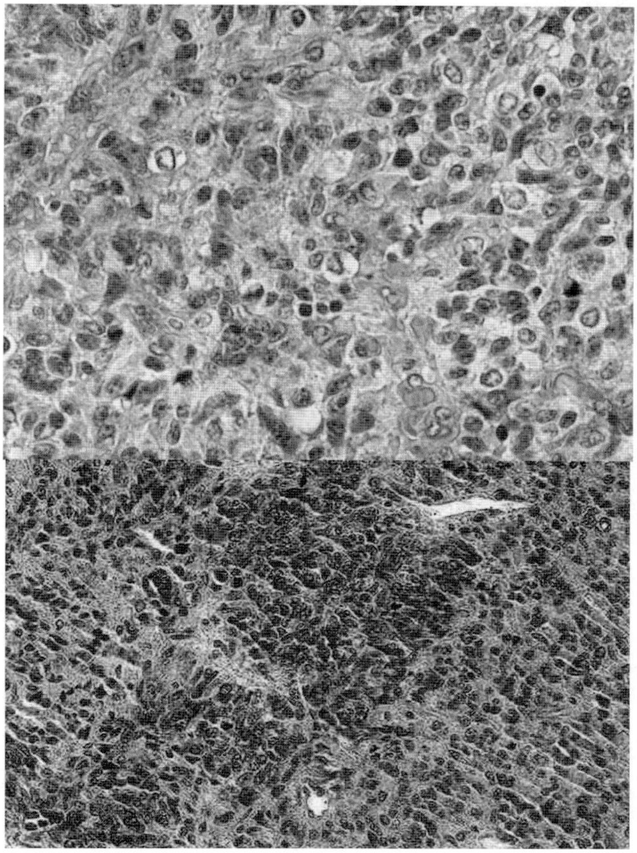

B

Figure 3. Anaplastic ependymoma. A. Incomplete pseudo-rosettes. H&E, x 400; B. Area with a high cell density. H&E, x 200

A

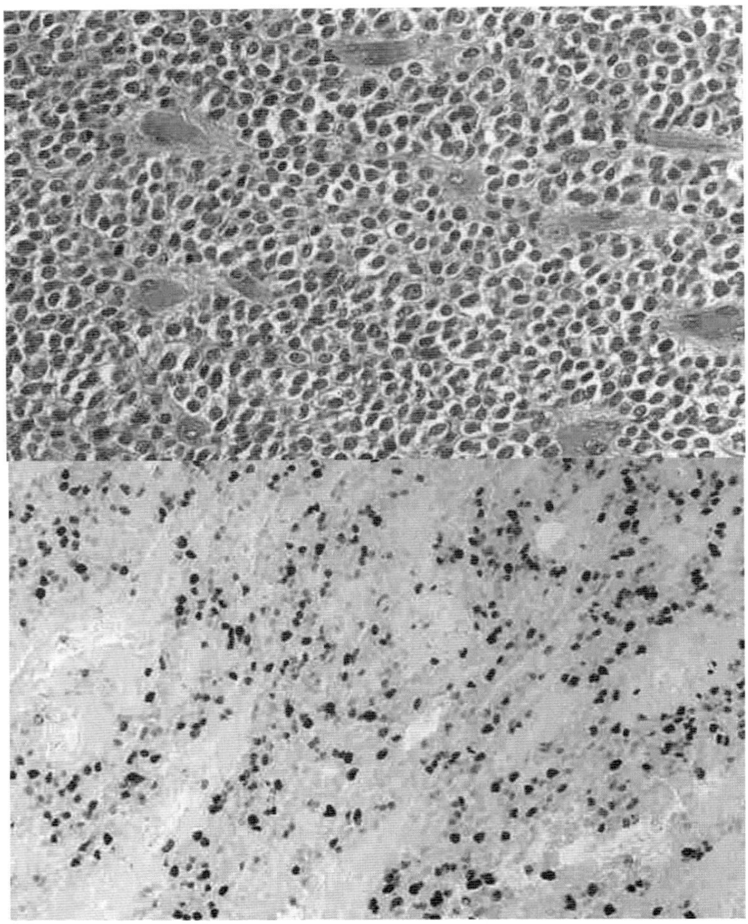

B

Figure 4. Ependymoma. A. Clear cell variant. H&E, x 400; B. High MIB. 1 LI in the anaplastic variant. DAB, x 400

4. SUBEPENDYMOMA

As mixopapillary ependymoma it is a grade I tumor. It grows attached to the ventricular walls of IVth, lateral, IIIrd ventricles or in cervical and dorsal spinal cord (Figure 5). Histologically, it is characterized by a lobular structure, hypocellularity, clusters of nuclei immersed in a fibrillary matrix (Figure 6). Frequently, astrocytes, resembling those of the subpendymal layer, can be found isolated or forming astrocytic areas (Figure 7). Mitoses are very rare. Cystic degeneration, microvascular proliferations and calcifications can be found as well as Rosenthal fibres and patchy necroses (Wiestler et al., 2000). The MIB-1 LI is very low (Rushing et al., 1997; Prayson and Suh, 1999).

Subependymoma can be symptomatic, but most tumors remain asymptomatic. The tumor when totally resected does not recur and does not require radiotherapy (Im et al., 2003).

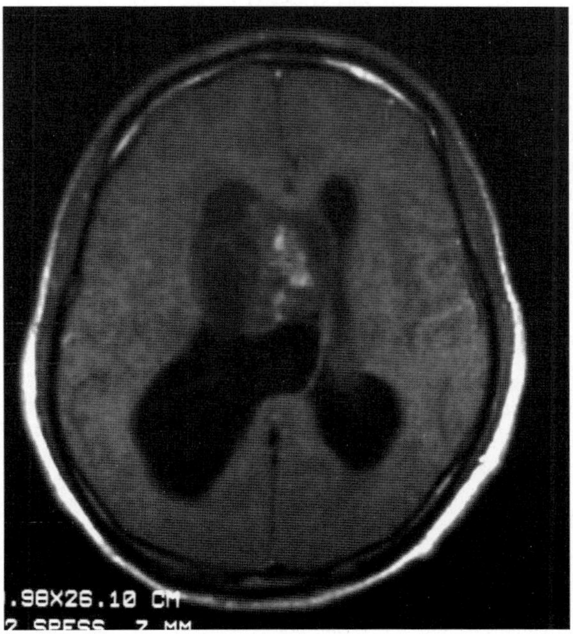

Figure 5. Sub-ependymoma, MRI, T1. From the Neuroradiology Unit, Dpt Neuroscience, University of Turin

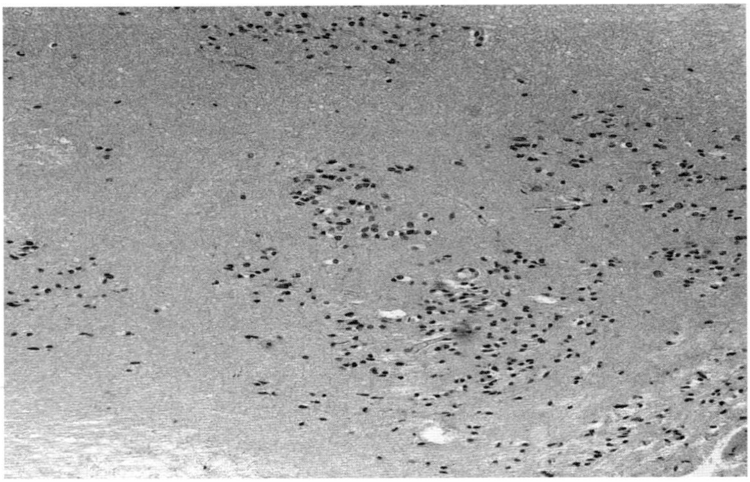

Figure 6. Subependymoma. Cell clusters in a fibrillary matrix, H&E, x 100

The main problem encountered in histological diagnosis is represented by the possible coexistence of an additional ependymoma which gives the tumor the dignity of grade II. Occasionally, nuclear pleomophism and hypercellularity can be found in some tumors, but they have not been considered responsible for the recurrence in two cases, since they were present also in nonrecurrent tumors (Im et al., 2003).

The coexistence of ependymomatous foci may go unrecognized in too small a sample, with the consequence of a missed indication to radiotherapy in a grade II tumor.

A

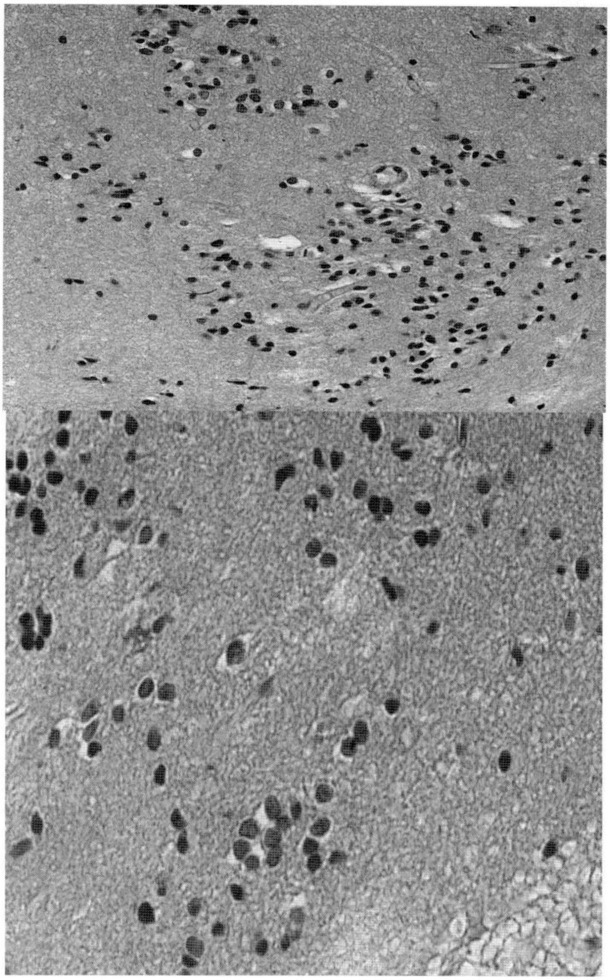

B

Figure 7. Subependymoma. A. Cells Cell clusters, H&E, x 200: B. Cell nests mimicking the subependymal ones, H&E, x 400

Chapter 8

NEURONAL AND MIXED GLIO-NEURAL TUMORS I

Immunohistochemistry greatly contributed to the nosographic definition of these tumors. The detection of neuron-specific and neuron-associated antigens led to the discovery of neuronal cells in many neuroectodermic tumors. This also made the diagnosis more difficult and sometimes erroneous, because of the possible uneven distribution and paucity of neuronal cells in tumors, especially when the diagnosis has to be made in very small specimens.

1. GANGLIOGLIOMA

It is a rare tumor with an incidence of 1.3%, affecting young patients with intractable focal epilepsy, mainly localized in temporal, followed by frontal and occipital lobes (Figure 1). It is a grade I tumor with an atypical variant grade II and an anaplastic variant grade III, less frequent in the temporal lobe. In a recent series, at 7 years follow-up only one out of 86 tumors showed recurrence (Blümke and Wiestler, 2002).

Frequently cystic and calcified, it is well circumscribed and hypointense at MRI in T1 weighted images and hyperintense in T2 images, with some degree of contrast enhancement.

Histologically the tumor shows a combination of neuronal and glial cell aspects in a typical two-phase way. When the neuronal phenotype prevails the lesions display the aspect of gangliocytoma or cortical dysplasias; when the glial phenotype prevails the lesions appear as astrocytomatous; and when clear cell aspects prevail the lesion comes nearer to oligodendrogliomas or dysembryoplastic neuroepithelial tumors. Neurons may show a cytoarchitectural derangement, be abnormally located or clustered or very large, even bi- or multinucleated, but sometimes they are not easily distinguishable from normal neurons of an infiltrated gray area (Figure 2). The glial component (Figure 4) shows fibrillary or pilocytic aspects (Wolf et al., 1994; Hirose et al., 1997), even with Rosenthal fibres, or oligodendroglial appearance. Papillary architectures until a papillary variant have been described (Komori et al., 1998; Prayson, 2000). Calcification, a dense capillary network and lymphocytic infiltrates may be present. The latter must be carefully distinguished from small neurocytic cells.

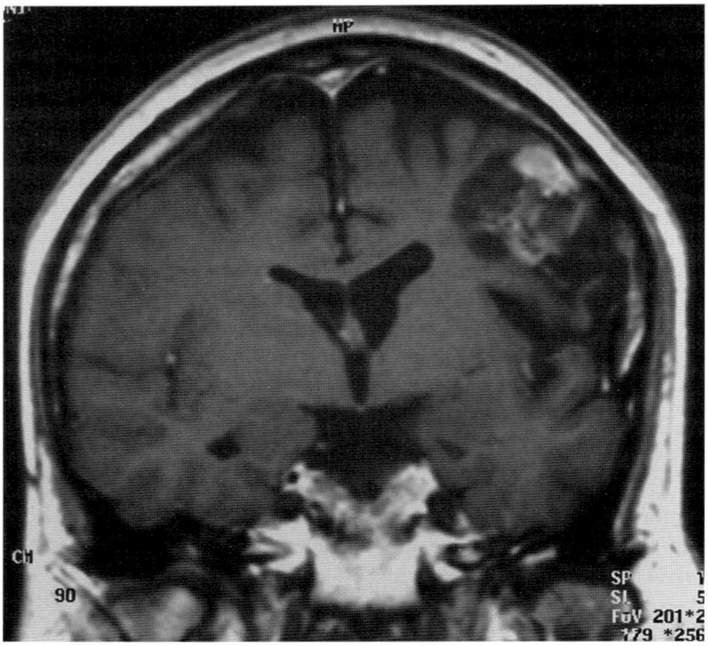

Figure 1. Ganglioglioma, MRI. From the Neuroradiology Unit, Dpt Neuroscience, University of Turin

GFAP staining, synaptophysin and antibodies against neurofilaments (Figure 3) and MAP2 help in the recognition of the glial and neuronal components respectively. Most tumors of the temporal lobe show a positive staining for CD34 in glial cells and maybe in neurons. CD34 is known as a marker of endothelial cells and also of early precursors in neural tube formation (Lin et al., 1995) and its positivity in gangliogliomas denounces the immature phenotype of the tumor (Blümke et al., 1999).

Ganglioglioma is a typical malformative tumor deriving from precursor cells. These can be found also in the adult brain hippocampus and generate both glial and neuronal cells. The molecular genetics of the tumor is very poor. Usually TP53 mutations or EGFR and PTEN alterations are lacking (von Deimling et al., 2000).

Malignant transformation is infrequent and it affects the glial component (Zentner et al., 1994). Out of 326 gangliogliomas, 30 tumors have been classified as atypical and 17 as anaplastic on the basis of the MIB-1 LI which was >5% and >10% respectively (Campos et al., 1994). The tumors may recur. The recognition of the tumor can be hindered by excessive modification of neuronal cells that become no more identifiable, so that the suspicion that the tumor can be a ganglioglioma may arise only from the intense contrast enhancement of sub-arachnoidal diffusions.

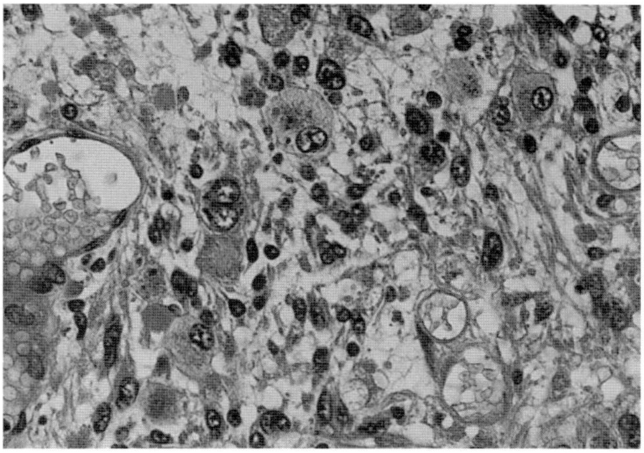

Figure 2. Ganglioglioma, H&E, x 400. From Dr. Bianca Pollo, Istituto Neurologico, Carlo Besta, Milan

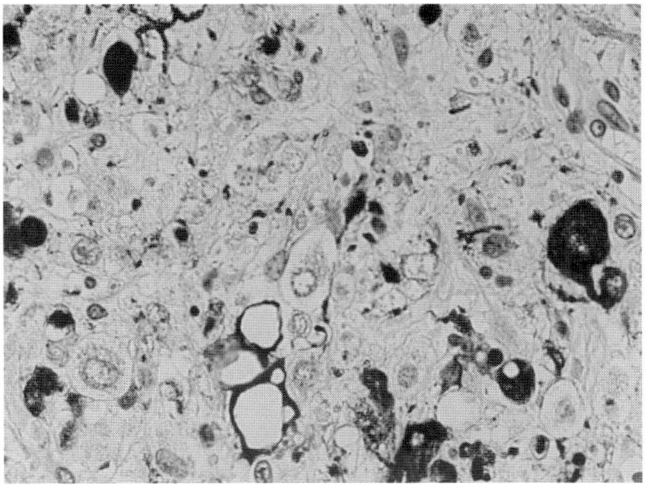

Figure 3. Ganglioglioma, NFs, Dab, x 400. From Dr. Bianca Pollo, Istituto Neurologico, Carlo Besta, Milan

The differential diagnosis of gangliogliomas is not easy, especially when it has to be carried out in small fragments. It must be made towards pilocytic and diffuse astrocytoma, oligodendroglioma, DNT, pleomorphic xanthoastrocytoma and cortical dysplasia (Blümke and Wiestler, 2002). Gangliocytoma or hamartoma of the hypothalamus and Lhermitte-Duclos disease can easily be recognized, also for their typical aspect and location. Gangliocytoma, moreover, at the neuronal extreme of the ganglioglioma spectrum, is so rare that it can be practically neglected. The differentiation from a diffuse astrocytoma can be really difficult, and particular carefulness is recommended when it is a temporal tumor in a young patient. (Blümke and Wiestler, 2002). The possible absence of neuronal cells may be misleading, as has been mentioned above, but also the opposite error is possible, i.e. interpreting as ganglioglioma a diffuse astrocytoma invading the cortex and with normal neurons entrapped in the tumor (Quinn, 1998). The neuronal component of gangliogliomas is usually recognized after a careful examination, in spite of the various neuronal alterations. Sometimes the latter are totally absent and the cytological aspect of the neurons is completely preserved; in this case, the neoplastic nature is deduced only from the abnormal clustering of the neurons. When the neurons are so deeply changed that they cannot be recognized any more and it is really difficult to distinguish the tumor from a diffuse astrocytoma, immunohistochemistry can still be of help. CD34 positive staining is of fundamental importance, at least for tumors located in the temporal lobe, as well as that of MAP2. However, with the latter marker the interpretation of a positive staining must also take into consideration the morphology of the cells: MAP2 is positive in glial precursor cells and in oligodendrocytomas and diffuse astrocytomas, even though in differently looking cells (Blümke et al., 2001), but it is negative in the glial component of ganglioglioma (Blümke and Wiestler, 2002).

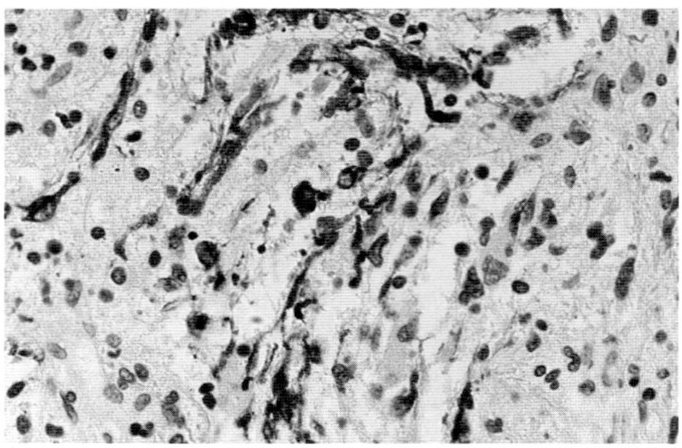

Figure 4. Ganglioglioma: GFAP-positive glia cells, Dab, x 400. From Dr. Bianca Pollo, Istituto Neurologico Carlo Besta, Milan

Oligodendroglioma, with which ganglioglioma may share a clear cell aspect, can be distinguished by its own histological features and also by MAP2 positive staining of its cells. The distinction of ganglioglioma from pleomorphic xanthoastrocytoma is not usually problematic, even though the two lesions have even been found in combination (Evans et al., 2000). On the contrary, difficulties may arise in the differential diagnosis of dysembryoplastic neuroepithelial tumors, especially when in a small fragment the gliomatous component of the latter prevails. The demonstration of glio-neuronal units is of great help as well as the negative staining for CD34. Finally, the differential diagnosis from cortical dysplasias may be really difficult, especially in laboratories where surgical material from temporal lobes of patients with intractable epilepsy does not concentrate and adequate experience is lacking.

Desmoplastic infantile ganglioglioma is a cystic tumor of infancy of the surface of frontal and parietal lobes. It has a dense fibrous aspect and variable small, round and neurocytc cells (VandenBerg et al., 1993). Its aspect must be kept in mind when diagnosing an infantile tumor on small tissue fragments.

Recently, under the eponym of "malignant glioneuronal tumors" neoplasias have been described, not containing ganglion cells, but differentiating along the neuronal line in scattered cells. They may resemble any known malignant glioma and show a histology of malignant gliomas, but contain cells co-expressing GFAP and NFP. The importance of identifying such tumors resides in the fact that gross total surgical resection proves to be curative in some cases or followed by a long-term follow-up (Varlet et al., 2004).

2. CENTRAL NEUROCYTOMA

Described for the first time by Hassoun et al. (1982, 1993), this is a rare tumor affecting mainly young patients, located in the lateral and third ventricles in the region of the foramen of Monro and with a good prognosis (Schmidt et al., 2004). The tumor is believed to derive from bipotential progenitor cells of the sub-ventricular residual matrix (von Deimling et al., 1991). In the typical location it appears in CT scans as a hyperdense mass, isointense in MRI, with moderate enhancement after gadolinium, containing calcification in 69% and cysts in 85% of the cases (Chang et al., 1993).

Histologically, it is characterized by round, isomorphic cells with features of neuronal differentiation and by fibrillary areas (Figure 5A). Irregular rosettes and pseudo-rosettes may be present as well as the "honeycomb" appearance resembling that of oligodendrogliomas. Also ganglioid cells and Homer-Wright rosettes can be observed (Von Deimling et al., 1990; Robbins et al., 1995). There is an intra-cytoplasmic positivity for synaptophysin (Figure 6A) and other neuronal antigens, such as NSE, class III β-tubulin, MAP2 (Figure 6B). Usually GFAP is negative with the possible exception of reactive astrocytes (Figure 5B). Nuclei have been found to be positive to the Hu antibody which identifies a family of RNA binding proteins limited to neurons (Gultekin et al., 1998). By electron microscopy meuronal structures may be made visible (Figure 7).

Anaplastic cases may exist with increased mitotic activity, vascular proliferations and necroses (von Deimling et al., 1990). The latter alone, however, do not indicate malignancy. The LIs of Ki-67, PCNA and AgNor are usually low; when Ki-67 LI is >2% may indicate shorter survivals (Söylemezoglu et al., 1997;

A

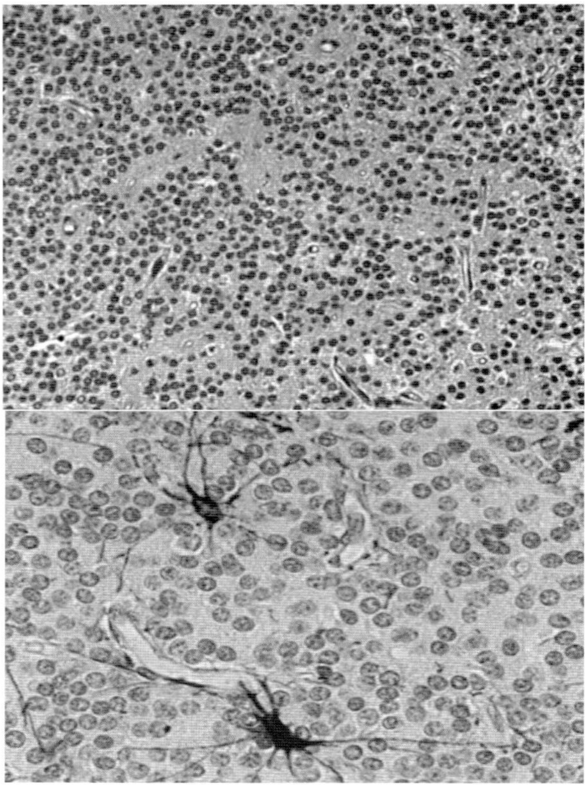

B

Figure 5. Neurocytoma. A. Neurocytic nuclei and nuclei-free areas, H&E, x 400; B. Reactive astrocytes. GFAP, DAB, x 400

Fujimaki et al., 1997; Mackenzie, 1999; Kuchiki et al., 2002). In six cases examined, p53 expression was very low, Bcl-2 never expressed and MIB.1 LI < 4.5% (Uro-Coste et al., 1999).

The prognosis is good after total removal, with shorter survivals after incomplete surgery.

The major problems with this tumor are the differential diagnoses towards oligodendroglioma, ependymoma, pineocytoma and the existence of extra-ventricular tumors. In all these conditions a correct diagnosis is mandatory for prediction of survival and the decision about radiotherapy. The diagnosis will be based on the histological aspect, the positivity for neuronal antigens and also on the ultrastructural aspect which will include microtubuli and vesicles, besides ER and synapses (Hassoun et al., 1993; Cenacchi et al., 1996). The nuclear positivity for the neuronal nuclear antigen NeuN may help (Söylemezoglu et al., 2003). For oligodendroglioma, losses of 1p and 19q as well as TP53 mutations can be used (Fujisawa et al., 2002).

Neurocytomas are benign tumors, but cases with malignant features or with a high MIB-1 LI have been reported (Eng et al., 1997; Fujimaki et al., 1997; Uro-Coste et al., 1999; Warton et al., 1998), whereas those with cranio-spinal dissemination are rather rare (Elek et al., 1999; Brandes et al., 2000; Takao et al., 2003), even though there are also cases with a classic histological aspect with diffusion in the subarachnoidal spaces (Eng et al., 1997), as well as cases with the same diffusion after radiotherapy (Tomura et al., 1997). The identification of the classic variant is very important, because radiotherapy is indicated in the anaplastic variant only, mainly identified by the occurrence of vascular proliferations, mitoses and necroses (Kuchiki et al., 2002). A meta-analysis demonstrated that total removal of the tumor gives good local control, whereas sub-total removal requires radiotherapy (Rades et al., 2003) and good results have been obtained after partial removal with LINAC (Martin et al., 2003) and a γ-knife (Hara et al., 2003). Data on chemotherapy are poor and inconsistent.

Extra-ventricular neurocytomas represent a special problem, because they show in a noncharacteristic location the same cytologic features of classic neurocytomas, with small and round cells, clear cytoplasm with halo, also associated with an astrocytic component. In both cases a ganglion cell component may be present (Giangaspero et al., 1997; Buccoliero et al., 2002). About 60 cases have been published to date (Buccoliero et al., 2002). In a series of 35 cases, 11 were "atypical" with microvascular proliferations, necroses and mitoses >3 for 10 HPF. As in the ventricular cases, those totally removed did not show recurrences, whereas the atypical or incompletely resected tumors recurred (Brat et al., 2001).

Particularly interesting are extra-ventricular tumors with neurocytic features, but showing 1p and 19q losses, which are not features of central neurocytoma. Two such cases have been reclassified as oligodendrogliomas (Fujisawa et al., 2002).

A B

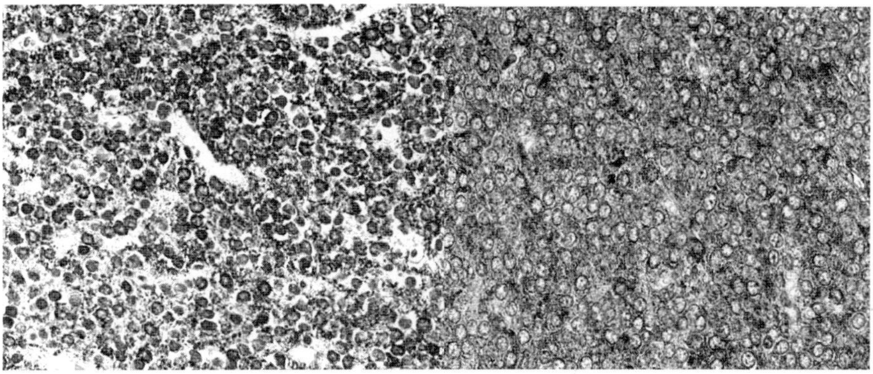

Figure 6. Neurocitoma. A. Synaptophysin, DAB, x 400; B. MAP2, DAB, x 400. From Dr. Bianca Pollo, Istituto Neurologico "Carlo Besta", Milan

Three other similar cases, showing in addition an infiltrative growth, have been classified as oligodendrogliomas with neurocytic differentiation (Perry et al., 2002). This diagnosis is in line with the positivity for neuronal markers repeatedly described in oligodendrogliomas (Ng et al., 1994; Patt et al., 1996; Wharton et al., 1998) and interpreted as representing possibly one end of a spectrum which includes oligodendrogliomas with neurocytic differentiation that would demonstrate the existence of a common oligodendroglial/neuronal precursor line. In another case with 1p and 19q losses and infiltrative growth of oligodendrogliomatous type the name of "atypical" neurocytoma was preserved (Mrak et al., 2004), because the use of genetic changes for a primary classification of tumors was considered less important than for sub-classifying histological defined entities for better prognosis and therapy (Reifenberger and Louis, 2003). In 2 of these cases, losses of 1p and 19q, usually absent in central neurocytomas, have been found.

All these features are not easily detectable in small samples of tumor tissue and they must be taken into account every time a diagnosis of neurocytoma occurs in a extra-ventricular tumor, all the more in that most tumors of this type do not behave like typical intra-ventricular neurocytomas.

Molecular genetics is rather poor: genomic alterations, mainly gains, were found by CGH in 60% of cases (Yin et al., 2000), but practically no EGFR amplification (Tong et al., 2000) or TP53 mutations (Fujisawa et al., 2002) were found.

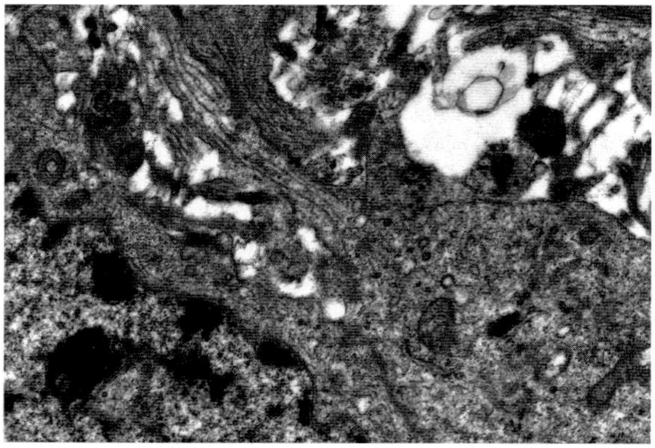

Figure 7. Central neurocytoma. Cytoplasm with a nucleus. In the central part there is a neurite with microtubules and upwards a synaptical structure with abundant vesicles. x 50,000. From Prof. Giovanna Cenacchi, Dpt of Radiologic and Histopathologic Sciences, Policlincio Sant'Orsola, Bologna.

2. GANGLIONEUROCYTOMA

In rare cases, the picture of neurocytoma may be enriched with the occurrence of an advanced neuronal differentiation and small neurons. The tumor occurs mostly in children and has a good prognosis. There is a positive staining for synaptophysis and NSE. It is still debated whether this tumor is a variant of central neurocytoma or a true tumor entity (Figarella-Branger et al., 2000). These tumors occur in children (Baehring et al., 2005), but also in the adult (Shin et al., 2002; Buhl et al., 2004) and in an extraventricular position (Funato et al., 1997; Biernat et al., 2000; Shin et al., 2002), for example in the spinal cord (Baehring et al., 2005).

The tumor must be kept separate from ganglioneuroblastoma, which was a central neuroblastoma with differentiated neurons (Horten and Rubinstein, 1976) included now in the category of PNET. In this tumor, among densely packed, hyperchromatic cells, clearly recognizable neurons can be seen (Figures 8, 9, 10).

4. DYSEMBRYOPLASTIC NEUROEPITHELIAL TUMOR (DNT)

This is still a controversial and ill-defined lesion of heterogeneous composition and with associated pathologies that are not easily recognizable, especially in small surgical samples. It is a multinodular cortical lesion found in patients with epilepsy often associated with abnormal neuronal migration (Daumas-Duport et al., 1988;

A

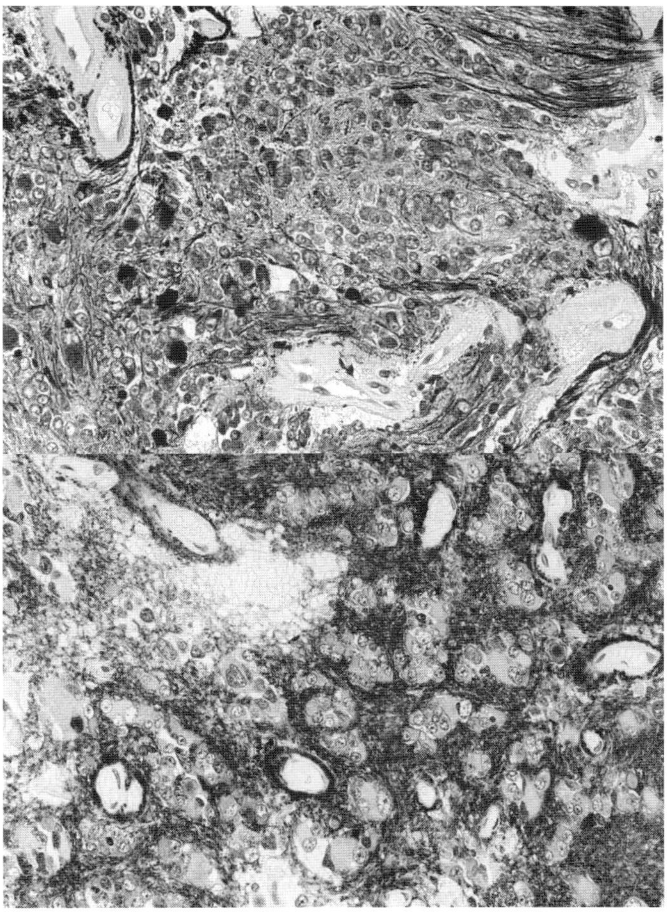

B

Figure 8. Ganglioneurocytoma. A. Neuronal cells positive for NFs, DAB, x 400; B. Synaptophysin-positive reticulum, DAB, x 400

A

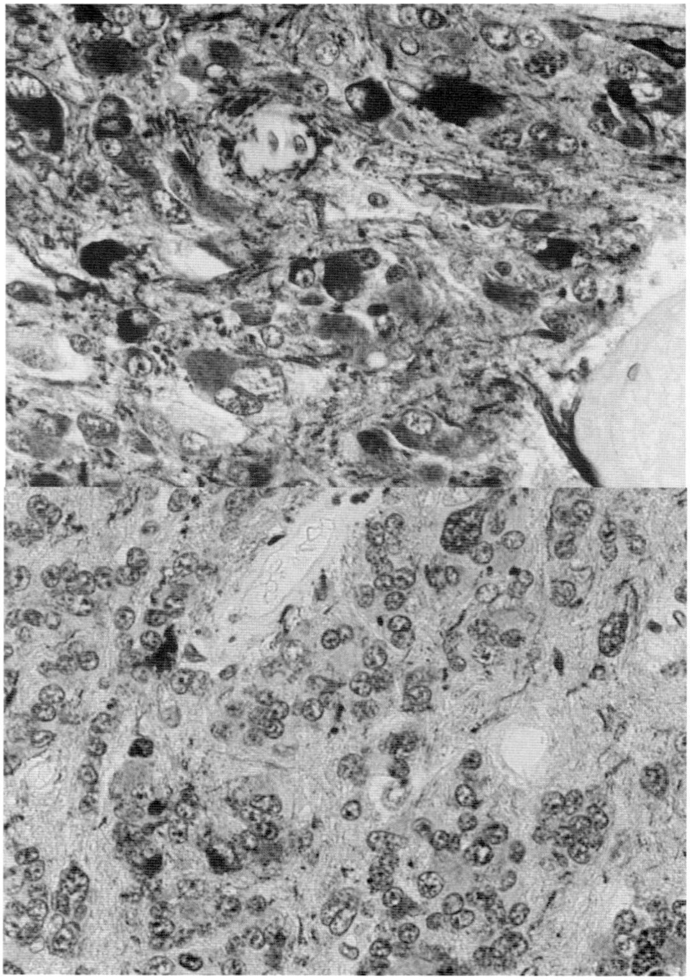

B

Figure 9. Ganglioneurocytoma. A. Neuronal cells positive for MAP2, DAB, x 400; B. Small neuroblastic cells are negative for NFs, DAB. x 400

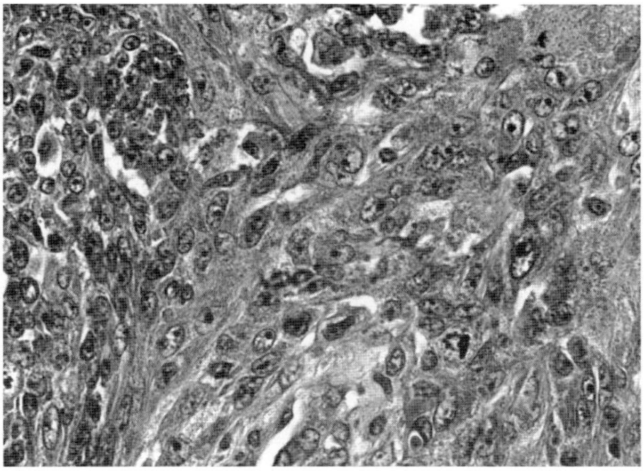

Figure 10. Ganglioneurocytoma. Neuroblastic cells differentiated into neurons, H&E, x 400

Wolf et al., 1995). The tumor is hypodense on CT and hypointense on T1- and hyperintense on T2-weighted images of MRI. There is no contrast enhancement (Figure 11).

Macroscopically, the lesion often appears as a translucent mass. Histologically, it is multi-lobulated and resembles an oligodendroglioma with abundant mucous degeneration. It shows migrational abnormalities of neurons in the adjacent cortex where they form cortical dysplasias. Simple and complex or non-specific forms of DNT have been distinguished. Typical of the simple forms are the “glioneuronal units” with a globular neuron associated with a double array of oligodendroglia-like cells that is lacking in the non-specific forms (Daumas-Duport et al., 1988; Daumas-Duport and Varlet, 2003). The isolated neuron immersed in a mixoid matrix, the “floating neuron” is characteristic (Figure 12), but frequently the neuron is absent or mature ganglion cells cannot be distinguished from preexisting ganglion cells in an infiltrative zone of oligodendrogliomatous type. The vascular pattern may be prominent, also with glomeruloid formations and anomalous or angiomatous vessels from which calcifications may start to form wide areas with densely packed concrements. An astrocytic component is variably present (Figure 13) and tumors may have gangliogliomatous aspects (Hirose et al., 1998). The glial component may show mitoses, necrosis and nuclear atypia, but Ki-67 LI is generally very low, $< 1\%$. An important negative feature is the lack of lymphocytic cuffings, typical of gangliogliomas (Daumas-Duport et al., 2000).

Cortical dysplasias contain unevenly distributed and dysmorphic neurons, whereas ectopic neurons can be found in the white matter. Dysplastic neurons have been accurately categorized and the three main anomalies are disorganization, cytomegaly and increased molecular layer neurons (Prayson and Frater, 2003) (Figures 13, 15).

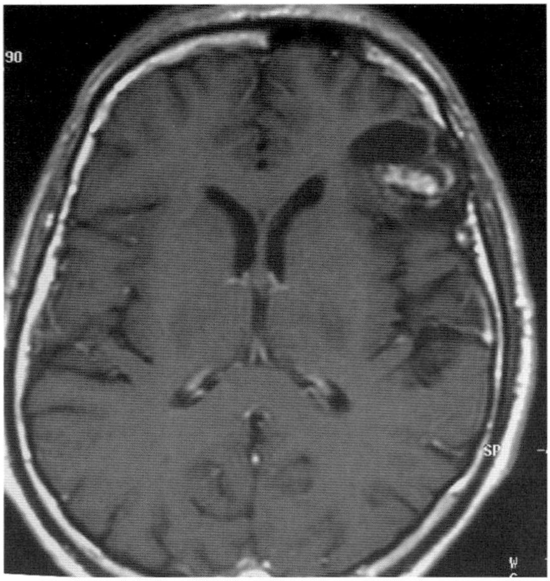

Figure 11. Frontal DNT, MRI. From the Neuroradiology Unit, Dpt. Neuroscience, University of Turin

In a series of 43 patients with chronic pharmacoresistant epilepsy, 24 had a circumscribed lesion of the temporal lobe, characterized by irregularly distributed and clustered cortical neurons, an increased number of GFAP-positive astrocytes (Figure 14A) and clusters of oligodendrocytes associated with immunoreactivity of the neuropil and cell membranes for E-NCAM. These hamartias were called glioneuronal malformations, similar to tuberous sclerosis, but without tuberous sclerosis features (Wolf et al., 1995). The E-NCAM-positivity could indicate cell immaturity and indirectly a migration derangement (Figure 14). The Ki-67 LI is regularly low in these lesions, with the exception of a few cases.

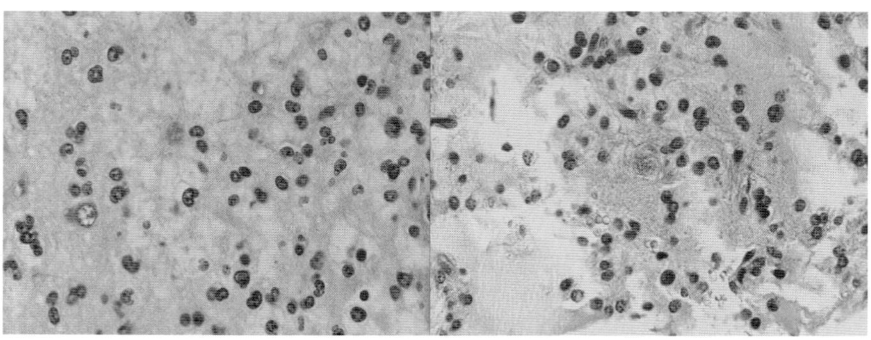

A B

Figure 12. DNT. A and B. Glio-neuronal units, H&E, x 400. From Dr. Bianca Pollo, Istituto Neurologico, Besta, Milan

In 14 cases of DNT, multidrug transporters such as P-glycoprotein or multridrug resistance associated protein were overexpressed, very likely playing a role in the resistance to anti-epileptic drugs (Vegelgesang et al., 2004).

Some interpretations have been put forward on the origin of these tumors and on their differentiation from oligodendrogliomas. In half the cases of oligodendroglioma, oligodendroglia-like cells contained NR1, a subunit of NMDA receptors, which is not neuron-specific, because it can be found also in astrocytes, or E-NCAM, but no NeuN. In DNT Neu-N and NR1-positive cells were present in 44% of cases, whereas E-NCAM-positivity was less frequent (Wolf et al., 1997). NR1 could indicate either a neuronal or a glial differentiation. However, the immunohistochemical profile of oligodendroglia-like cells in DNT and of oligodendrogliomas largely overlap, leaving the distinction very hard to make and suggesting an early neuronal differentiation (Wolf et al., 1997). It has been shown, however, that these cells transcribe myelin genes and are oligodendrocytes (Wong et al., 1999) and cannot be of help in distinguishing the tumor from oligodendroglioma, because in both tumors small dark neurons occur.

The occurrence of neuronal cells positive for MAP2 and nestin-positive cells of neuronal morphology with cells co-expressing both antigens, suggests the origin from pluripotent neuroepithelial cells (Duggal et al., 2003).

DNT are benign lesions which do not recur, even after partial resection (Daumas-Duport et al., 2000). Therefore, they must be distinguished from other tumors, such as oligodendrogliomas and pilocytic astrocytomas and this is not an easy task, especially when only small fragments of tissue are at one's disposal. The glioneuronal units may be lacking, at least in the biopsy, and the multilocular character of the lesion may go unrecognized in biopsies. Also cortical dysplasias can be missed or even simulated by normal neurons in badly oriented small samples (Daumas-Duport and Varlet, 2003). Some clinico-radiological features may greatly help, such as epileptic seizures and cortical location of the lesion on CT and MRI, but histologically, the main difficulty resides in the occurrence of areas with abundant mucin production and confluent cysts, not forgetting that true oligodendrogliomatous areas may occur in DNT.

The other features of DNT, i.e. cells with neuronal characteristics, must be interpreted very carefully. An important question is represented by areas with mitoses, nuclear atypia and increased MIB-1 LI which could suggest the diagnosis of an anaplastic tumor, if the DNT nature of the lesion is not recognized. These areas may be very small and missed or even be the prominent feature when only small fragments are examined. Finally, DNT in extracortical locations may not be easily recognized (Leung et al., 1994; Cervera-Pierot et al., 1997; Baisden et al., 2001). Worth mentioning is that in many instances, the main feature of the tumor remains the "floating neuron".

5. ROSETTE-FORMING GLIONEURAL TUMOR OF THE FOURTH VENTRICLE

Eleven cases have recently been described of this tumor type involving the IVth ventricle, vermis and acqueduct of young subjects (Komori et al., 2002). At neuroimaging it appears as an irregularly enhancing mass often associated with hydrocephalus or multicentric. Histologically, it shows two cell components. One is given by neurocytic cells forming rosettes and pseudorosettes around central

A

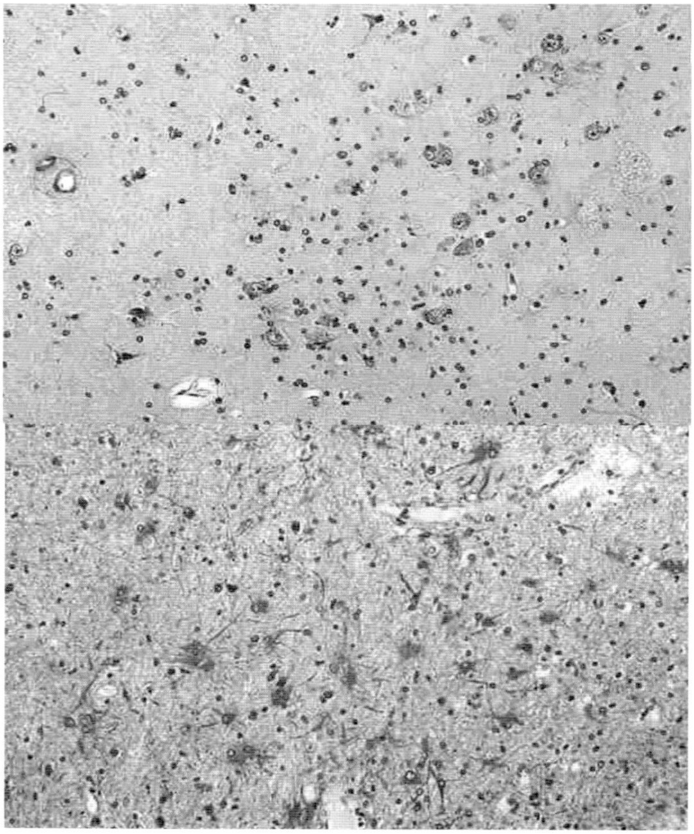

B

Figure 13. Dysembryoplastic neuroepithelial tumor. A. Clusters of oligodendrocytes and displaced neurons, H&E, 200; B. Astrocytic focus, H&E, x 200

A

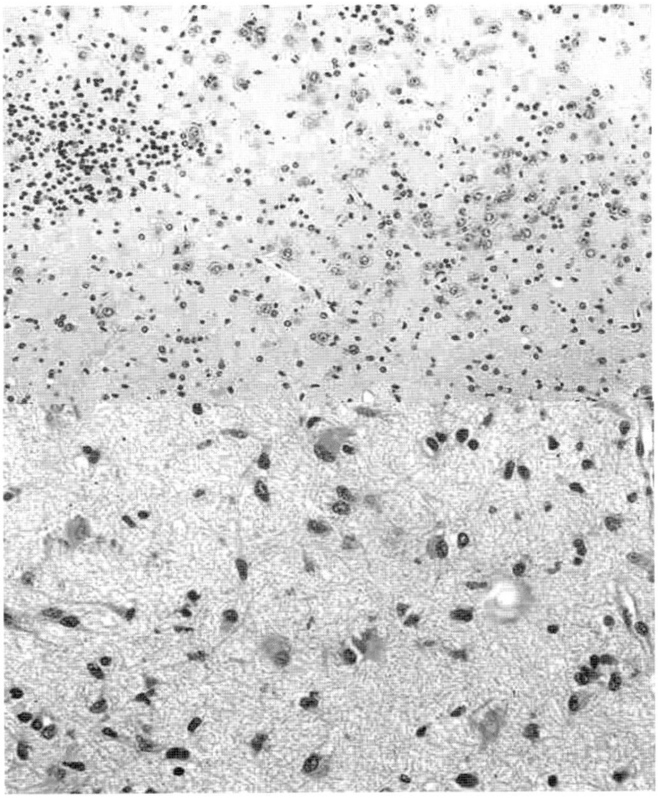

B

Figure 14. Dysembryoplastic neuroepithelial tumor. A. Clusters of GFAP-positive astrocytes, DAB, x 400; B. Astrocytomatous areas, H&E, x 400

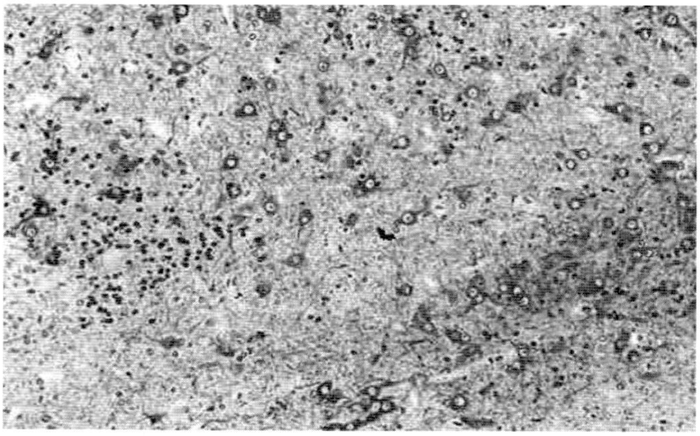

Figure 15. Dysembryoplastic neuroepithelial tumor. Cortical dysplasia with foci of displaced neurons, H&E, x 400

capillaries, with round nuclei and delicate cytoplasmic processes, in a loose texture (Figure 16). The second component is given by spindle or stellate astrocytes in a fibrillary background, associated with oligodendroglia-like cells with cytoplasmic halos. Rosenthal fibres and hyaline bodies can be found as well as ganglion cells. Synaptophysin and MAP-2 are positive in the first component and GFAP and S-100 in the second one. Gangliod cells are positive for neurofilaments and synaptophysin.

The tumor resembles DNT and similar cases have been described in the cerebellum under this eponym (Kuchelmeister et al., 1995); however, there are biological and histological differences that confer this tumor the dignity of a tumor entity, because its phenotypic features do not depend uniquely on the location in the cerebellum (Komori et al., 2002). The lesion is benign and recurrences after surgery are not known, however it is considered less as a hamartoma than a low-grade glioma.

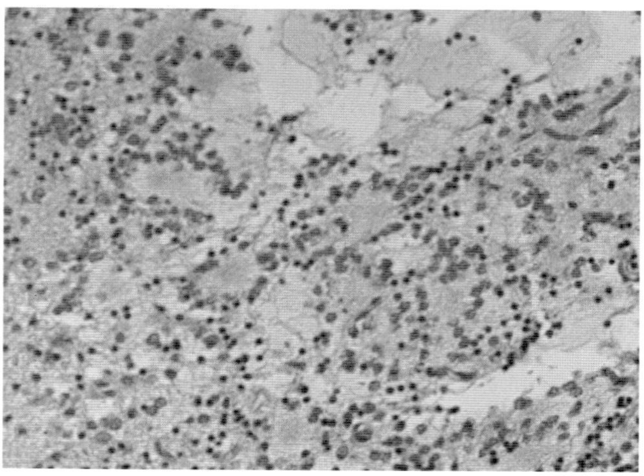

Figure 16. Rosette-forming glioneural tumor. Pseudo-rosettes, H&E, x 200

A recent case has been described (Preusser et al., 2003) and one case has been seen by us in the last year.

6. LIPONEUROCYTOMA

This is a recently recognized cerebellar tumor entity which must be differentiated and kept separate from medulloblastoma, because of its favourable outcome. Histologically, it is composed of small, round neoplastic cells, with a clear neurocytic differentiation and with few mitotic figures. Scattered there are accumulations of adipocytes (Figure 17). Medulloblastomas with simply lipidized cells and no clear neurocytic differentiation should be included in this tumor category (Giordana et al., 2000). The MIB.1 LI is 1%–6%. Tumor cells express synaptophysin, NSE and MAP-2 (Kleihues et al., 2000). GFAP is also focally expressed. The first case was reported 25 years ago (Bechtel et al., 1978) and since then other cases have been described under different names. The tumor has its own molecular genetics characteristics: TP53 mutations in 20% of cases, no mutation of PTCH or β-catenin or isochromosome 17q, typical of medulloblastoma. Very recently, 24 samples of the tumor from 20 patients, published by different authors, were studied. Cluster analysis of cDNA put the tumor closer to central neurocytoma, from which it differs for TP53 mutations, than to medulloblastoma (Horstmann et al., 2004). The follow-up of the cases ranged from 56 months to 16 years with 5 tumors recurring after 5–12 years.

Given the cerebellar location, the recognition of this tumor and its distinction from medulloblastoma is of paramount importance, because it does not need heavy treatments. Problems arise when tumors with the features described, but with no adipocytic cells, are encountered. What are they? Their nature will be further discussed in the next tumor type.

A

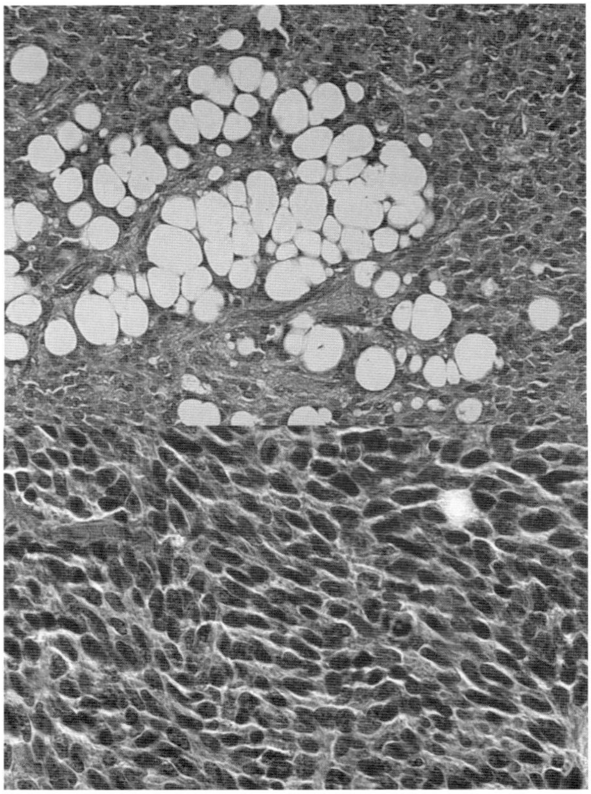

B

Figure 17. Liponeurocytoma. A. Lipidized adipocytes, H&E, x 200; B. Parenchyme of the tumor, H&E, x 400

Chapter 9

NEURONAL AND MIXED GLIO-NEURAL TUMORS II

1. MEDULLOBLASTOMA

The tumor has been put into the category of embryonal tumors of the CNS and it is the commonest malignant CNS tumor of childhood with an age peak at eight years, but well represented in adulthood in which 30% of tumors occur. Medulloblastoma is a malignant tumor, but with a variable and often good response to radio- and chemotherapy. After surgery and radio- chemotherapy in standard-risk patients, survival at 5 years is 70%, whereas in high-risk patients, i.e. with age < 3 years or metastasis at presentation or residual tumor after surgery, survival at 5 years is 25% (Ellison, 2002).

Histologically the classical tumor is composed of densely packed cells with a round or oval elongated nucleus and scanty cytoplasm (Figure 1A). Neuroblastic rosettes are frequently found, whereas ganglion cells are occasionally seen. Neuronal differentiation can be indicated by ultrastructure or by positive staining for neurofilaments and synaptophysin in cells without a clear neuronal phenotype (Figure 2A, B). Glial differentiation is much rarer, but GFAP-positive glial cells of reactive nature are frequently found. Mitoses are usually frequent, but sometimes very few and not rarely they are even difficult to find. Also frequent are apoptotic nuclei, especially in highly proliferating areas (Schiffer et al., 1994). Nuclei are mostly isomorphic, but nuclear pleomorphism and atypical mitoses are not infrequent (Figure 1B). Less frequently, microvascular proliferations, calcifications and necroses can be found.

After a long debate that occurred twenty years ago, medulloblastoma was included in the category of PNET which grouped cerebellar and hemispheric tumors with the same histological aspect and biological behavior (Rorke, 1983). The discussion on the rationale of this inclusion concerned mainly the histogenesis of the tumor and its differentiation capacity (Schiffer, 1997); and an abundant literature was produced extending to the rare supra-tentorial tumors the same molecular genetics of the more frequent infra-tentorial PNETs. Recently, however, dissimilarities have been emphasized as well.

A

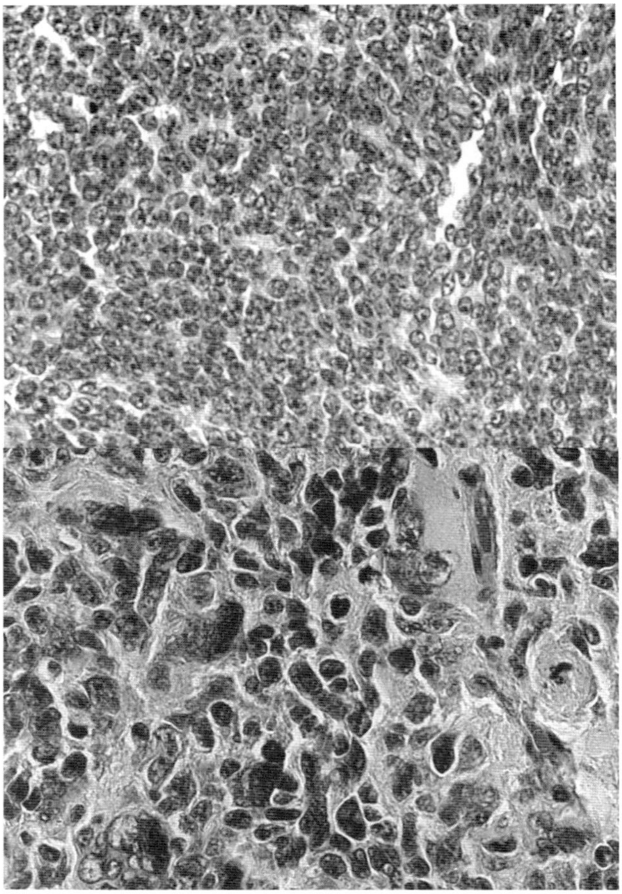

B

Figure 1. Medulloblastoma. A. Densely packed isomorphic nuclei, H&E, x 400; B. Pleomorphic nuclei, H&E, x 400

On CT Scan and MRI medullastomas appear as solid and diffusely contrast-enhancing masses.

2. HISTOGENESIS OF MEDULLOBLASTOMA

The knowledge of the different steps of medulloblastoma development may be of help in recognizing in surgical tumor samples morpho-histochemical and genetic

A

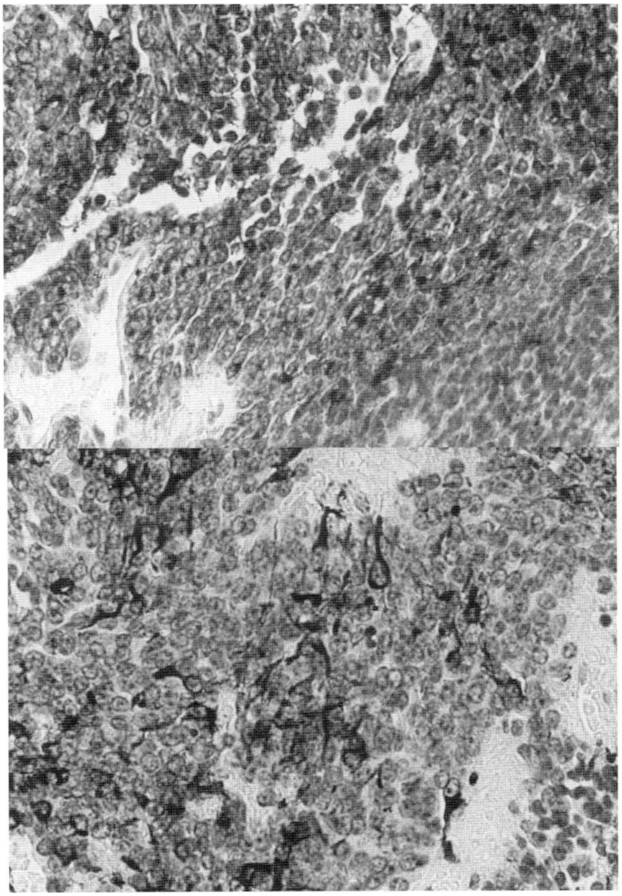

B

Figure 2. Medulloblastoma. A. Synaptophysin ; B. Neurofilaments. DAB, x 400 features useful for prognosis. Classically, medulloblastoma has been regarded as developing from pluripotential cells of cerebellar development (Rubinstein, 1985), i.e. from posterior medullary velum. Some hypotheses have been formulated on the origin of the tumor. The most accredited one is that medulloblastoma develops from the external granular layer of the cerebellum, produced by migration of undifferentiated cells from the roof of the IV ventricle. The major support for this theory is that proliferating precursor neurons of the external layer are under the control of Sonic Hedgehog whose receptor PTCH is mutated in many tumors. This hypothesis can explain the wide neuronal differentiation of medulloblastomas, but not the glial one, because it is not certain that glial cells originate from the external granular layer (Swarz and Del Cerro, 1977). The hypothesis that medulloblastoma, as supratentorial PNET, arises from subependymal matrix cells is contradicted by the molecular genetics of the two PNET types. There is another hypothesis (Katsetos et al., 1994–96) suggesting different origins of the tumor on the basis of staining for Calbindin-D28k and for class III β-tubulin. The former is a neuronal calcium-binding protein of the ventricular matrix, which is not expressed in the external granular layer, whereas the second one is expressed in both locations.

3. VARIANTS

The most important one is *desmoplastic medulloblastoma* characterized by nodules in which tumor cells are more rarefied, with isomorphic nuclei, surrounded by cells with a higher density, more pleomorphic nuclei and producing densely packed reticulin fibres (Figure 3A). Since nodules are the main characteristics of this tumor it has been proposed to call the tumor a "nodular" variant (Eberhardt and Burger, 2003). There is some confusion with the terms "nodularity" and "desmoplasia": the first term refers to nodules and the second to the distribution of tumor cells in rows or chords with intervening abundant reticulin fibres (Figure 3B). In some important findings not all nodular lesions are desmoplastic and not all desmoplastic tumors are nodular (Eberhart et al., 2002). The cells of the nodules are positive for synaptophysin indicating neuronal differentiation, whereas outside the nodules cells show higher Ki-67 and PCNA LIs and are frequently GFAP-positive. This variant accounts for 15% of medulloblastomas and seems to be more frequent in adults. Since nodularity is quantitatively variable, it has been graded, but no specification has been given about the nodularity cut-off between classic and desmoplastic medulloblastoma, whereas extensive nodularity was found to be typical of the following variant (Eberhart et al., 2002).

When the nodules are elongated, contain more neuropil and more marked neuronal differentiation with a neurocytic aspect of the cells, until the occurrence of mature neurons, the picture is that of a variant once known as "cerebellar neuroblastoma" (Shin et al., 1978; Pearl and Taikei, 1981; Yagishita et al., 1982). This variant has recently been called "*medulloblastoma with extensive nodularity and advanced neuronal differentiation*" (Giangaspero et al., 1999) and it is more frequent in children < 3 years of age. On MRI the tumor assumes a "grape-like" aspect.

A

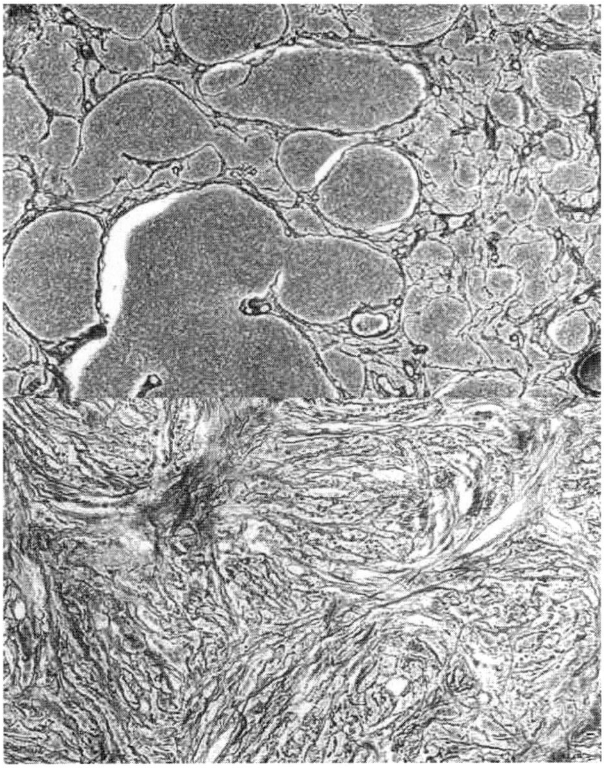

B

Figure 3. Medulloblastoma. Nodules of different sizes, x 200; B. Desmoplasia by rows of tumor cells and bundles of reticulin, x 400, Gomori for reticulin

Another variant described as "*large cell medulloblastoma*" (Giangaspero et al., 1992), is characterized by cells with more abundant cytoplasm, pleomorphic nuclei and prominent nucleoli, circumscribed necroses and abundant apoptotic figures (Figure 4). Foci with strong nuclear atypia, but lacking prominent nucleoli, were found either in association with the above described aspect or alone in tumors with similar aggressive biological behavior and this variant was called *large cell/anaplastic medulloblastoma* (LC/A) (Brown et al., 2000) which would represent not more than 4% of medulloblastomas. Interestingly, foci of this type were found in a series of 7 tumors with other characteristics, for example in desmoplastic or classic tumors or myomedulloblastomas (Leonard et al., 2001). Anaplasia in medulloblastomas was graded on the basis of nuclear atypia and of the number of mitoses and the category of LC/A was considered as severely anaplastic. In the same series anaplastic foci were found in other variants, such as desmoplastic medulloblastoma (Eberhart et al., 2002). Still in another series of 250 tumors, anaplastic tumors were kept separate from large cell tumors, representing respectively 17% and 2% (McManamy et al., 2003) and the latter are still to be kept separate from the "atypical theratoid/rhabdoid" tumors of children less than 2 years of age (Eberhart and Burger, 2003).

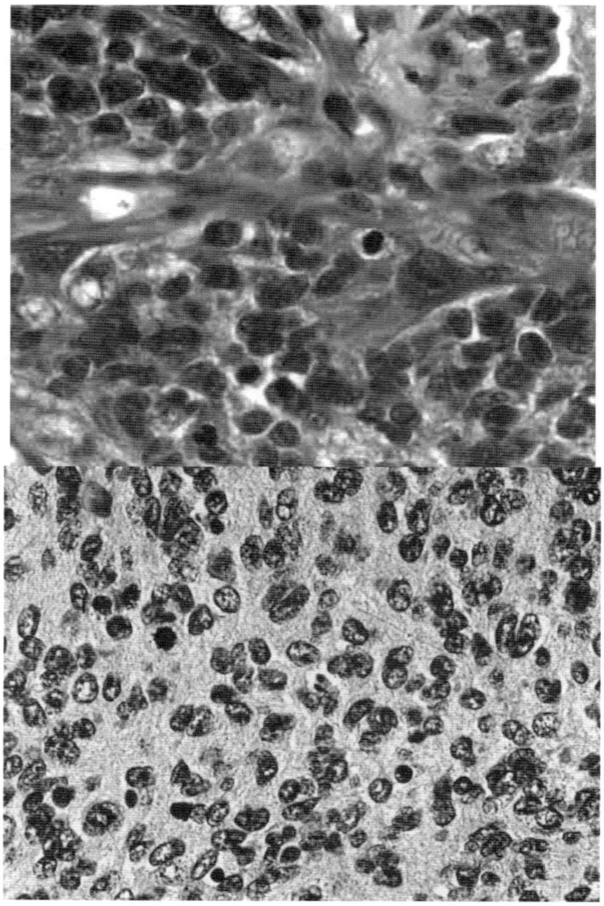

Figure 4. Large cell medulloblastoma, H&E, x 400

There are other tumor types, such as "*melanotic medulloblastoma*" and "*myomedulloblastoma*" (Smith and Davidson, 1984; Schiffer et al., 1992) to which the dignity of a variant is no more recognized, because the first tumor is simply a medulloblastoma with cells rich in melanin and the second one is really rare.

4. METASTASIS

Medulloblastomas often spread through CSF, but rarely disseminate to systemic sites. Some series of metastasizing tumors had been published (Campbell et al., 1984; Sure et al., 1995) and facilitation to metastasis was attributed to ventriculoperitoneal shunting, a common procedure in the past (Kessler et al., 1975), even though this was not a constant finding (Spencer et al., 1984). Bone marrow, soft tissue, lymphnodes and lung are the preferred sites. Metastases can be present at initial diagnosis or appear later after treatment; anaplasia is more common in metastasizing tumors than overall; and it can be detected either in the initial tumor and in metastases (Eberhart et al., 2003). Survival time of cases with metastases is shorter than overall.

5. MOLECULAR GENETICS

A great amount of data are available, but a definitive systematization is still lacking. A shared observation by CGH and FISH is that chromosome gains and losses are more common in large cell/anaplastic medulloblastomas, associated with c-myc and N-myc amplification and 17p losses (Brown et al., 2000; Eberhart et al., 2002). The most frequent alteration, present in 30–45% of cases, is LOH of 17p, whereas TP53 mutations are found in a low percentage of cases. Also frequent, in 20–40% of cases, are losses of 1p and 10q, whereas a novel gene DMBT1 has been described in a small subset of tumors (Mollenhauer et al., 1997).

Two developmental mechanisms promoting tumor growth have been identified. That of Gorlin's syndrom is based on PTCH encoding for a receptor of the Hedgehog family (Hh) of signaling proteins, which is a transmembrane protein repressing the signaling if not bound by an Hh ligand or when the gene is mutated. In the first case granule cell proliferation is promoted. PTCH or downstream molecules of the cascade, for example SUFU, are frequently mutated (Pietsch et al., 1997; Taylor et al., 2002), especially in the desmoplastic variant (Pietsch et al., 1997; Pomeroy et al., 2002). Shh signaling drives the expression of mRNA encoding N-myc protein, essential for expanding cerebellar granule neuron precursors. N-myc is stabilized by PI3K which inhibits GSK3 phosphorylation (Kenney et al., 2004). A cooperation of IGF with Shh/PTCH in producing medulloblastomas has also been described (Rao et al., 2004). Mice injected with Shh and c-myc developed medulloblastomas from nestin-expressing neural progenitor cells (Rao et al., 2003). An interesting observation is that during the development of granule cells, Bmi1 is expressed in the external granular layer promoting cell proliferation and is over-expressed in tumors where it can be an alternative or an additive to PTCH mutation (Leung et al., 2004).

Another developmental mechanism is that of Turcot's syndrome, represented by mutation of the APC gene, a component of the Wnt signaling cascade, activated by binding of Wnt ligands to the receptor Frizzled. The signals transferred to the nucleus act through APC, GSK3β, axin and β-catenin. Normally, β-catenin is phosphorylated by GSK3β and degraded into the proteasome; but binding of Wnt to Frizzled inhibits such phosphorylation; and β-catenin, translocated to the nucleus, activates a series of genes conditioning proliferation and differentiation of cells, such as c-myc, Cyclin D1 (Pomeroy and Sturla, 2004). β-catenin has been demonstrated in sporadic medulloblastomas (Huang et al., 2000; Yokota et al., 2002).

MATH1, essential for the development of the cerebellar granule cells regulating the components of the Notch signaling pathway (Gazit et al., 2004), PEDF encoding for an anti-angiogenic factor (Dawson et al., 1999) and BIRC5 encoding for survivin (Altieri, 2003) play a role in the development of the cerebellum, but are not significantly expressed in the adult cerebellum. High levels of MATH1 were found in hemispheric and in adult tumors and not in those of the vermis (Salsano et al., 2004), confirming their origin from the external granular layer and that of childhood and vermis tumors from the ventricular matrix (Katsetos et al., 1995). Other previous data, however, indicated that MATH1 was up-regulated in pediatric patients (Lee et al., 2003). High levels of PEDF were found in 30% of medulloblastomas, all showing high levels of MATH1, and BIRC5 was over-expressed in all the tumors, as is known for every kind of tumor (Salsano et al., 2004).

Other molecular genetic findings important in the pathogenesis of the tumor (Waha et al., 2004) are: the occurrence in >50% medulloblastomas of proteins of neural transcription factor PAX6 or mRNA of PAX5 (Giangaspero et al., 2000), the epigenetic silencing of the HIC-1 gene by methylation and the down-regulation of *trkC* by microarray gene profiling in tumors with poor prognosis (Pomeroy et al., 2002), whereas it is related to good outcome and differentiation of tumors. In tumors with metastasis at presentation, PDGFRA and members of the Ras/MAPK pathway are up-regulated in comparison with control tumors (MacDonald et al., 2001). Finally, in many medulloblastomas, α- and β-synuclein have been found in the cytoplasms, especially in pale islands, but not γ-synuclein (Fung et al., 2003).

6. PROGNOSTIC FACTORS

These were rather elusive and basically represented by M stage (Chang et al., 1969), location, surgery, age, radiotherapy and chemotherapy (Jenkin et al., 2000). They became more refined, with the improvement of our knowledge of the pathogenesis and molecular genetics of the tumor. Surgery, radiotherapy and chemotherapy led to 50–70% five years survival both in children (Giordana et al., 1995; Packer, 1999) and in the adult (Brandes et al., 1998), being the three main negative clinical factors age < 3 years, metastases at presentation and partial surgical resection (Packer, 1999). The identification of prognostic factors must be pursued in histology, proliferation capacity, differentiation/undifferentiation and the variants, but it is not easy. As for undifferentiation/differentiation, for example, no definite conclusion has been reached, partly because of the difficulties in assessing differentiation. Observations that differentiated tumors had longer survivals (Packer et al., 1984) than undifferentiated tumors were in contrast with other observations which demonstrated the opposite (Caputy et al., 1987). Glial differentiation was shown to be associated with favorable (Goldberg-Stem et al., 1991) or unfavorable (Janss et al., 1996) outcome of tumors and it was not prognostic (Coffin et al., 1990).

Proliferation capacity has long been considered from the prognostic point of view. Medulloblastoma is a malignant tumor with a Ki-67 LI >20%, with a range between 15% and 50%, but with a large regional variability. BrdU LI >20% was found to correlate with worse prognosis (Ito et al., 1992), but MIB-1 LI was not found to correlate with survival, both in adult and children tumors (Schiffer et al., 1995; Giordana et al., 1995). On the contrary, in pediatric medulloblastomas, MI was shown to correlate with survival (Gilbertson et al., 1997). Apoptosis (Figure 5), considered as the main cause of cell loss and important for establishing tumor capacity of growth, showed a positive (Korshunov et al., 2002) or negative (Haslam et al., 1998) correlation with shorter survivals. Intuitively correlated with longer survival, it is an event related with mitosis and, therefore, indicating proliferation, beside being elicited by different pathways (Schiffer et al., 2003), so that contrasting results are not surprising. In a series of 181 classic and desmoplastic medulloblastomas of children there were a higher MIB-1 LI, a lower AI and a lower AI/LI than in adult tumors, negating a biological difference between the two age group tumors (Sarkar et al., 2002).

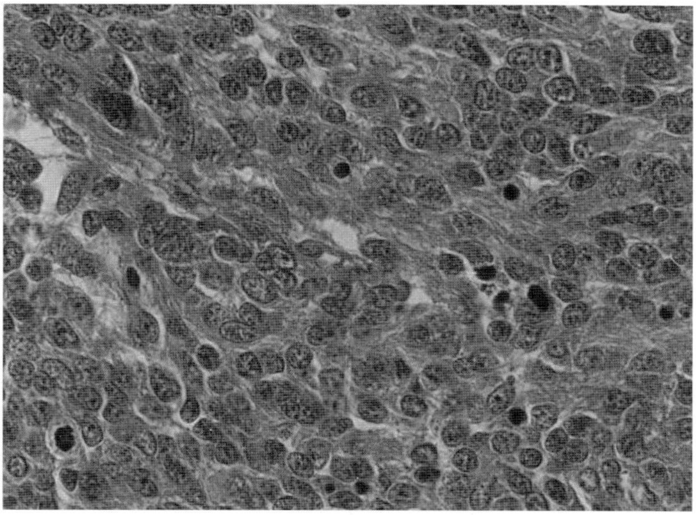

Figure 5. Apoptotic nuclei, H&E, x 400

The variants themselves represent prognostic factors and it must be acknowledged that a histological progression over time does occur in medulloblastoma. Desmoplastic medulloblastomas were regarded as showing a better prognosis than classic tumors, but their definition has changed in the last few years and it does not include any more tumors characterized only by a great amount of collagen and reticulin fibres. The diagnosis of desmoplastic medulloblastoma requires nodularity, even though the extent of this feature has not been specified. The grade of nodularity correlates with better survival and TTP, but not in a statistically significant way (Jenkin et al., 2000), whereas desmoplasia did not correlate (Eberhart et al., 2002) or showed a negative correlation in the adult (Giordana et al., 1995-97).

Medulloblastomas with *extensive nodularity and advanced neuronal differentiation* show longer survival, but the cases till now observed are too few for a significant evaluation. On the contrary, anaplasia correlates with shorter survivals and so does its extent (Eberhart et al., 2002), but most contributions have not been conclusive. The single items composing anaplasia, i.e. Ki-67 LI, mitotic or other indexes showed variable, mostly negative, correlation with survival (Giordana et al., 1997; Ellison, 2002), even though MIB-1 LI and PCNA LI appeared to be higher in adult than in children tumors (Giordana et al., 1997). Neuronal or glial differentiation showed variable correlation with survival, but with no conclusive achievements.

A new chapter has opened concerning the prognosis of medulloblastoma by Myc transcription factors, but expression of c-myc and increased TrkC levels have been found associated with less (Herms et al., 2000) or more (Grotzer et al., 2000) favourable outcome. Cytogenetics and CGH demonstrated that a 10–15% of medulloblastomas show amplification of *c-myc* or *N-myc* (Fruhwald et al., 2000; Gilhuis et al., 2001) and the bad prognosis of large cell/anaplastic medulloblastoma is associated with myc amplification (Jay et al., 1999; Brown et al., 2000). By *in situ* hybridization, c-myc mRNA was found to be significantly associated with shorter survival in 31% of medulloblastoma and with anaplasia, likely regulated by Wnt signaling and Mxi-1 mutation, whereas low N-myc and high TrkC expression were not significantly associated with longer survival (Eberhardt et al., 2004). In other findings, TRK and N-myc or C-myc were not prognostic, whereas ERBB2 was associated with a poor prognosis, especially in LC/A tumors (Gajjar et al., 2004). Loss on 17p was found to possibly correlate with survival (Ellison, 2003). Microarray profiling techniques are now greatly contributing to stratification and prognosis of medulloblastomas, but the results are not yet fully in agreement. In one finding, unfavorable prognosis was associated with STK15, stathmin 1 and Cyclin D1, but not with Cyclin B1, MYC, RAS, p53 (Neben et al., 2004). Combining gene expression profiles and clinical parameters for risk stratification it has been found that the former can predict medulloblastoma outcome independent of clinical variables (Fernandez Teijeiro et al., 2004).

7. RATIONAL BASES FOR THERAPIES

Beside radiotherapy and chemotherapy and the molecular bases of their rationale, focused on TP53, there is today a great interest in the possibilities of interfering with genes/proteins of the various molecular pathways at work in tumor development. Most of them derive from experiments on *in vitro* cultures of medulloblastoma cells or their precursors. As an important example, there is the possibility of specifically inhibiting REST/NSRF which is a neural silencer element regulated by wnt cascade (Willert et al., 2002) and expressed in medulloblastoma cell lines (Lawinger et al., 2000). Specific inhibitors may block PDGFRA and MAPK which enhance migration of tumor cells *in vitro* as well as ERBB2, clinically correlating with poor outcome and metastasis (Hernan et al., 2003) and promoting and activating MAPK, which in turn is associated with IGF, inversely correlated with TrkC. There are trials to antagonize these factors (Pomeroy et al., 2004).

8. PROBLEMS IN DIAGNOSIS USING SURGICAL SAMPLES

Treatment of medulloblastomas consists of surgery, radio- and chemotherapy. The clinical diagnosis is not difficult for the characteristics on CT and MRI and histological diagnosis do not present particular problems, unless there are transitional aspects. The main question for histological diagnosis is whether factors requiring variations in the treatment or influencing prognosis exist which affect the variant type. Most frequently the question concerns the desmoplastic variant, because this is not always so sharply definable. The occurrence of nodules, the so-called pale islands, is mandatory, but not that of hyper-production of reticulin, as could happen in the diffusion to the sub-arachnoidal space. Nodules can be easily recognizable, but sometimes they are not distinguishable from areas of neuropil with rarefied cells, free of reticulin. Sometimes they are immersed in a densely packed area of reticulin fibres, which by themselves do not authorize the diagnosis of desmoplastic medulloblastoma. Another point for debate is the focality of nodules with a double consequence. On the one hand, a desmoplastic focus can be missed when the surgical sample is too small. On the other hand, in a standard surgical sample one wonders whether the finding of a single or of a few desmoplastic foci is enough to make the diagnosis of desmoplastic medulloblastoma. In a large series of tumors, this problem has been solved by distinguishing desmoplasia, represented by rows of tumor cells alternating with bands of collagen and reticulin, from nodularity which means nodules: the two are not synonyms (Eberhart et al., 2002). On the scale of nodularity it is not known where the cut-off point is for calling a tumor desmoplastic. The confusion can be greater when the surgical sample is made almost entirely by the subarachnoidal growth of the tumor, with intense production of reticulin.

Large cell medulloblastomas fall into the category of anaplastic tumors of which they represent the highest degree (Eberhart et al., 2002). Regardless of the regional variability both of large cells and anaplasia which in small samples may be missed, there is no clear-cut separation of the two aspects from classic medulloblastoma if not in the extreme expressions. In the grading of anaplasia *large cell* tumors have been identified with the highest degree (Eberhart et al., 2002), but no indication is available about a cut-off point between slight and no anaplasia, i.e. between anaplastic and nonanaplastic tumors. "Cell wrapping", similar to that found in other tumors, seems to be nothing more than a sign of anaplasia.

Teratoid/rhabdoid tumors of children less than 2 years of age must be kept separate. On the other hand, this can be confirmed by molecular genetics with chromosome 22 losses and mutation of INI1 gene.

Finally, the possibility of a histological progression over time with the transformation of a non-anaplastic tumor into an anaplastic one must be acknowledged, even though rare, and this implies the occurrence of foci of high-grade histological features (Eberhardt et al., 2002; Perry et al., 2002), also independently of the variant.

9. DIFFERENTIAL DIAGNOSIS WITH OTHER EMBRYONAL TUMORS

Atypical teratoid/rhabdoid tumors. These are typical of children less than 2 years of age and are mainly localized in the cerebellum. Histologically they have a heterogenous aspect, being composed of undifferentiated cells, rhabdoid cells with round cytoplasm of epithelioid aspect, neuronal and glial cells. There is a positive staining for vimentin, often for smooth-muscle actin, sometimes for cytokeratin and EMA. The differential diagnosis must be carried out with medulloblastoma and is not easy by histological criteria alone. Detection of mutations of INI1/SNF5 gene (Biegel et al., 1999) may help in the diagnosis and it is a mandatory test in the protocol of Children's Oncology Group (COG), also since this variant may undergo a different therapy. A study of COG is under progress with the goal of identifying specific targets for therapies (Pomeroy et al., 2004).

Supra-tentorial PNET. Recognized by Hart and Earle (1973) and defined by Rorke (1981), these occur within the first 10 years of age, and are often cystic and necrotic. Morphologically they are similar to medulloblastoma (Figure 6) (McNeil et al., 2002), and may show differentiation toward the neuronal, glial or mesenchymal line. They do not share with medulloblastoma the same genetic alterations (Vogel and Fuller, 2003). Generally, they are kept separate from infra-tentorial medulloblastomas, also for different responses to therapies. Only rarely and in particular locations, e.g. the spinal cord, does the differential diagnosis have to be done with peripheral PNET.

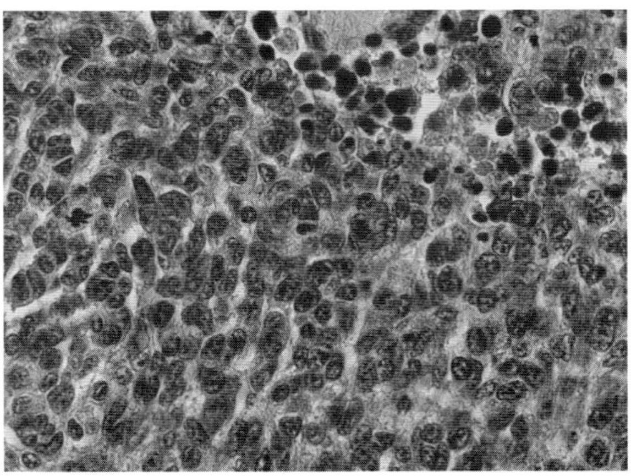

Figure 6. Supra-tentorial PNET. Densely packed hyperchromatic cells, H&E, x 400

Pinealoblastoma. It occurs mainly in children, but also in adults. It is initially located in the pineal region and histologically is composed of undifferentiated cells. Occasional neuroblastic rosettes and Flexner-Wintersteiner rosettes can be found. Necrosis and mitoses in variable amounts can occur. The outcome is poor. Differentiation from medulloblastoma is really difficult, especially when typical features of one or other of the tumors are lacking.

Primary rhabdomyosarcoma. They arise from the meninges, especially from the infra-tentorial ones, and are very malignant tumors, being fatal within two years. Histologically the tumors contain characteristically elongated cells with cross-striation that, however, is not mandatory. When absent, the diagnosis is based on muscle differentiation antigens such as desmin, and myogenin. Molecular genetics shows abnormalities on chromosome 11p15.

Ependymoblastoma. It is a rare tumor of children of 1 or 2 years of age. It must not be confused with anaplastic ependymoma. It shows typical rosettes containing mitoses and lacks pseudo-rosettes. It is a malignant tumor.

Chapter 10

PECULIAR TUMORS

1. ASTROBLASTOMA

This is a rare tumor affecting young subjects and located mainly in the hemispheres, histologically characterized by perivascular GFAP-positive astrocytes with processes radiating on a central vessel (Lantos and Rosenblum, 2000). It has not yet achieved a nosographic position, because it has not yet been clearly defined. Histological patterns like that described, can be found in gliomas of different types or they may characterize the entire neoplasia. Also the malignancy grade of the tumor is still uncertain, because astroblastic aspects can be found in benign and in malignant gliomas.

Few systemic studies have been dedicated to astroblatoma. In a series of 23 cases the tumor was observed to be well demarcated with perivascular pseudo-rosettes showing a prominent hyalinization. The tumor could be of low-grade with long survival or high-grade with recurrence and short TTP (Bonnin and Rubinstein, 1989), as confirmed later (Tiessen et al., 1998). In the most recently published series of 20 cases (Brat et al., 2000) the tumor appeared circumscribed on MRI, with a dominant solid enhancing component, and characterized by GFAP-, vimentin-, S-100-positive perivascular pseudo-rosettes with prominent hyalinization. Well differentiated forms show a MIB-1 LI <5%, whereas malignant forms, besides anaplastic nuclei, microvascular proliferations and necroses, show 4–15 mitoses/10 HPF and MIB-1 LI mean >15%.

From the nosographic point of view, astroblastoma has been considered a tumor *per se* and focal expressions of similar histological patterns, as seen in gliomas of different type and grade, do not indicate astroblastoma.

The differential diagnosis of the tumor must be carried out mainly towards ependymoma. As a matter of fact, there are similarities between the two tumor types, the most important of which is the perivascular orientation of tumor cells, the occurrence of intermediate filaments in astroblastoma and also of microvilli. On the other hand, cilia and inter-cellular junctions are so rare that astroblastoma cells show an intermediate differentiation between ependymoma and astrocytoma and have been interpreted as deriving from tanycytes (Rubinstein and Herman, 1989). Both tumors affect young people, are circumscribed, but astroblastoma has no ventricular or para-ventricular location. Histologically, the main difference is the lack of fibrillarity in astroblastoma which, on the contrary, characterizes ependymomas (Brat et al., 2000). Genetic alterations consisted of losses of chromosomes 10, 21 and 22 (Jay et al., 1993). Gains on 20q and 19 have mainly been demonstrated, as opposed to those found in astrocytomas or ependymomas (Brat et al., 2000).

Histology is a predictive factor. Low-grade tumors have long-term survivals after total surgical resection, whereas malignant tumors recur and progress either after total or partial resection and even after radiotherapy (Brat et al., 2000).

In small specimens the main pitfall is represented not by the possibility that a focal astroblastic expression is not included in the section, but by the opposite, i.e. that all the area examined is covered by an astroblastic focus of a glioma that shows different histological aspects in non examined areas. The misdiagnosis, however, is not a treatment affecting one, because the malignancy grade is established on the occurrence of the known criteria.

2. CAPILLARY HAEMANGIOBLASTOMA

Capillary haemangioblastoma is a grade I tumor associated with Von Hippel-Lindau disease in 25% of cases. It may have any location, but in sporadic form it occurs predominantly in the cerebellum. Histologically, it is characterized by two distinct components: vacuolated, lipid-containing cells (stromal cells) and a network of fine capillary (Figure 1). Intra-tumor hemorrhages and cysts frequently occur. The tumor is well demarcated and with a low proliferation rate.

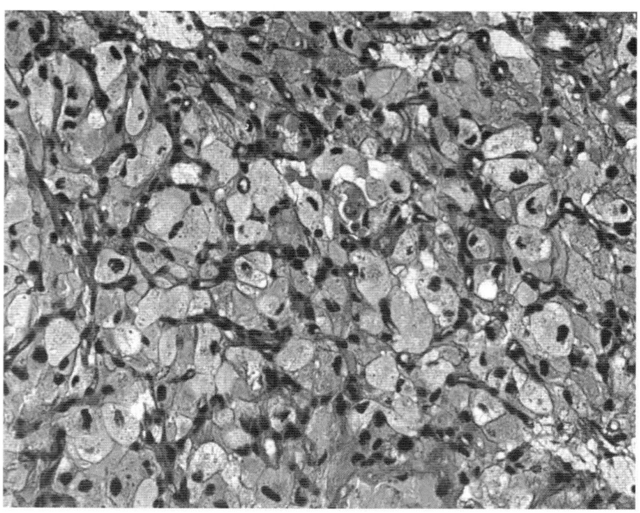

Figure 1. Hemangioblastoma, H&E, x 400

The main pathogenetic problem is represented by the origin of stromal cells. They have been supposed to derive from angiogenic (Lach et al., 1999) or undifferentiated mesenchyme (Frank et al., 1989), neuroendocrine cells (Becker et al., 1989), neuroectoderm (Kepes et al., 1979), but nothing definite has been established. The problem is made more difficult by the frequent finding of GFAP-positive cells of astrocytic type (Figure 2) which have been considered as reactive astrocytes entrapped in the tumor (Kepes, 1979; Schiffer et al., 1983) or as stromal elements that have taken up GFAP from reactive astrocytes (Deck and Rubinstein, 1981). The stromal component may be conspicuous (McComb et al., 1987), without reaching the picture of a mixed tumor (Bonnin et al., 1983). GFAP-positive staining has also been interpreted as a glial differentiation of stromal cells (Ishizawa et al., 2004). There are observations on the occurrence in stromal cells of other markers which are typical of neuroectodermal cells: for example, Ezrin (Bohling et al., 1996; Ishizawa et al., 2004), vimentin (Schiffer, 1997; Ishikawa et al., 2004), CD56 (NCAM), CD57 (Leu-7), CD99, NSE, VEGF and Flk-1 (Hatva et al., 1996; Machein and Plate, 2000). This wide positivity of neuroectodermal markers could indicate a differentiation of stromal cells with respect to neuroectoderm (Ishizawa et al., 2004).

Cases have been reported of clear-cell ependymomas mimicking capillary haemangioblastoma with GFAP-positivity, perivascular pseudo-rosettes and microvilli under electron microscopy and abundant capillaries (Kawano et al., 1999). This interpretation has been criticized and the tumors have been considered as real haemangioblastomas (Burger, 1999). However, other cases of haemangioblastomas expressing microscopic, immunohistochemical and ultrastrucutral features of ependymoma have been reported (Hishizawa et al., 2004). These cases, sharing many antigenic properties with neuroectoderm, have been interpreted in the framework of the capacity of neuroectodermal differentiation of stromal cells. This possibility has been confirmed by another case of hemangioblastoma with repeated recurrences, in a patient with characteristics of Von Hippel-Lindau disease, with increasing areas of glial differentiation interpreted as deriving from GFAP-expressing stromal cells. The observation could hint at a glial histogenesis of haemangioblastoma (Adams and Hilton, 2002).

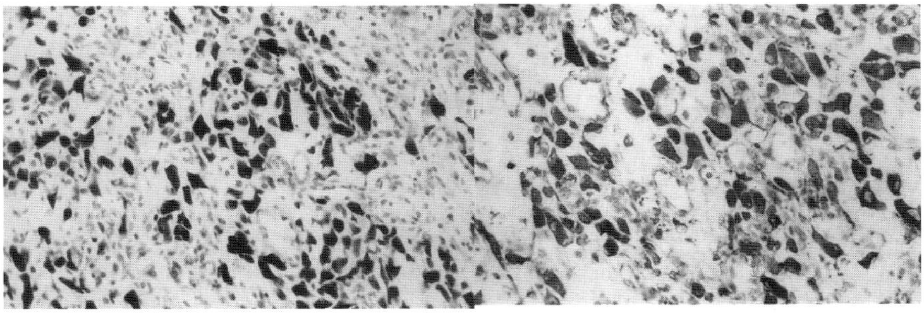

Figure 2. Hemangioblastoma. A. GFAP-positive cells, DAB, x 400; B. Vimentin-positive cells, DAB, x 400. From. D. Schiffer, 1997

3. GLIOMATOSIS CEREBRI

Described by Nevin (1938), it has been defined as a diffuse glial tumor infiltrating the nervous system, affecting more than two lobes and extending to infratentorial structures. It is a grade III lesion (Lantos and Bruner, 2000). In order, the following regions are affected: brain hemispheres, mesencephalon, pons, thalamus, basal ganglia, cerebellum and bulb. On MRI it shows enlarged nervous structures, hyperintense on T2 weighted images, with no contrast enhancement and a low relative cerebral blood volume, in accordance with the absence of vascular proliferation in the tumor (Yang et al., 2002) (Figure 3). At autopsy the involved structures appear enlarged.

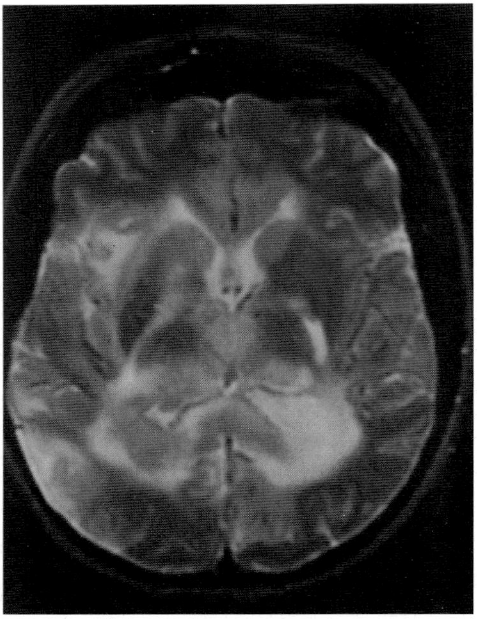

Figure 3. Gliomatosis cerebri, MRI, T2-weighted image. From the Neuroradiological Unit, Dpt Neuroscience, University of Turin

Histologically it is composed of elongated glial cells of the astrocytic type, with oval nuclei, which can be distributed in rows parallel to those of myelin fibres (Figure 4A). In some cases there are also neoplastic masses. Focally, the histological aspect may be that of astrocytoma, oligoastrocytoma (Figure 4B), anaplastic astrocytoma or oligoastrocytoma, or the tumor may be prevailingly of oligodendroglioma-tous type (Balko et al., 1992). MIB-1 LI is usually low, but it varies according to the examined areas.

The histogenesis of the lesion is still a matter for discussion. It may be conceived of as a common glioma, but purely infiltrative or as a different tumor entity or of oligoclonal origin or as a collision glioma. Median survival is 38 months (Kim et al., 1998).

Molecular genetics: TP53 mutations have been found in 3 out of 7 cases, whereas PTEN mutations and EGFR over-expression have been found in 1 case only (Herrlinger et al., 2002). In one case, the study of TP53 mutations in the

A

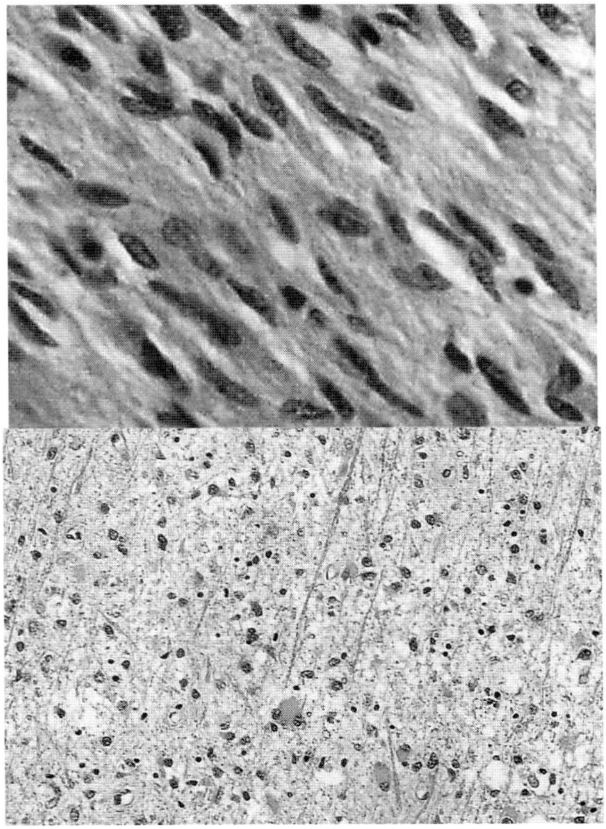

B

Figure 4. Gliomatosis cerebri. A. Elongated astrocytic cells among myelin fibres. H&E, x 400; B. Oligoastrocytomatous aspects, H&E, x 200

different areas of the brain led to the conclusion of a clonal proliferation and therefore it confirmed the nature of an infiltrative low grade glioma (Kros et al., 2002). In a series of 18 cases, TP53 mutations were found in two cases only, at the sites known to be altered in malignant astrocytomas; and in another case different TP53 alterations have been found in different tumor areas, so that polyclonality could not be ruled out, whereas no mutation of PTEN was found (Mawrin et al., 2003). Nine cases were recently examined and the rare genetic alterations found seemed to reflect tumor progression associated with astrocytoma transformation, even within a single case (Mawrin et al., 2005). Radiotherapy and chemotherapy by PCV seem to be indicated.

The histological recognition of the tumor in small biopsies can be difficult. The clinical diagnosis of the tumor largely depends on neuro-imaging and the histological diagnosis of biopsies depends on the area where they have been taken from. Since the histological pattern found may belong to different neoplastic lesions, the diagnosis will depend largely on the aspects of neuro-imaging, whereas the identification of the histological grade will be reliable only if it is grade III or grade IV.

4. CHORDOID GLIOMA OF THE IIIrd VENTRICLE

Recently described (Brat et al., 1998), this is an infrequent lesion of the floor of the third ventricle. Not more than 29 cases have been described till now (Sato et al., 2003). Histologically it is composed of clusters or cords of polygonal cells immersed in a PAS-positive mucinous matrix. The cells are positive for GFAP, vimentin and CD34. MIB-1 is <1%. The nature of the tumor has been discussed and the ependymal (Cenacchi et al., 2001; Pasquier et al., 2002) or tanycytic (Sato et al., 2003) origin have been suggested, the latter supported by location, GFAP-positive staining, and ultrastructure (Sato et al., 2003).

Chapter 11

CELL MIGRATION AND INVASION

1. GENERAL REMARKS

One of the factors which prevent malignant gliomas from being cured is their extensive diffusion into the surrounding brain tissue. Motility of glioma cells, demonstrated both *in vivo* and *in vitro*, and invasion are the products of events involving extracellular matrix (ECM) (Pilkington, 1994), adhesion molecules such as cadherins, integrins and chemoattractants and properties of the cells themselves. Fundamentally, cell motility increases with malignancy (Chicoine and Silbergeld, 1995) and all its mechanisms are at the basis of glioma spreading. The growth of gliomas is generally attributed to cell proliferation which conditions invasiveness, but not enough attention has been paid to cell motility. In a mathematical model it has been calculated that the velocity of expansion is linear with time and varies from about 4 mm/year for low-grade gliomas and 3 mm/month for high grade gliomas (Swanson et al., 2003).

2. MECHANISMS OF MIGRATION AND INVASION

There are excellent less recent and recent reviews on these mechanisms involving ECM, proteolytic activity of tumor cells, cell adhesion molecules and regulating factors. Moreover, different experimental tumor models are available. First of all it is really important to know that neoplastic glial cell motility is dependent upon dynamic remodeling of the actin cytoskeleton, that vimentin characterizes developing and poorly differentiated glial cells such as nestin and that these are typical of developing neuroectodermal cells. Co-expression of nestin and vimentin serves as a marker of enhanced motility and invasion in gliomas and GFAP has the opposite significance (Bolteus et al., 2001). The components of ECM, i.e. laminin, fibronectin, collagen IV, tenascin and vitronectin interact with invading glioma cells as permissive substrates (Tysnes and Mahesparan, 2001).

Tenascin is highly concentrated around hyperplastic vessels in gliomas (Zagzag et al., 1995) and enhances migration of endothelial cells and phosphorylation of FAK (focal adhesion kinase) which interacts with integrin-β1 mediating tenascin C signaling

(Plopper et al., 1995). It is over-expressed in invasive gliomas (Mariani et al., 2001). The interest in tenascin has recently increased, because specific anti-tenascin antibodies labeled with I^{131} are being used for therapy (Bigner et al., 1995; Goetz et al., 2003).

The proteins of the matrix must be disrupted by proteases or protease-activators such as the zinc-dependent enzymes metalloproteinases (MMPs), classified as collagenases, gelatinases and stromelysin and secreted as proenzymes, which are in balance with their inhibitors or tissue inhibitors of metalloproteinases (TIMPs). Urokinase-type plasminogen activator (uPA) binds to its receptor converting plasminogen to plasmin that degrades fibrin, laminin, fibronectin and proteoglycan. Among cysteine proteinases, cathepsin B must be borne in mind. Cell adhesion to ECM is favored by integrins, composed by transmembrane glycoprotein units of which β1 is the critical one. Integrins variously occur on glioma cells both in cell lines and biopsies and can be considered as the rungs of a ladder on which the cells are attached (Tysnes and Mahesparan, 2001). The *αvβ3* complex can recognize several ligands, such as laminin, fibronectin, vitronectin and C-tenascin and it can play a role in the angiogenesis activating VEGFR-2. Finally, other molecules must be considered: cadherins are cell-cell β-adhesion molecules regulating adhesion that can be influenced by NCAM. Hyaluronates are important in many inter-cellular processes, whereas CD44 is the principal receptor of hyaluronan, and inhibits adhesion of glioma cells to fibronectin, laminin, vitronectin and collagen I. It is present in glioblastomas and shows numerous isoforms derived from alternative splicing the functions of which are still unclear. Its blocking reduces invasion (Bolteus et al., 2001).

Many factors intervene in regulating cell migration and invasion, first of all EGF. *In vitro*, cells with strong expression of EGFR are more stimulated to migrate than those with lower expression (Tysnes et al., 1997). In this regard the observation that highly amplified cells for EGFR are found at the invading edge of the tumor rather than at the solid tumor centres acquires a specific significance (Okada et al., 2003). Another very important factor in the regulation of tumor glial cell motility is PTEN: Its phosphatase-independent domains reduce the invasive potential of glioma cells, distinct from the PKB/Akt pathway (Maier et al., 1999). There is a complicated integrin-mediated signaling to which kinases such as FAK and ILK belong. FAK seems to be necessary for integrin-mediated motility (Sieg et al., 2000) and it is in the focus of a very complicated circuit (Günther et al., 2003). Integrins also play a role in cell growth and proliferation. Other factors involved in controlling cell motility are SF/HGF (scatter factor/hepatocyte growth factor) (Lamszus et al., 1999), TGF-β1 (Merzak et al., 1995). The regulation of the entire process of cell invasion is really not so simple (Demuth and Berens, 2004).

3. PATHOLOGICAL FINDINGS

Many neuropathological contributions during the last decades outlined how gliomas spread in the brain; and a systematic study of the spreading in one hundred autopsy cases of glioblastomas and astrocytomas has been carried out in our Institution (Schiffer, 1986). Knowledge about the spreading modalities of gliomas is important especially when a pathologic evaluation has to be done on small surgical samples with the purpose of recognizing the tumor type from its spreading modalities or to establish whether a tissue sample does or does not contain glioma cells. Gliomas may spread in the homolateral hemisphere or to the contralateral hemisphere, mainly along the long axis of short and long fibre bundles. Typical is the diffusion to the contralateral hemisphere through the corpus callosum and lamina terminalis. Fibre bundles may also represent an obstacle to diffusion, when they are reached by tumor cells along their short axis. Each tumor location has preferential pathways: fronto-parietal tumors may spare the temporal lobe; and the opposite occurs with occipital tumors. Temporal tumors may spread toward the hypothalamus and low midline structures or the temporal stem. Interestingly, glioblastomas with evident astrocytic areas, very likely remnants of a previous astrocytoma, spread more frequently to the same hemisphere, whereas glioblastomas, very likely of primary origin, spread through the corpus callosum (Figure 1).

Spreading modalities	%	Tumor type
Homolateral diffusion	44	1 > 3 > 2
Contralateral diffusion	56	3 > 2 > 1
Sub-arachnoidal (intracranial) diffusion	25	2 > 3 > 1
Sub-pial diffusion	9	2 > 1 > 3
Corpus callosum	35	3 > 2 > 1
Septum pellucidum – fornix	18	3 > 2 > 1
Infiltration > 2 cm from tumor edge	22	2 > 3 > 1
Seeding on ventricular walls	6	3 > 2
Multicentric growth	8	2 > 3 > 1
Necrotic tumor with no regrowth	10	1 0 0

1 = tumors with evident astrocytic character; 2 = tumors with diffuse anaplastic aspect; 3 = tumors with mixed aspects

Figure 1. Spreading modalities of glioblastomas

A

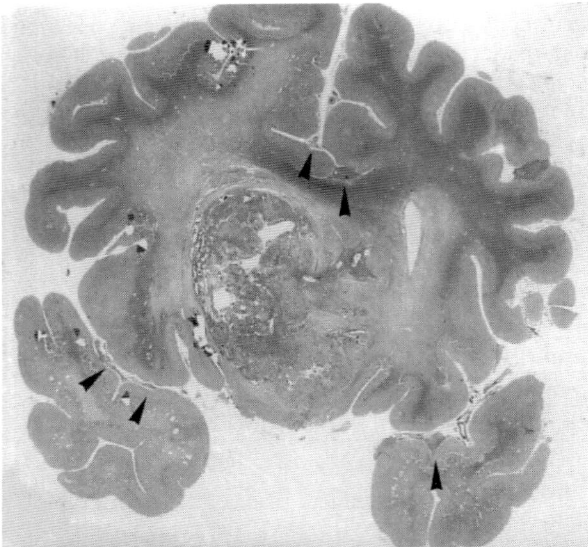

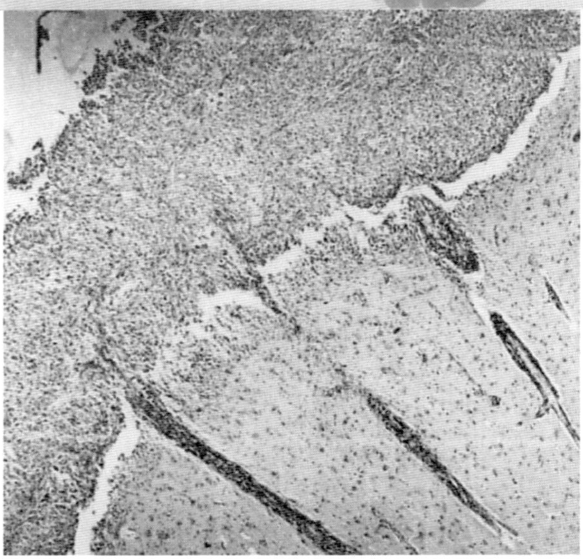

B

Figure 2. Glioblastoma. A. Sub-arachnoidal seeding; B. Invasion of the cortex through penetrating vessels, H&E, x 100

A

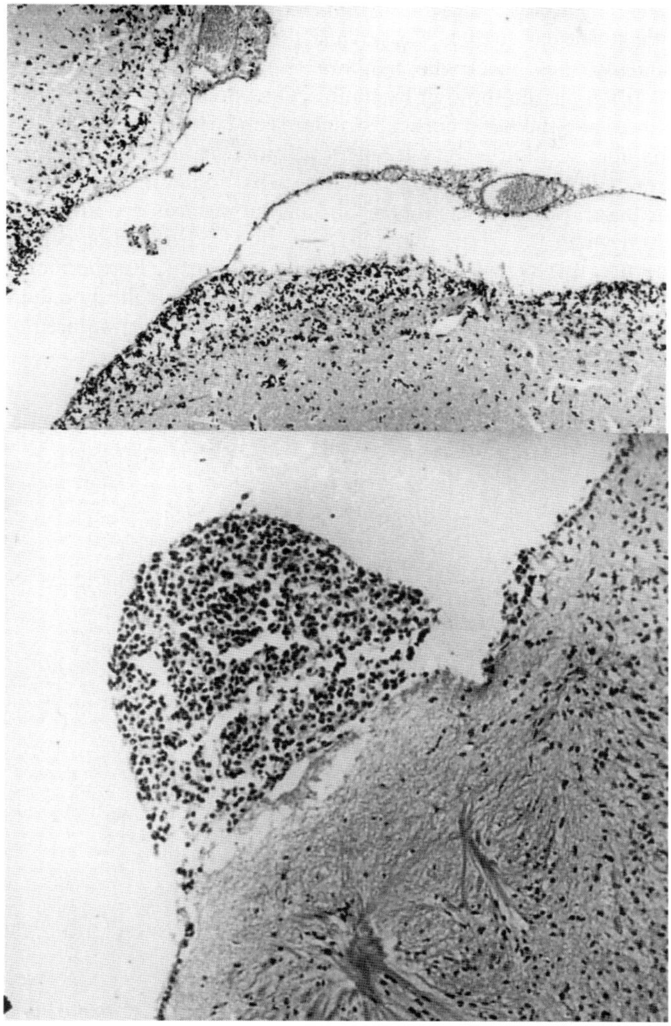

B

Figure 3. Glioblastoma. A. Sub-pial invasion; B Seeding on the ventricular wall, H&E, x 100

The cerebral cortex is invaded from the tumor in the white matter, either with or without a perineuronal satellitosis, or from a sub-pial infiltration (Figure 3A) from a tumor that invaded the opposite gyrus, or from cells coming down from sub-arachnoidal seedings along penetrating vessels (Figure 2A, B). Basal ganglia are invaded by local tumors, which also invade the corpus callosum, or by adjacent tumors. Frequenly they reach the temporal stem or the hypothalamus. Septum pellucidum is often passed through by tumor cells which establish a traffic between the hypothalamus and the basal cortical structures and corpus callosum.

Sub-arachnoidal seeding is very frequent (Nishio et al., 1982; Rosenblum, 1995) and it is represented by small clusters of tumor cells, sometimes visible to the naked eye. Anterior basal, posterior cerebellar and lateral cisterns are involved and even sagittal scissura when the gyrus cynguli is invaded. When tumor cells invade the underlying cortex, this shows a remarkably intense gliosis. Furthermore spreading in the ventricular system is frequent (Figure 3B). Tumor cells collect on the ventricular surface and adhere to it where ependymal cells are lacking on an area of pilocytic gliosis.

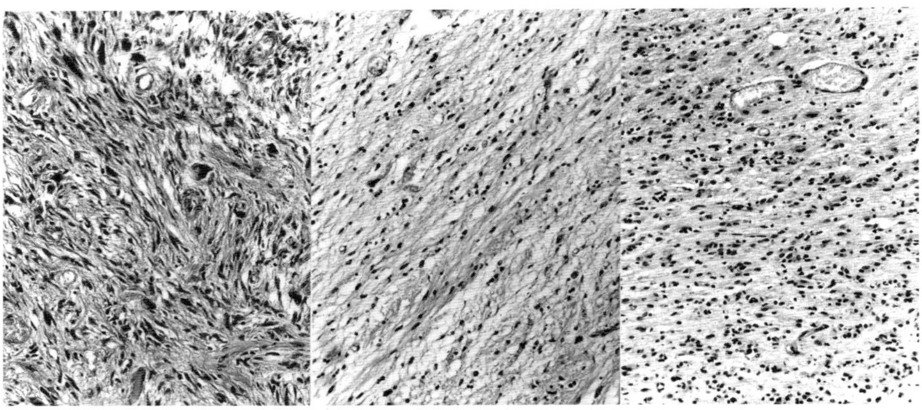

A B C

Figure 4. Glioblastoma. A. Solid tumor; B, C. Periferal rarefied areas. H&E, x 200

The most important points of the tumor spreading are the existence of a gradient of tumor cell density towards normal tissue and how far from the tumor edge neoplastic cells can be found and recognized (Figure 4A, B, C). These data are of paramount importance for tumor resection and for the planning of post-surgical irradiation. Classically 2 cm distance from the tumor edge is considered the safety limit (Burger et al., 1988). Given the great practical weight that this point has and the frequency by which infiltrated cortex represents the entire surgical specimen given to the pathologist, it is useful to go into details. However, the discussion of this important point requires previous consideration of what kind of relationship exists, if any, between cell invasion and cell proliferation, since there is *in vitro* evidence suggesting that the two events may be antithetic (Pilkington, 1992; Merzak et al., 1995) and examples of infiltrating, but non-proliferating tumor cells are known (Darlymple et al., 1994).

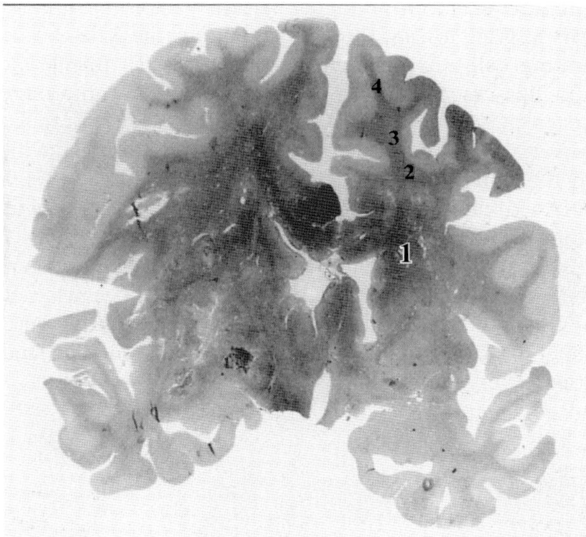

Figure 5. Glioblastoma. Gradient of tumor cell density toward normal tissue, H&E

Cell Count (mean number of nuclei per μm^2)	
1. Normal white matter	96±10
2. Tumor peripheral area	213±36
3. Infiltrated area	171±13
4. Apparently normal area	148±8
5. Edematous infiltrating area	55±8

Between the solid tumor and the cortex there is a cell density gradient (Figure 5) more frequently than between a solid tumor and the white matter where the border is sharp. There is also a gradient for mitoses and nuclei stained for proliferation markers, such as MIB.1. When the border is clear-cut the gradient is also sharp. In this case, a tumor peripheral ring with high mitotic density or MIB.1 LI occurs. In long fibre bundles nuclei positive for MIB.1 or PCNA can be found in an otherwise completely normal structure (Figure 6C, D).

The occurrence of isolated tumor cells can be ascertained by this method (Burger et al., 1986; Schiffer et al., 1997) or by stereotactic (Kelly et al., 1987) or systematic topographic studies (Burger et al., 1988; Burger and Kleihues, 1989). They can also be deduced from cell counting in the peri-tumor tissue, if cell concentration is above a certain value (Schiffer et al., 1997) (Figure 5). It may happen that in infiltrated cortex where tumor cells migrate toward the cortical surface the MIB 1 LI is very low as well as in migrating cells that accumulate under the pia membrane in the outer cortical layers. In this case there is a dissociation between the migratory and the proliferation capacity (Schiffer et al., 1997). *In vitro*, the two properties appear as mutually exclusive: cells expressing A2B5, i.e. gangliosides, which are highly expressed during development in migratory cells (Small et al., 1987), are not labeled by BrdU or PCNA (Pilkington, 1992–94). Also TGF-β1 intervenes modulating cell adhesion and stimulating ECM formation (Pilkington et al., 1994). Isolated tumor cells and solid tumor cells seem to be under different genetic control (Liotta and Stetler-Stevenson, 1991). This finding is very important, because radiotherapy and certain forms of chemotherapy are likely to be scarcely effective on these cells. On the contrary, the proliferation rate of subarachnoidal seedings and of the cells invading the cortex is very high.

A B

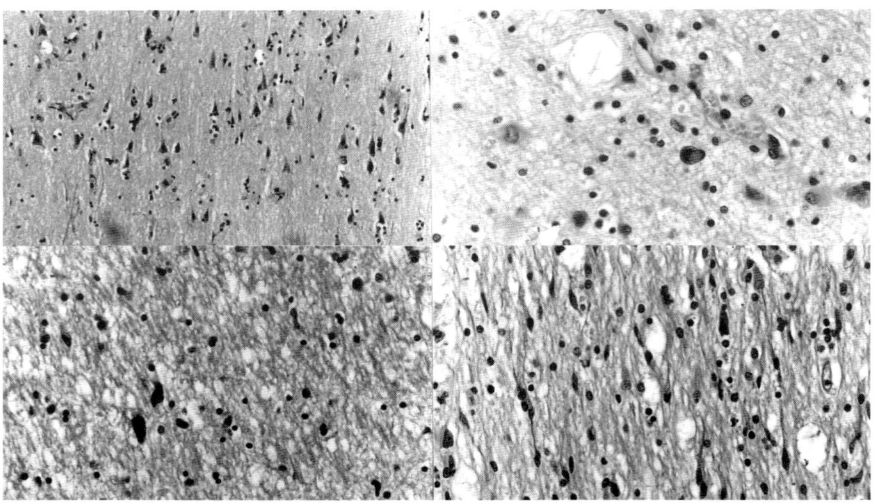

C D

Figure 6. Glioblastoma. A. Uncertain cortical infiltration; B. Abnormal nuclei in peritumoral white matter after irradiation, H&E , x 200; C, D. PCNA positive nuclei in almost normal white matter, DAB, x 400

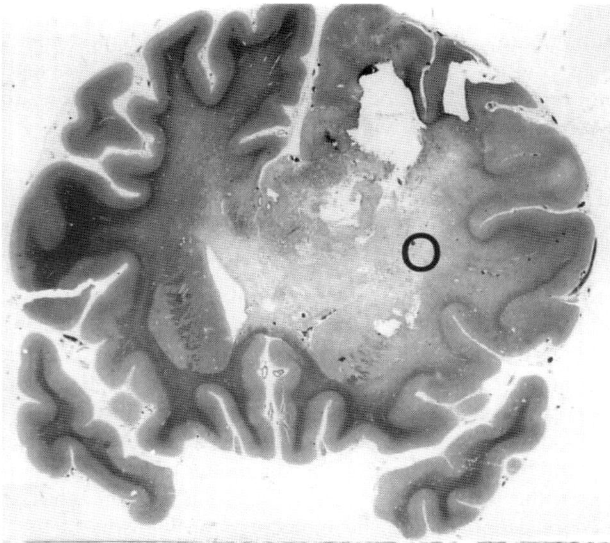

Figure 7. Radionecrosis. Luxol Fast Blue B

A

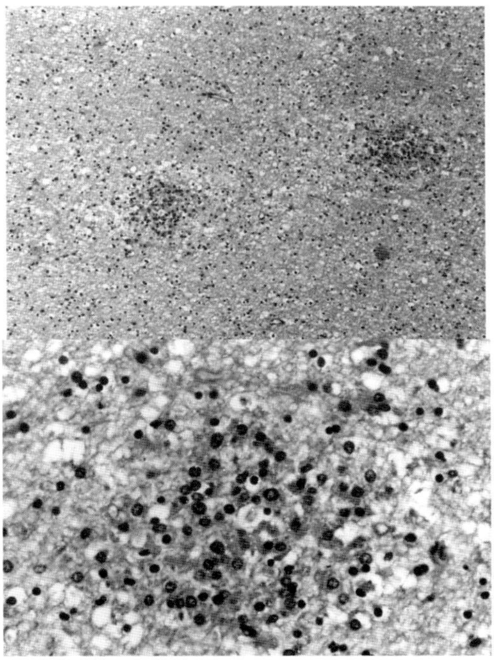

B

Figure 8. Glioblastoma after irradiation. Radionecrosis A. Two cell foci in the circle of Fig. 7, H&E, x 200; B. At a higher magnification, H&E, x 400

Very interesting are the regrowth modalities (Schiffer et al., 1982) of malignant gliomas after radiotherapy. They are not discussed in this chapter and radionecrosis is just referred to (Figure 7). Particularly important are the frequent findings of abnormal, pleomorphic nuclei around the tumor after irradiation (Figure 6A, B) and the occurrence of tumor foci in the radionecrosis (Figure 8).

4. THERAPEUTIC CONSIDERATIONS

Knowing the pathological aspects of cell migration and invasion of gliomas has a practical importance in the recognition of tumor type and grade from events associated with malignancy or with specific tumor types. This is especially useful when the tissue sent to the pathologist does not come from a full solid tumor or it is sent just for detecting a macroscopically undetected tumor invasion. It is important also as a tool for evaluating experimental procedures against migration and invasion, both in humans and in animal models. As a matter of fact, cell migration and invasion have been the target of a series of therapeutic approaches and for these attempts to go more deeply inside the molecular mechanisms at the basis of the different steps leading to migration and invasion represents one of the future guidelines to halt glioma growth.

The possibility that therapies enhance invasion in malignant gliomas has also been considered. Irradiation may activate wt p53 and increase MMP-2 expression (Bian and Sun, 1997) through Bcl-2 activation which, regulating cell-cell interaction, includes integrin-dependent regulation of cell adhesion via r-ras (Reed, 1997; Zhang et al., 1996). Sub-lethal radiotherapy doses promote migration and invasiveness of the glioma cells that become responsible for local relapse of the tumor (Wild-Bode et al., 2001) .

Chapter 12

APOPTOSIS

1. DEFINITION AND REGULATION

Cell cycle time, growth fraction, tumor doubling time and cell loss regulate tumor growth which can be indexed by the balance between cell loss and cell proliferation. Cell loss is due to necrosis and apoptosis. Necrosis is a sudden event killing at the same time more cells through cell membrane damage, energy depletion and inflammatory response. Apoptosis is a "programmed cell death" with nuclear changes, preserved cytoplasmic organelles, new gene transcription and complicated regulatory pathways (Hengartner, 2000). In gliomas, it is considered the major source of cell loss and, unlike neurodegenerative diseases where apoptosis represents an ominous sign as it is regarded as the way neurons die, in gliomas as in other tumors it is interpreted as the opposite of cell proliferation. Therefore, in gliomas it is studied and discussed from two points of view: as a possible prognostic factor and as the expression of a complicated molecular network to which pathways to cell death and to cell proliferation belong.

2. REGULATORY CIRCUIT OF APOPTOSIS

Apoptotic nuclei can be recognized in tissues because of chromatin compacting, splitting and leaning against the nuclear membrane, followed by the formation of apoptotic bodies which are then phagocyted by macrophages. The entire process lasts not more than few hours. DNA is split by a DNAse in fragments of 180 bases. Inserting nucleotides labeled by digoxygenin in the DNA breaks and using polymerase or nucleotydil terminal transferase represents the immunohistochemical way of demonstrating apoptosis, called TUNEL. The same basis supports the DNA "laddering" on gel electrophoresis which is the biochemical way of demonstration (Wyllie et al., 1999).

The regulation of apoptosis is extremely complicated and in mammals it can be initiated by three distinct pathways (Figure 1). The intrinsic (Green and Reed, 1998) or transcriptional way via mitochondria is centered on p53 which, activated by DNA damage signaling, induces cell cycle arrest through p21, DNA repair through PARP (Rich et al., 2000) and apoptosis through Bax transportation to mitochondria which release AIF (apoptotic inducing factor), Smac/DIABLO and cytochrome c that through Apaf-1, procaspase-9 lead to activation and cleavage of caspase-3, the last step before DNA breaking. Caspase-3 cleaves the inhibitor of caspase-activated

DNAse (ICAD) and activates CAD followed by DNA fragmentation (Hengartner, 2000). In this system Bcl-2 has an anti-apoptotic function activating the transcription factor NFkB (Zhivotosky et al., 1997) which transfers to the nucleus when released by IkBα that keeps it in the cytoplasm.

There is also an extrinsic (Ashkenazi et al., 1999) or receptorial pathway which can be triggered by ligation of death receptors such as Fas or CD95 or APO-1, TNF or TRAIL receptors and proceeding through pro-caspase-8 DISC, inhibited by FLIP, activates caspase-3. The triggers belong to the super-family TNF, such as APO-2 (TRAIL- TNF-related apoptosis inducing legand) acting with the agonistic receptors DR4 and DR5 and through the death domains FADD and TRADD. Cross-talk between the intrinsic and extrinsic pathways can occur. Activated caspase-8, alternatively to the activation of caspase-3, can cleave Bid (Bcl-2 inhibitory BH3-domain-containing protein) to activate the intrinsic pathway (Gross et al., 1999). Truncated Bid translocates to mitochondria activating their pathway through the release of cytochrome c, AIF, SMAC/Diablo, IAP (Song et al., 2003). Proteasome inhibitors induce Fas-mediated apoptosis by c-Myc accumulation and subsequent induction of FasL in human glioma cells (Tani et al., 2001) or p53/p21-independent apoptosis (Wagenknecht et al., 1999) or they involve the processing of multiple caspases and cytochrome c release (Wagenknecht et al., 2000).

Also PI3 kinase-Akt pathway intervenes in regulating apoptosis, controlled in turn by Ras and by PTEN.

There are also a ceramide pathway downstream CD95, c.Jun/JUNK or IkBα (Alkalay et al., 1995) and the granzyme B pathway. Besides p53, Bax, PARP, caspases, also the function of NFkB must be considered (Amirlak and Couldwell, 2003).

The regulation of apoptosis is very complicated, because it is at the crossroads of different pathways, including the Ras-MAPK one leading to cell proliferation. IAP (inhibitory of apoptosis proteins) produced by mitochondria and inhibiting caspase3, including Survivin particularly important in tumors (Li et al., 1998), must be mentioned, in the effort to prevent apoptosis. IAP can be inhibited in turn by SMAC/Diablo produced by mitochondria themselves (Wu et al., 2000).

An inconsistency seems to exist between the TNFα related TRAIL induced apoptosis and apoptosis induced by inhibitor of the ubiquitin-proteasome system. It has been shown that in human astrocytoma CRT-MG cells inhibition of ubiquitin-proteasome system enhances TRAIL-mediated apoptosis, because it blocks degradation of caspase-8 and -3. Proteasome inhibitors suppress TRAIL-mediated activation of NFkB, whereas NFkB inhibition alone is not sufficient to enhance TRAIL-mediated cell death. The ubiquitin-proteasome system plays a role in the anti-apoptotic surveillance (Kim et al., 2004).

Poly(ADP-ribose)polymerase (PARP) is an enzyme located in the nucleus and activated in response to DNA damage for its repair (Oliver et al., 1998) after radiation or chemotherapy (Wang et al., 1998). The intact molecule prevents apoptosis keeping endonuclease inactive (Rice et al., 1992). PARP is cleaved by caspase and its cleavage can be assumed as a marker of early apoptosis (Bursztajn et al., 2000). When over-activated, it consumes ATP leading the cell to necrosis. (Ha et al., 1999). PARP and its cleaved fragments can be recognized by monoclonal

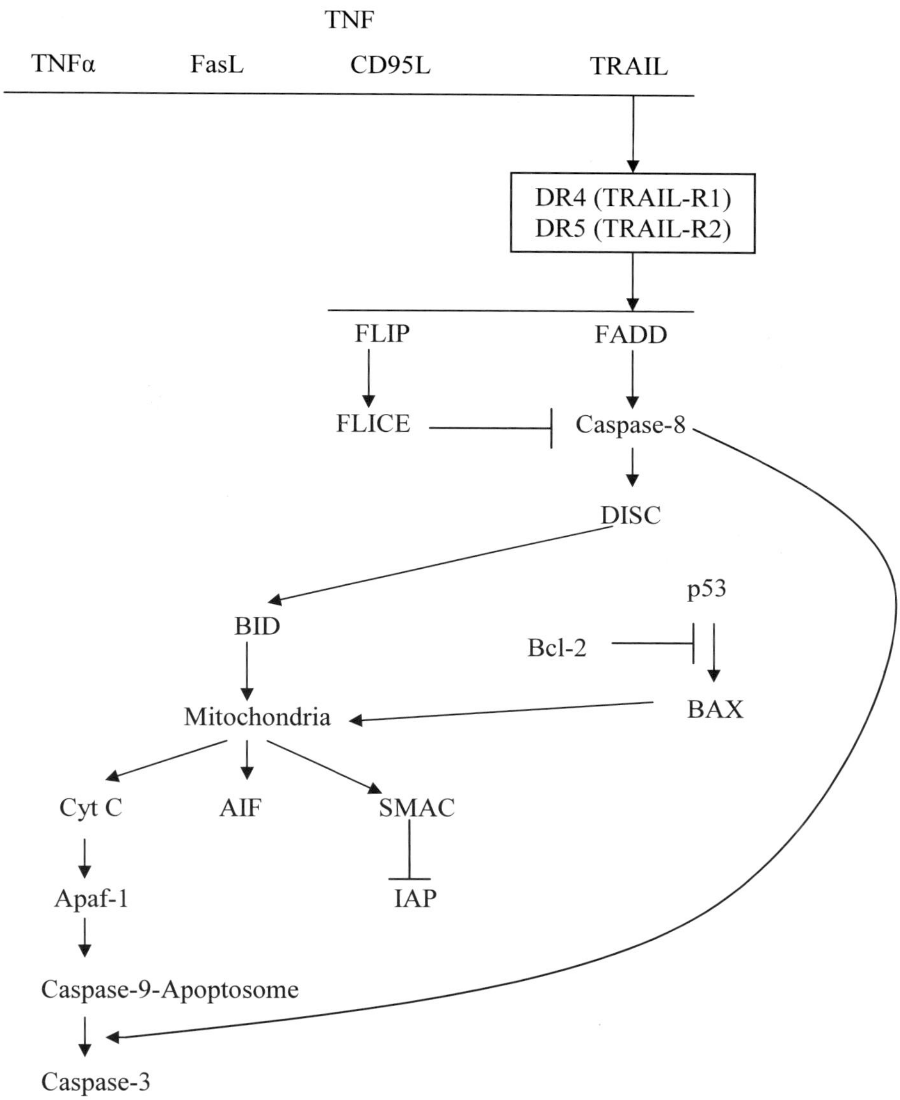

Figure 1. Apoptotic pathways

antibodies (Duriez et al., 1997). In medulloblastomas cleaved PARP co-localizes with cleaved caspase-3 (Puig et al., 2001).

Interesting observations are available on genes/proteins of the apoptotic pathways which can be evaluated in relation to tumor pathogenesis or therapies. For example, LOH of 12q22 where Apaf-1, the major p53-mediated apoptosis effector, is located has been found in 42% of glioblastomas to reduced expression of its mRNA and protein and with no relation to TP53 mutations and EGFR amplification. This means that the abrogation of Apaf-1 and of p53 mediated apoptosis can play a role in the tumorigenesis of glioblastoma (Watanabe et al., 2003).

3. ASSESSMENT OF APOPTOSIS IN TISSUES

The assessment of apoptosis in tissues is based on nuclear morphology, positive TUNEL and DNA laddering on gel electrophoresis. The direct demonstration has been long discussed and good reliability has been recognized for TUNEL. It should however, be critically applied, because different staining intensities of nuclei and the staining of nonapoptotic nuclei can be observed. Once the problem is resolved whether in apoptosis single or double DNA strand breaks take place and whether they can be distinguished using polymerase or nucleotydil transferase (Iseki, 1986; Wijsman et al., 1993; Gold et al., 1993,94; Gorczyca et al., 1993; Migheli et al., 1994; Mundle e Raza, 1995), it is ascertained that TUNEL can show up DNA breaks which do not belong to apoptosis, but to necrosis, DNA duplication, gene transcription and even post-mortem autolysis (Figure 2). Through a cut-off of the different nuclear reaction intensities, an intense positive staining due to apoptosis has been distinguished from a weak reaction due to DNA duplication which could acquire a progressive indication of proliferation (Rhodes, 1998). It could indicate also a single DNA strand break (Sgonck and Wick, 1994), or cells which survived apoptosis (Arends et al., 1994; Tomlinson and Bodmer, 1995). It is known that DNA breaks are associated with the accumulation of mutations and that tumor cells which escaped apoptosis live with a broken DNA that can accumulate mutations favoring tumor progression. On the other hand, tumor cells after exposure to radiation or chemotherapy can repair themselves, thus escaping apoptosis (Amirlak and Couldwell, 2003).

It may also be that apoptosis or its initial stages escape the demonstration for the short duration of the process and that early stages can be detected by single strand DNA antibodies (Frankfurt et al., 1997; Korkolopoulou et al., 2001). The best way to assess apoptosis in tissues is the association of morphology with TUNEL.

Recently a more objective and comprehensive method of quantitatively analyzing apoptosis and, at the same time, associated events is laser scanning cytometry (Amirlack and Couldwell, 2003). This could overcome some limitations of traditional microscopic study, including the short duration of the process.

Indirect demonstrations of apoptosis are those based on the many proteins/genes involved in the pathways to apoptosis or in their regulation. Caspase-3 is the most important one, because it is the last step in the cascade to apoptosis. However, independently of its being really the no-return point in the way to apoptosis, it can be positive in the cytoplasm, in the nucleus or in both, or even to be negative once apoptosis is completed. Its positive staining corresponds, but does not overlap with TUNEL positivity (Schiffer et al., 2001).

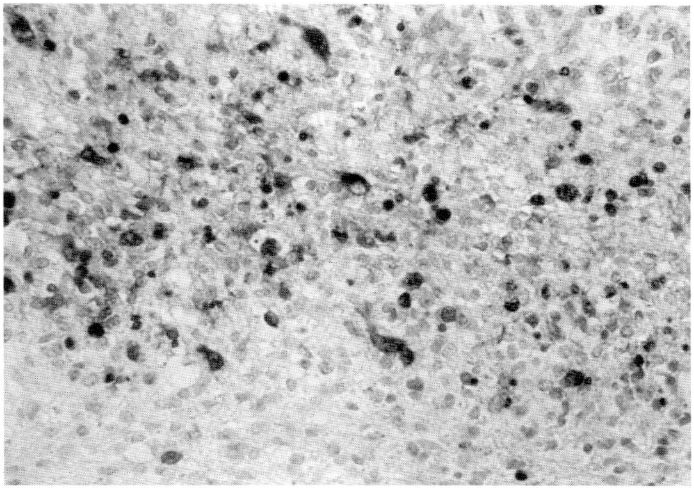

Figure 2. Perinecrotic pseudo-palisading of glioblastoma. TUNEL-positive apoptotic and necrotic nuclei, x 400

4. APOPTOSIS IN BRAIN TUMORS

4.1 Direct demonstrations of apoptotic nuclei

The study of apoptosis in brain tumors has been stimulated by three main possibilities: apoptosis can represent cell loss thus indicating regression and better prognosis; its failure could be responsible for tumor development; its induction in tumor cells could be instrumental to therapies. The first demonstration of apoptosis in brain tumors was given by us in medulloblastoma (Schiffer et al., 1994). The so-called "lymphocyte-like nuclei" of this tumor (Zülch, 1956), already recognized as the remnants of pathologic mitoses from which nuclei do not recover, and containing denaturated DNA by fluorescence procedures (Schiffer et al., 1966), were demonstrated to be actually apoptotic (Schiffer et al., 1994). Since the first demonstrations in astrocytic gliomas (Schiffer et al., 1995; Ellison et al., 1995; Nakagawa et al., 1995), many studies revealed the relationship of apoptosis with cell proliferation for the increasing frequency of apoptotic nuclei going from astrocytoma to glioblastoma (Patsouris et al., 1996; Tachibana et al., 1996; Kordek et al., 1996; Kochi, 1997; Korshunov et al., 1999; Heesters et al., 1999; Haapasalo et al., 1999; Delgado et al., 1999; Mizoguchi et al., 2000). In gliomas and glioblastomas, it was shown then that apoptosis was in relation to shorter survivals (Rhodes, 1998), with the exception of only one observation (Korshunov et al., 1999). Recent studies demonstrated that both apoptosis and cell death/cell proliferation ratio are associated with patient survival and can be used for patient stratification for treatment evaluation (Kuriyama et al., 2002).

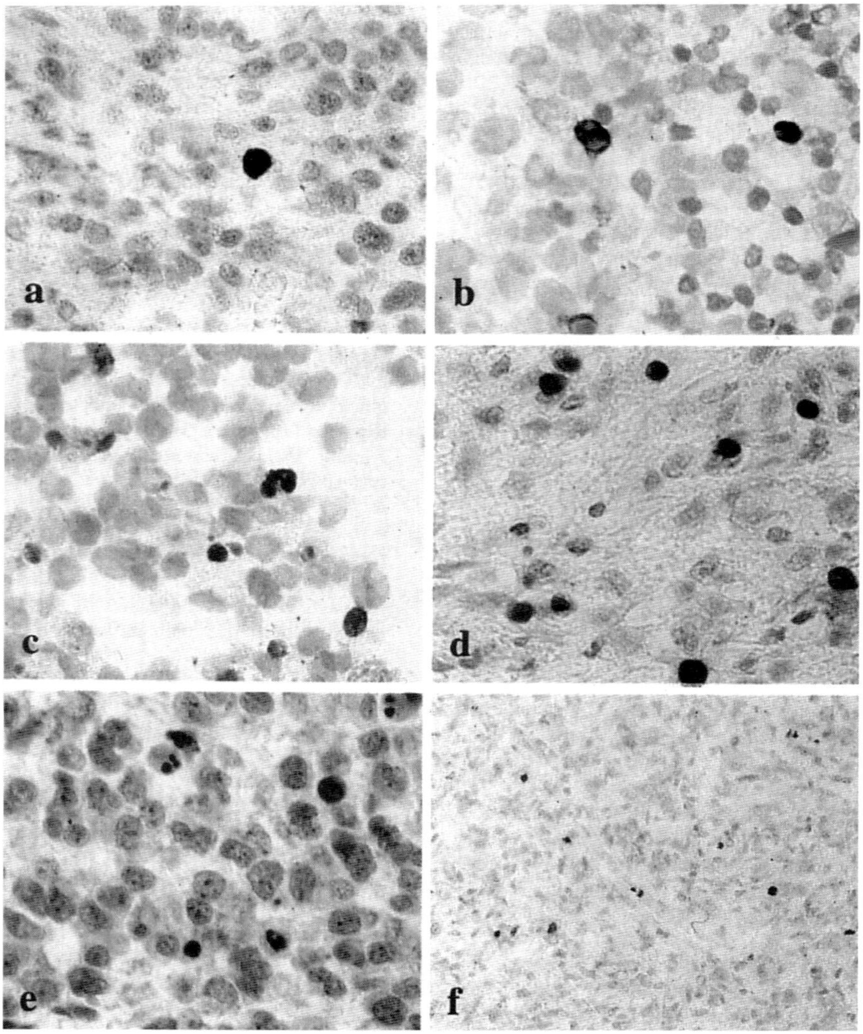

Figure 3. Glioblastoma. a) Activated caspase-3, DAB x 1000; b) Activated caspase-3 positive in the cytoplasm, DAB, x 1000 ; c) caspase-3-positive apoptotic bodies, DAB, x 1000; d) Apoptotic nuclei in proliferative area, TUNEL, x 1000; e) Nuclei positive for Jun, x 1000; f) nuclei positive for JUNK, DAB, x 200. From Schiffer et al., 2002

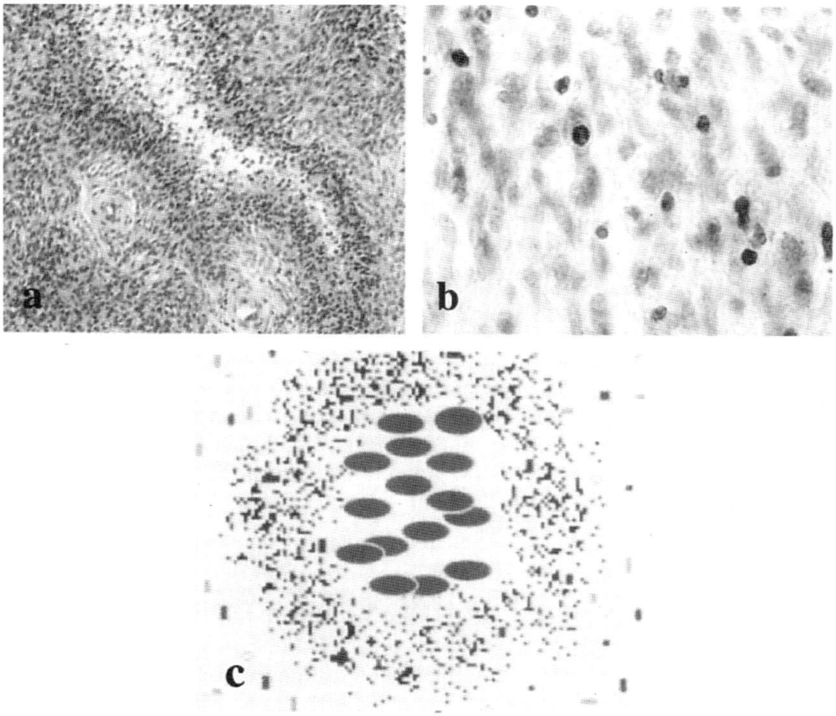

Figure 4. A. Perinecrotic pseudo-palisading of glioblastoma, H&E, x 200; B. Apoptotic nuclei in the palisade, TUNEL, x 400; C. Scheme of a necrosis with pseudo-palisades. Necrotic nuclei are in the necrosis and apoptotic nuclei are in the palisades. From Schiffer et al., 2002

A more detailed analysis of apoptosis in glioblastomas demonstrated that apoptotic nuclei could be found either in perinecrotic pseudo-palisadings and in proliferating areas, coexisting with necrotic nuclei in the former and showing a linear correlation with mitoses in the latter (Schiffer et al., 2001) (Figures 3, 4). Maybe the pathway to apoptosis is the external or receptorial one in the former where it is triggered by hypoxia and the internal or transcriptional one in the latter focused on p53. The hypothesis was that the small circumscribed necroses originated from tumor proliferating centers, where necrosis develops from the very high cell density, because of the imbalance between quick tumor cell proliferation and much slower of endothelial cells (Schiffer, 1997). As a matter of fact Fas/Apo-1 was found to be expressed mainly in cells around large necroses (Tachibana et al., 1996), not linked to p53 status (Tohma et al., 1998). Recently an interesting hypothesis has been put forward concerning necrosis in glioblastoma. The pathway to necrosis could involve interactions between *Ras* and *Akt* pathways and the switch from apoptosis to necrosis by TNF-procoagulation activity (Raza et al., 2002). In this regard two remarks are necessary. The location of apoptotic nuclei in perinecrotic pseudo-palisadings of glioblastomas, where they are in relation to hypoxia should confer on them a regressive and prognostically favourable implication, but since circumscribed necroses are typical of malignant tumors, they lose this significance (Figure 4). At the same time it can be said that the location of apoptotic nuclei in proliferating areas, produced by the internal pathway linked to the cell cycle, does not exclude the possibility that they are, on the contrary, in relation to hypoxia, because the imbalance between tumor cell and endothelial cell proliferation rates could start just in proliferating areas.

In recurrent astrocytic tumors, AI did not show any role in predicting the interval before recurrence or malignant progression (Ralte et al., 2001).

In oligodendroglioma the AI (apoptotic index) is higher than in astrocytic tumors and it increases with anaplasia to the point of being a prognostic factor (Schiffer et al., 1997a, b) and it correlates with topoisomerase IIα expression (Miettinen et al., 2000). In medulloblastoma, apoptotic nuclei are more frequent in nodules (Eberhart et al., 2001) and correlate with shorter survival (Haslam et al., 1998). A high AI has still been found in central neuroblastomas, lymphomas, metastases, PNETs. It correlates roughly with cell proliferation with no specific factor (Grotzer et al., 2001) or is related to therapies (Székessy and Stoltenburg-Didinger, 2001).

In the conclusions of their review on apoptosis in brain tumors, Steinbach and Weller (2002) infer that it is largely due to hypoxia, increases with malignancy, but is of no value for specific prognosis.

5. INVOLVEMENT OF PATHWAYS TO APOPTOSIS

Many investigations have been dedicated to genes/proteins involved in the pathways to and in the regulation of apoptosis, often contradicting one another. The most studied has been Bcl-2, because of its anti-apoptotic function. However, the network of pro- and anti-apoptotic factors in tumors is so complicated that it is difficult to recognize to individual factors a certain pro- or anti-apoptotic function, because of their active role in other molecular pathways. For example it is of great importance to remember that Bcl-2 plays the role of a constitutive protein in neuroepithelial cells. Some studies did not show any correlation of Bcl-2 with malignancy in gliomas (Nakasu et al., 1994; Schiffer et al., 1996; Krajewski et al., 1997) or with apoptosis in glioblastoma (Takekawa et al., 1999) or, together with TP53, p21 and CD95, with survival (Kraus et al., 2001). Besides negative investigations, others showed a correlation of Bcl-2 with survival in anaplastic astrocytomas, but not in glioblastomas (Martin et al., 2001). Contrasting results have been obtained for Bcl-2 and Bax between immunohistochemistry and immunoblots in low-grade and high-grade gliomas. In the former Bcl-2 showed a low level by immunoblots and a high level by immunohistochemistry in low-grade gliomas, and the opposite in high-grade gliomas and the same was found for Bax. The two proteins appeared to be regulated at different levels (Martin et al., 2001). Moreover, recently it has been demonstrated that patients bearing gliomas with an N-terminal truncated form of Bax, called Baxψ, a more powerful inducer of apoptosis, had longer survival (Cartron et al., 2002). Many observations are available on the effect of over-expression of Bcl-2. In glioblastoma cell lines it reduces TRAIL-induced cleavage of caspase-8 and Bid, blocks cleavage of caspase-9, -7 and -3 and of XIAP conferring resistance against TRAIL (Fulda et al., 2002). In malignant gliomas calpain mRNA was increased, as well as the expression of caspase-9 and -3 and PARP (Ray et al., 2002).

In oligodendrogliomas, Bcl-2 was found to show an increased expression with anaplasia (Deininger et al., 1999), whereas in other studies no correlation was found between Bcl-2 and AI (Delgado et al., 1999; Tews, 1999) or Fas/FasL (Frankel et al., 1999). In medulloblastoma a loss of caspase-8 gene expression (Zuzak et al., 2002) and shorter survival in cases with MDM2 amplification, probably due to p53 inactivation (Giordana et al., 2002), were found. Survivin has been found expressed in gliomas, meningiomas and schwannomas (Katoh et al., 2003), whereas in our hands it was positive in metastases, and in malignant gliomas (Schiffer et al., 2002). Its mRNA and a spliced variant were found higher in malignant than in benign gliomas (Yamada et al., 2003).

APO2/TRAIL has been variably demonstrated in a series of neuroectodermal tumors with a distribution pattern parallelling that of GFAP. It was negative, however, in neoplastic oligodendrocytes (Nakamura et al., 2000).

It is known that PDGFR over-expression is a marker of embryonal glia cells and may accompany astrocytoma development which has been hypothesized to be associated with an apoptosis failure due to inactivation of p53 (Louis, 1997). As a matter of fact, a higher AI was found in astrocytomas which did not transform at recurrence than in those that transformed into anaplastic tumors (Schiffer et al., 2002) (Figure 5). In this regard it is interesting that mouse neural precursor cells exposed to transplacental ethyl-nitrosourea show apoptosis and caspase-3 activation and develop more high grade glial tumors if p53 is inactivated (Leonard et al., 2001).

NOT TRANSFORMED ASTROCYTOMAS		TRANSFORMED ASTROCYTOMAS	
MI	AI	MI	AI
0.39%	0.94%	0.00%	0.00%
0.22%	1.55%	0.13%	0.00%
0.00%	0.22%	0.00%	0.20%
0.00%	0.67%	0.00%	0.05%
		0.37%	2.70%
		0.07%	0.84%
		0.00%	0.47%
0.15%	0.84%	0.08%	0.61%

Figure 5. Percentage of apoptotic nuclei calculated at the first surgery in astrocytomas which remain unchanged or transformed into anaplastic astrocytoma at a second surgery. Schiffer et al., 2002

6. *IN VITRO* CULTURES AND CELL LINES

Many studies have been carried out *in vitro*, because it is very easy to induce apoptosis in cells, to follow the process and to induce it by drugs with an anti-tumor activity. Several proteins involved in apoptotic regulation or belonging to apoptotic pathways have been tested and the information obtained contributed to the theoretical schemes of apoptosis. For example, in cell lines of rat malignant gliomas and in glioblastoma cell lines, Fas expression was found to parallel apoptosis and survival (Frankel et al., 2001). Particularly useful have been monolayer and cell suspension systems. A long list of experiments have been made (Schiffer et al., 2002). There is only one limitation in the use of *in vitro* observations: great caution must be taken when transferring their conclusions to *in vivo*, because the micro-environment of the tumor mass cannot be reproduced and no nutrient gradient and necrosis model can be realized. Only spheroid systems of culture allowed a central area of cell death with peripheral apoptotic nuclei so that a relationship between apoptosis and development of necrosis and the role of the energy status of the cell could be studied (Bell et al., 2001).

A very interesting observation is that TRAIL induces apoptosis even without any increase of DR5 by wild-type TP53; and in resistant glioma cell lines this is achieved by pretreatment with some chemotherapeutics, for example camptothecin, etoposide, cisplatin. Caspase-8 inhibitor FLIP is down-regulated whereas the pro-apoptotic Bak is up-regulated (Song et al., 2003). In cultures of human neuroblastoma and oligodendroglioma inhibition of survivin with antisense oligonucleotides induces cell death without caspase activation and PARP cleavage, but with AIF nuclear translocation and XIAP increase (Shankar et al., 2001). In malignant glioma cell lines TRAIL-induced DISC was demonstrated together with caspase-8 and -10 with recruitment of cFLIP (Xiao et al., 2002) of which different isoforms have been isolated and up-regulated by Calcium/Calmodulin-dependent protein kinase II (Yang et al., 2003).

Supernatants of lypopolysaccharide stimulated macrophages increase apoptosis in cell lines from ENU induced glioblastomas through Fas/FasL and the caspase-3 system. The supernatants contained elevated levels of TNFα and IFN-γ (Chen et al., 2002). Apoptosis turned out to be induced by death-receptor dependent and independent (Bcl-2 family, Bax) pathways (Chen et al., 2003). This property of macrophages adds to others already known such as those affecting vascularization, growth rate and stroma formation.

Interestingly, in transfection of caspase-3 into cultures of C6 glioma and immortalized cell lines of rat brain capillary endothelial cells, apoptosis is induced, demonstrated by TUNEL (Zassler et al., 2005). Of course, the method used for protein delivery is toxic for non-tumor cells.

7. EFFECTS OF THERAPIES ON APOPTOSIS

There are a considerable number of observations, both *in vitro* and *in vivo,* that the anti-tumor effects of radio- and chemotherapy on tumor cells and tumors are mediated by apoptosis (Schiffer et al., 2002), because it mediates therapy-induced cytotoxicity in response to drug treatment, irradiation and cytokines. Spontaneous or induced drug resistance is the main obstacle to apoptosis-mediated therapies. Either death receptors or mitochondria may be the entry sites of apoptotic pathways. Stimulation of death receptors of the TNF super-family, through FADD reaches caspase-8 and the caspase cascade, whereas mitochondria stimulation goes to the latter through cytochrome c, Apaf-1 and caspase-9. A cross-talk between the two pathways occurs through Bid (Kroemer and Reed, 2000), so that CD95 can activate caspase-3 directly through caspase-8 or through mitochondria. It is known that TP53 status influences the effect of therapies and Bcl-2 contributes to the resistance of tumor cells to the therapies (Strik et al., 1999; Deininger et al., 2000) and its over-expression can block both caspase-8 and -3. Another member of the TNF super-family, TRAIL, utilizes the same way as CD95, and has the advantage that it does not cause toxicity in animals and appears as a promising tool for inducing apoptosis (Roth and Weller, 1999). It has been shown that CD95 and APO2L/TRAIL induced apoptosis in malignant glioma cell lines is under the modulatory effects of EGFR, inhibited or amplified (Steinbach et al., 2002). Apoptosis induced by TRAIL can be inhibited by over-expression of Bcl-2 (Fulda et al., 2002). Another interesting observation which may indicate a potential therapy is that the apoptotic DNA endonuclease (DNase-γ) transfer into human glioma cell lines induces cell death accompanied by DNA fragmentation (Saito et al., 2003).

Epithelial neoplastic cells usually do not undergo apoptosis after irradiation; probably they respond with a second type of programmed cell death leading to an increased autophagy with early destruction of the cytoplasm preceding or without nuclear collapse (Zakeri et al., 1995). It has been demonstrated that acidic vesicular organelles are formed in the cytoplasm which may represent a defense mechanism with sequestration of intracellular toxins (Paglin, 2001). In irradiated glioblastoma cell lines, a transient increase of p21 and p27 has been shown in those sensitive to irradiation and a decrease in the resistant ones, whereas instead of apoptosis autophagic vauoles occurred regardless of the radio-sensitivity (Xao et al., 2003).

The discovery that many drugs were inducers of apoptosis raised great hope for better therapies of tumors. However, it has also been discovered that defects in apoptosis could be at the basis of tumorigenesis and drug resistance and therefore of the failure of chemotherapy. As a matter of fact, the use of chemotherapy against tumors is based on the induction of apoptosis by drugs that specifically act on some steps of normal apoptotic pathways, but with tumors this cannot be the case, because genetic or epigenetic events alter the pathways generating drug resistance. Dealing with this complicated matter, it must be considered that apoptotic pathways are so deeply interlaced with all the molecular machinery of the cell that its failure or disablement has been regarded as even having a role in tumorigenesis. During embryogenesis, an excessive stimulation of glial cells, for example by over-expression of PDGFR, may lead to overcoming growth control (Heldin, 1992) and with intact sensing mechanisms this leading to apoptosis (Lowe and Ruley, 1993), unless the latter are abrogated, for example, by inactivation of TP53. We have seen that in II grade astrocytomas recurring as III grade astrocytomas the AI was lower than in those recurring as II grade tumors (Schiffer et al., 2002). There is also the possibility that genotoxic agents induce further genetic mutations with damage without death (Johnstone et al., 2002). In the demonstration of apoptosis by TUNEL, beside apoptotic nuclei which associate positive staining with the typical morphological aspect, there are less intensely stained nuclei because of single DNA strand breaks, necrosis, gene transcription etc. (Iseki et al., 1986; Migheli, 1994) that can escape apoptosis. Cells living with a broken DNA strand accumulate mutations, so that apoptosis failure may favour tumor progression (Arends et al., 1994; Tomlinson and Bodmer, 1995).

A long list of initiators, regulators and executioners of apoptosis during the normal process, tumorigenesis and drug resistance has been made and the frequency by which intrinsic pathway to apoptosis is deregulated in tumors, more frequently than the extrinsic one, has been noted, but with emphasis on the escape of tumor cells to apoptosis, including that induced by death-receptors (Johnstone et al., 2002). The drug resistance originates from the failure of the apoptosis that has been induced by the drugs themselves. For a more rational approach to cancer therapy in the future, restoration of apoptosis in tumors seems to be an indispensable step, preferring drugs which silence apoptosis instead of inducing alterations in its pathways or attacking pro-survival pathways.

8. CONCLUSIONS

Even though apoptosis, representing the major source of cell loss, balances with cell proliferation in tumors for characterizing their growth speed, it also parallels tumor malignancy at the phenotypic level, without reaching the dignity of a prognostic factor, unless under special conditions. With tissues not only precautions must be taken in assessing apoptosis, but also the AI is hardly usable as a prognostic factor. The regulation of apoptosis is very complicated, but since there is little doubt that apoptosis can be the final product of therapies and manipulations on tumors, its study will provide further useful information towards counteracting tumor growth. In every day practice it is highly unlikely that the AI of a tumor has to be used for assessing the malignancy grade. Apoptosis can be taken into account just because it is often associated with a high proliferation rate. In a recent review its increase is generally associated with an adverse outcome, including genetic changes of TP53 (Kostantinidou et al., 2005). In statistical evaluations of survivals or of PFS it is often taken into account in the possible prognostic factors, even though it very rarely comes out to be one of them.

Chapter 13

THE UBIQUITIN-PROTEASOME SYSTEM

In higher eukaryotic cells, the proteasome is implicated in the ATP/ubiquitin-dependent proteolysis of most nuclear and cytosolic proteins and short-lived proteins involved in many molecular cell mechanisms: cell division and cell cycle regulation, differentiation and development, transcription, apoptosis regulation, cell surface receptor modulation, DNA repair, stress responses, immune and inflammatory responses etc. (Ciechanover and Iwai, 2004). Proteins such as p53, p27/Kip.1, cyclins and their kinases, IkBα, caspases etc, that have been encountered in every above-mentioned chapter and that play very important roles in the process of progression or regression are sent to the proteasome at the end of molecular processes of activation/inhibition. On the other hand, it has been shown that specific and peptide aldehyde inhibitors of proteasome, such as lactacystsin, LLnL, MG132, induce apoptosis by activation of caspase-3 in glioma cells (Wagenknecht et al., 1999, 2000; Tani et al., 2001). A highly specific inhibitor of the proteasome is PS341 that shows anti-tumor activity. It arrests the cell cycle in the G2/M phase with the increase of p21, p27/Kip.1 and cyclin B1, inducing apoptosis by stabilizing IkBα through the inhibition of chimotryptic activity of the 26S proteasome and decreasing the nuclear activity of NFkB in glioblastomas (Yin et al., 2004). PS-341 may facilitate the apoptosis induction of TNFα and TRAIL.

The regulation of the ubiquitin-proteasome pathway in malignancies is thus of paramount importance and can be accomplished at the level of ubiquitination or of proteasome activity. NFkB is a hyperactive pathway in glioblastomas (Hayashi et al., 2002) and its activity correlates with the growth rate (Nagai et al., 2002). The activation of the pathway depends on the ubiquitination and degradation of IkBα and if the latter is stabilized, because of the inhibition of chymotryptic activity of proteasome 26S, the nuclear activity of NFkB decreases. Conversely, activation of NFkB inhibits apoptosis. TRAIL and TNFα activate the caspase cascade, but at the same time stimulate NFkB (Franco et al., 2001; Dai et al., 2003). If the proteasome is inhibited, the stimulation of NFkB is blocked, but not apoptosis, and this explains why PS-341 enhances TNFα and TRAIL cell death (An et al., 2003).

The 26S proteasome is a 1500–2000 kD multi-subunit particle composed of two multimeric proteins. One with proteolytic activity is defined as the catalytic core or 20S proteasome and the other protein with regulatory function is called the 19S regulator or PA700. The 20S proteasome exists also as an independent unit, and has a cylindrical aspect with four rings each formed by 7 homologous but distinct α and β subunits (Figure 1). In eukaryotic cells the 20S proteasome has a number of peptidase activities capable of cleaving fluorigenic peptides and represented by chimothypsin-like, trypsin-like and peptidyl glutamyl peptide hydrolyzing (PGPH) activity. Entering into the 26S proteasome, 3 classes of enzymes are required for the ubiquitination of proteins: E1 or ubiquitin activating enzyme, E2 or ubiquitin conjugating enzyme and E3 or ubiquitin-protein ligase. SCF complexes (Skip 1, Cul-1, F-box protein) ensure a specific recognition and ubiquitination of different substrates through different F-box proteins. For example, the S-phase kinase-associated protein (Skp2) belong to the F-box family protein and is required for G1-S transition, targeting p27/Kip.1 for ubiquitination (Adams, 2003).

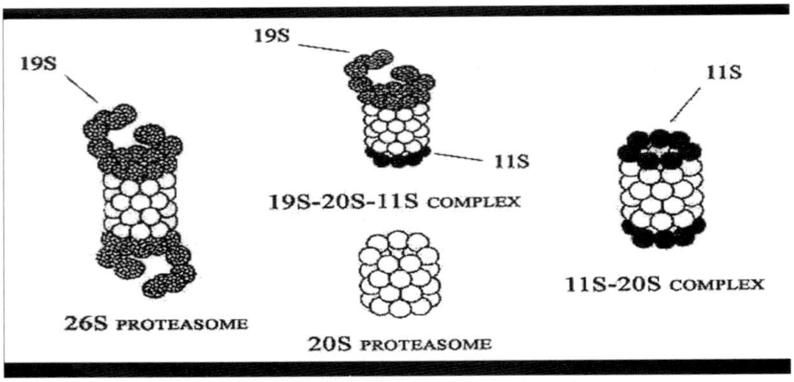

Figure 1. Scheme of the proteasome. From Piccinini et al., 2003

One of the most important problems, not yet resolved, is the high specificity and selectivity of the ubiquitin system towards the numerous substrates. They must be "recognized" first, before entering into the proteasome for degradation and this is determined by two protein groups: E3s and modifying enzymes and ancillary proteins or molecular chaperones or by other mechanisms. The involvement of the ubiquitin-proteasome system in neuro-degenerative diseases, documented by accumulation of ubiquitin-conjugates and ubiquitin associated inclusion bodies, today plays a very important role in the pathogenesis of these diseases (Ciechanover and Brundin, 2003). In the same way it plays a role in malignancies; and in those of the nervous system one need only mention the important function of the proteins listed above. In general it can be said that cancers can result from stabilization of oncoproteins and destabilization of tumor suppressor gene products; and some of them are substrates of the ubiquitin-proteasomal system (Ciechanover and Iwai, 2004).

Stimulated by IFN-γ or cytokines the 20S proteasome is transformed into an immunoproteasome; and its constitutive subunits β1, β2 and β5 are replaced by inducible subunits called LMP2, MECL-1 and LMP7, with a change in the peptidase activity and the formation of peptides more suitable for MHC class 1 presentation, since they bear at their terminus basic and hydrophobic aminoacids (Stohwasser et al., 2000). The 20S proteasome and the activator are modified in the passage from primary mouse microglia to activated microglia (Stohwasser et al., 2000).

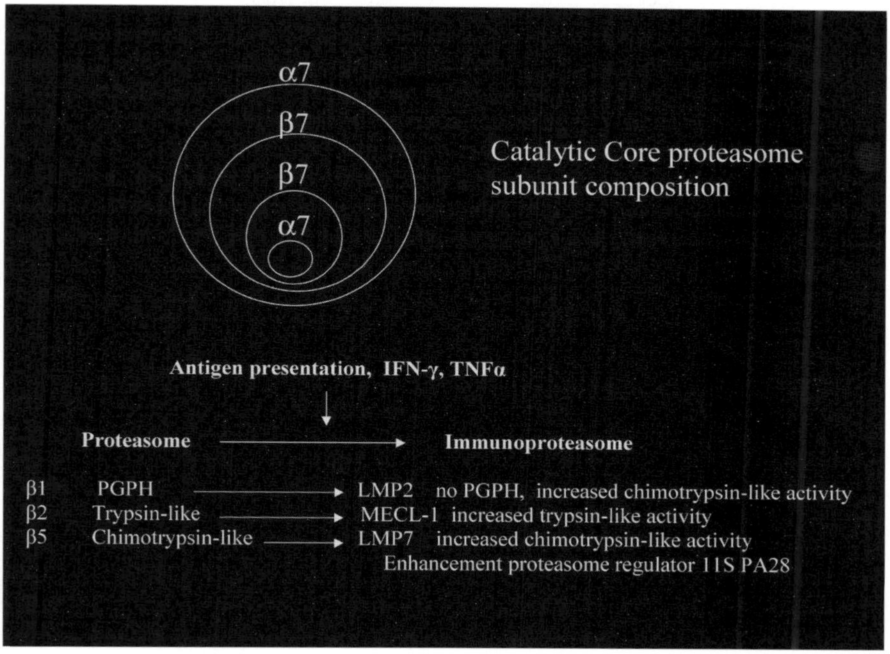

Figure 2. Constitutive and inducible sub-units. From Piccinini et al., 2005

Proteasomes 26S and 20S have been purified in glioblastomas and evaluated for peptidase activities and subunit composition. By comparison with controls, an increased expression of inducible subunits and a reduction of the peptidase activities have been found. In spite of the abundance of activated microglia and of macrophages and lymphocytes in glioblastomas, involved in the immune response and controlled by IFN-γ and TNFα, it is likely that the alterations mentioned previously can be attributed to tumor cells. LMP2, MECL-1 and LMP7 were found to be increased with unvaried trypsin-like and decreased PGPH and chimotrypsin-like activities (Figures 2, 3, 4) (Piccinini et al., 2005). The precise meaning of these alterations still escapes us, even though they should denounce an altered immunological attitude of glioblastomas. By immunohistochemistry the 20S core subunit can be detected in most cytoplasms (Figure 5).

It is useful to recall that a particular immunological situation is realized at the periphery of the tumors where Th cells are orientated toward Th2 regulatory phenotypes instead of Th1; and this is one of the mechanisms of their immune defects (Hao et al., 2002; Akasaki et al., 2004). A relationship between the increasing cyclin D1 LI and decreasing p27/Kip.1 LI and the modification of the proteasome 20 S could not be demonstrated (Schiffer et al., 2004).

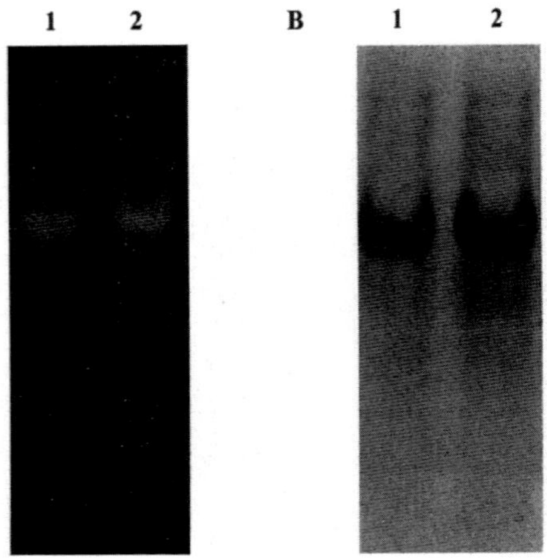

Figure 3. Purified 20S proteasome in glioblastoma (lane 1) and control (lane 2)

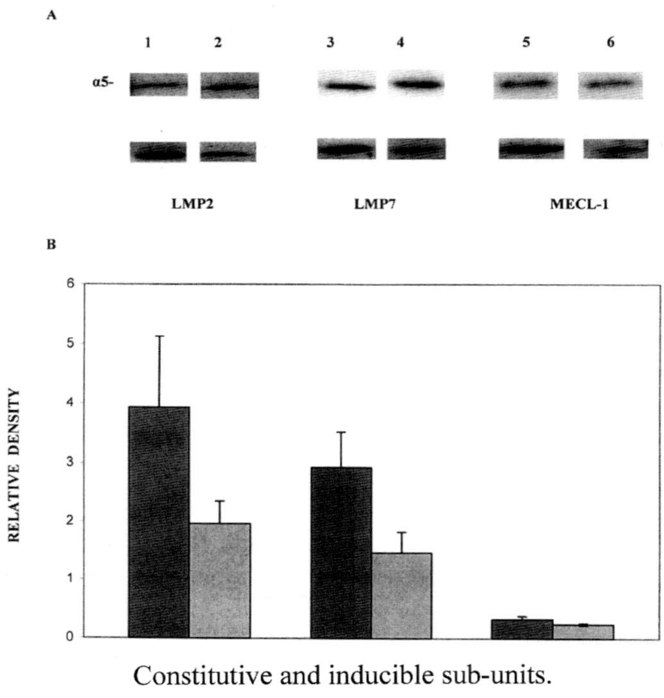

Figure 4. Panel A Glioblastoma: lanes 1, 3, 5; control: lanes 2, 4, 6. LMP2: lanes 1, 2; MECL-1: lanes 3, 4; LMP7: lanes 4, 5. Panel B Densitometric analysis. From Piccinini et al., 2005

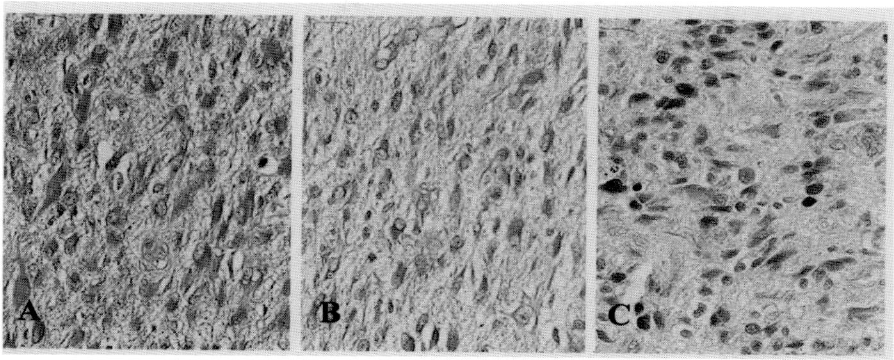

Figure 5. Glioblastoma. A. 20S Proteasome; B. LMP7; C LMP2. DAB, x 400. From Piccinini et al., 2005

Chapter 14

ANGIOGENESIS

1. THE PROCESS OF ANGIOGENESIS

Angiogenesis includes endothelial sprouting, i.e. endothelial cell proliferation and migration, tube formation and intussusception (Patan, 2000); and it is a key feature for high-grade gliomas. New capillaries sprout from preexisting host vessels and single endothelial cells form new vessels. From experiments with human glioma spheroid implants, a supplementary mechanism would be migration of individual endothelial cells (Goldbrunner et al., 1999). Classically, in glioblastoma different forms of new vessel formation can be recognized: endothelial hyperplasia, microvascular proliferation, increased vessel density and glomeruloid formation. In spite of the various models proposed, *in vitro* and *in vivo,* none is without limitations in comparison with the human vascular pattern development.

Angiogenesis in gliomas has a genetic and a hypoxic regulation. A key molecule at the basis of the process is the hypoxia inducible factor 1 (HIF-1) consisting of an α- and a β-subunit, which can be detected in perinecrotic pseudo-palisadings of glioblastoma and in the front of the invading tumor: its expression correlates with the tumor grade (Zagzag et al., 2000). The formation of hypoxic areas in tumors are thus very important for initiating angiogenesis; and in glioblastomas these are represented by circumscribed necroses which can develop from the imbalance between quick tumor cell proliferation and poor local vascular supply (Schiffer et al., 1989) or by occlusion of existing vessels that leads to apoptosis of endothelial cells, vessel collapse and tissue hypoxia (Holash et al., 1999). In normoxic conditions HIF-1 is degraded in the ubiquitin-proteasome system. In hypoxic conditions, it binds to hypoxia-response elements (HREs) and activates VEGF gene transcription, responsive genes for angiogenesis, tumor cell survival and invasion. The formation of the active HIF-1 depends on the co-activator p300/CBP which can be inhibited by the factor inhibiting FIH. HIF-1 expression can be genetically modulated and is enhanced when PTEN is inactivated (Zundel et al., 2000).

The molecular mechanisms (Figure 1) of angiogenesis in brain tumors are based on a series of factors, among which are bFGF, TGFβ1, EGF, PDGF (Dunn et al., 2000) and mainly VEGF and Ang-1 and Ang-2 must be mentioned. VEGF is produced by tumor cells and has a specific mitogenic function on endothelial cells, mediated by the tyrosine kinase receptors VEGFR-1 and VEGFR-2. The latter are highly expressed

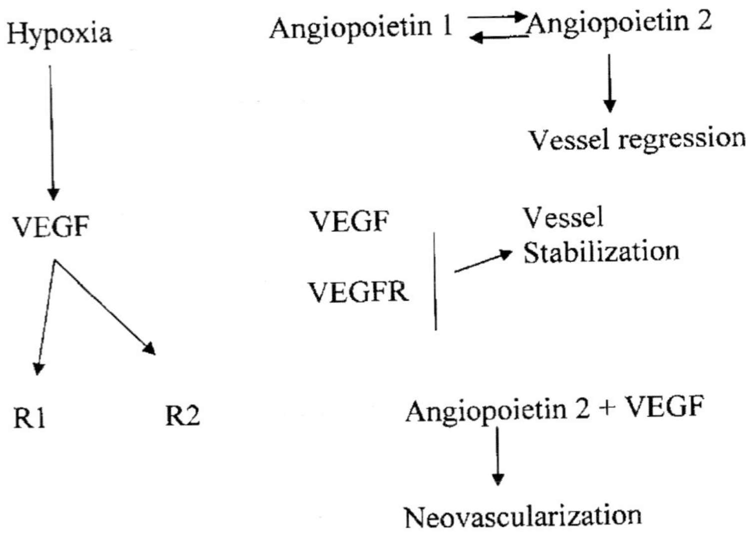

Figure 1. Pathways to angiogenesis

during development (Plate, 1999); and are regulated by oxygen tension: for example they are up-regulated by hypoxia (Damert et al., 1997). The family of angiopoietins are ligands of the tyrosine kinase receptors Tie-1 and Tie-2 expressed by endothelial cells. Ang-1 and Ang-2 are ligands of Tie-2 and they play a role in the interaction of endothelial cells with other vascular cells such as pericytes and smooth muscle cells (Yankopoulus et al., 2000). Ang 1 phosphorylates Tie-1; its overexpression increases vascularization and antagonizes apoptosis of endothelial cells *in vitro;* whereas Ang-2 antagonizes Ang-1 and produces vascular deficits during development. In opposition to the classical two phases of tumor growth, avascular and vascular, a theory of two phases, both vascular, has been proposed (Zagzag et al., 2000). In tumor implantation experiments, in the first phase tumor cells go on the native vessels, whose cells undergo apoptosis preceded by Ang-2 expression with the necrosis of tumor cells which in turn induces angiogenesis. Ang-2 would activate pericytes/smooth muscle cells (Zagzag et al., 1999). An important role in angiogenesis is played by HIF-1 located in perinecrotic pseudo-palisadings of glioblastoma, modulated by tumor oxygenation. It correlates with the degree of anaplasia and vascularization and activates VEGF gene transcription (Zagzag et al., 2000).

The progression from low-grade astrocytomas, which use pre-existing vessels, to highly vascularized glioblastoma, must be realized through an "angiogenic switch" (Machein and Plate, 2000), where VEGF up-regulation and neovascularization are

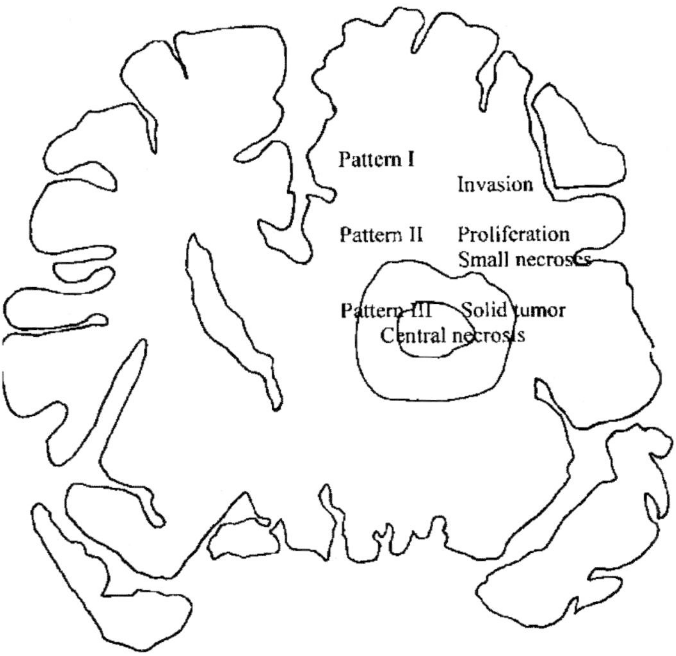

Figure 2. Phases and patterns of vascularization in glioblastoma

induced by hypoxia of circumscribed necroses (Plate et al., 1992). Ang-1 is poorly expressed in low-grade astrocytomas (Audero et al., 2001) and increase in pseudo-palisadings of glioblastoma (Ding et al., 2001); whereas Ang-2 is expressed in malignant gliomas both in tumor and endothelial cells, together with Tie-2.

The process of angiogenesis, besides growth factors and their receptors, requires a multitude of other factors such as adhesion molecules, β-catenin, regulators of extra-cellular matrix remodeling etc. (Lopes, 2003), among which tenascin must not be forgotten, because it is found at the peak of angiogenesis and disappears when angiogenesis had ceased (Zagzag and Capo, 2002).

In malignant gliomas angiogenesis can be modulated by genetic alterations. The activation of Akt regulates VEGF induction in hypoxic conditions, PTEN can regulate Thrombospondin-1, the activation of EGFR can increase VEGF secretion by increasing HIP-1 levels, TP53 mutations can increase bFGF, p53 can down-regulate VEGF etc. (Kaur et al., 2004).

An important remark on the origin of the new endothelial cells is that recently endothelial precursors and hematopoietic cells have been found to contribute to neo-vascularization in early phases of tumor growth. Although bone marrow cells contribute to microglia in the brain, endothelial precursors and hematopoietic cells have not been found to substantially contribute to angiogenesis in murine brain tumors (Machein et al., 2003).

2. DESCRIPTIVE ANALYSIS OF GLIOMA VASCULATURE

The vascular tree in gliomas when they transform into glioblastomas undergoes tremendous and widespread changes so that none of the different aspects it shows at the end of the angiogenetic process can be detected as a whole in surgical samples. It is therefore useful to categorize all the vascular productive changes in GBM and provide an interpretation of their distribution in space and time so that the limited aspects observed in the surgical sample can be correctly interpreted. In our observations of 1400 surgical samples and 100 autopsy cases of glioblastoma studied by the "whole mounting technique", three patterns have been recognized that can be encountered one after the other going from the peripheral peritumor normal nervous tissue to the necrotic centre of the tumor, or going from initial tumor lesions to a developed tumor composed of a necrotic hypodense centre and a peripheral enhancing ring (Figure 2).

A

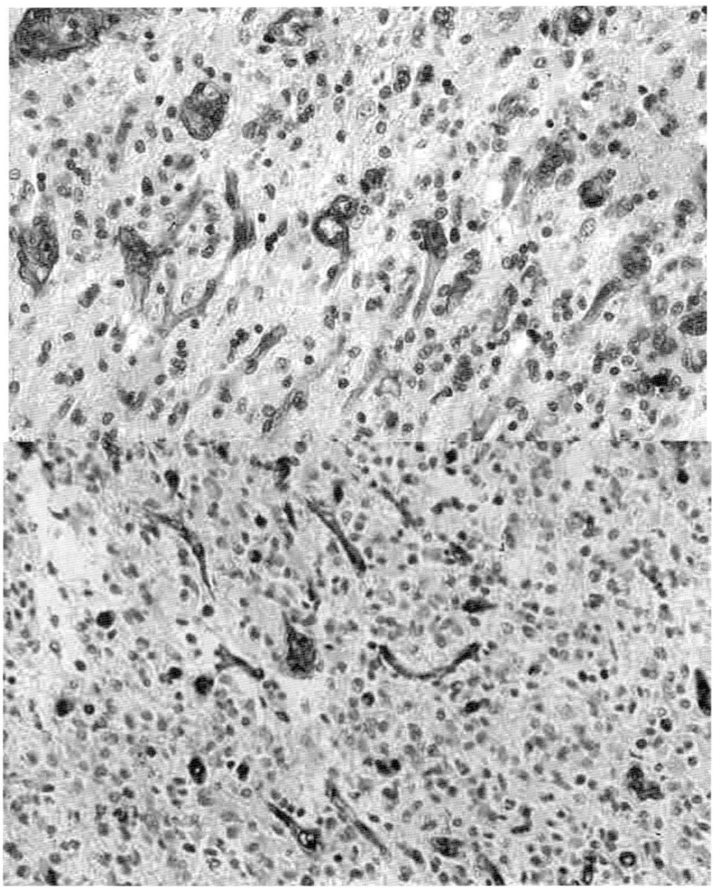

B

Figure 3. Glioblastoma. A. Endothelial proliferation and formation of buds, PAS Hematox; B. id. Factor VIIIRAg, x 400

The first pattern is found in infiltrated tissue and is given by endothelial buds, capillary with endothelial hyperplasia and new capillaries. Endothelial cells are recognizable for their CD31 and FVIII/Rag positive staining (Figure 3). More deeply, small vessels with more than one peri-luminal ring of cells can be found. The cells outside the peri-luminal ones are positive for α-sm-actin and vimentin.

The second pattern is found, also admixed with the first one, more deeply; and is composed of larger vessels with the same characteristics of vessel formations composed of many lumina, as multiple excavations in a multicellular glomerate or as many vessels with proliferated walls lumped together (Figures 4, 5A).

A

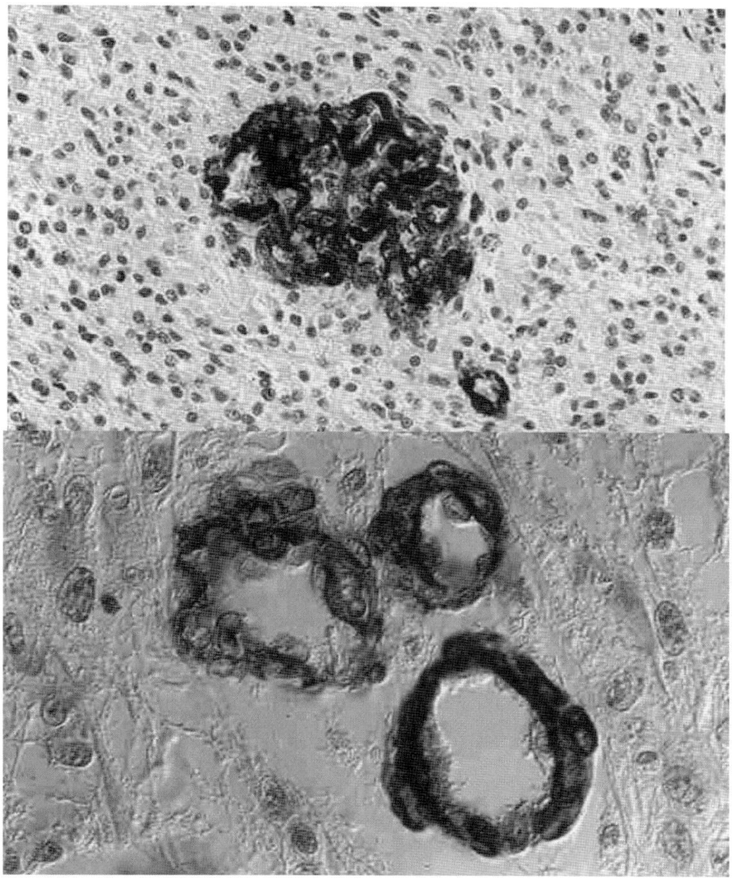

B

Figure 4. Glioblastoma A. α-sm-actin positive cells around lumina; B. id in larger vessels. DAB, x 400

A

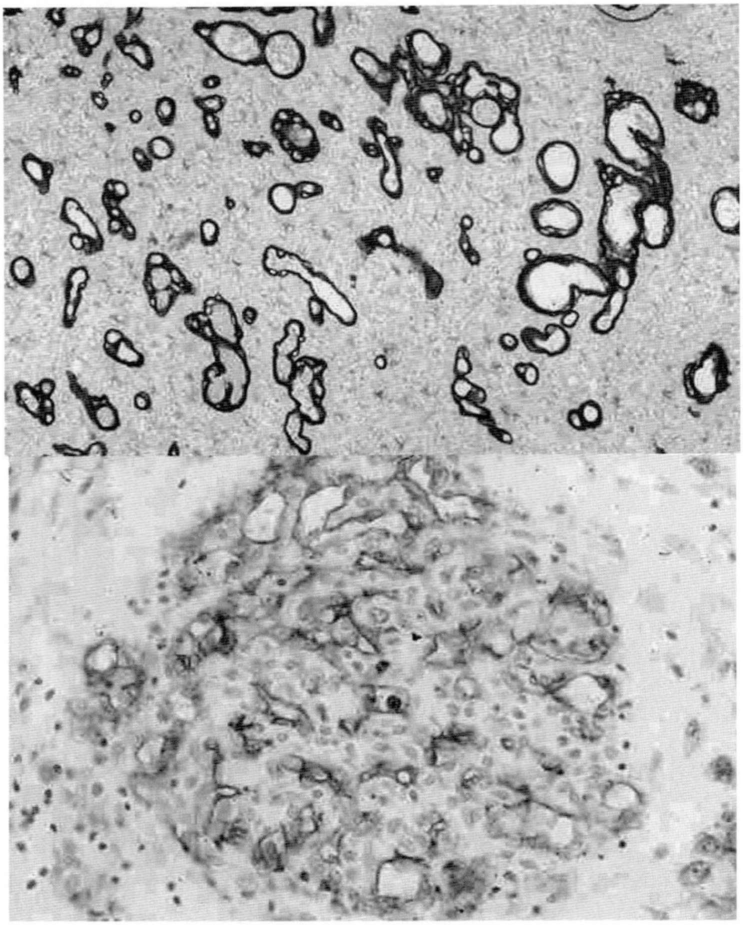

B

Figure 5. Glioblastoma A. Dilated vessels of the II pattern laminin, DAB; B. Glomeruloid multi-canal formation. CD.31, DAB, x 400

The third pattern is given by classic glomeruloid formations with proliferated vimentin- or α-sm-actin-positive cells (Figures 5B, 8). Circumscribed necroses are located between the second and third pattern and around them no capillary is found, in accordance with our interpretation of circumscribed necroses as arising from proliferating centres where endothelial proliferation for angiogenesis is delayed in comparison with quick tumor cell proliferation (Figures 6, 7).

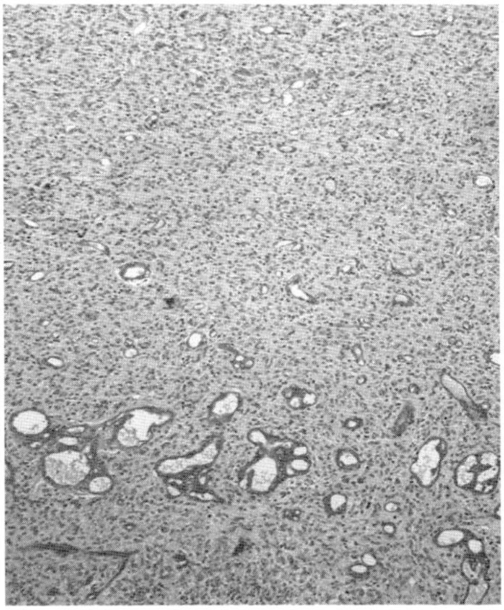

Figure 6. Glioblastoma. Microvascular proliferation in the deep cortical layers in a completely invaded cortex, H&E, x 200

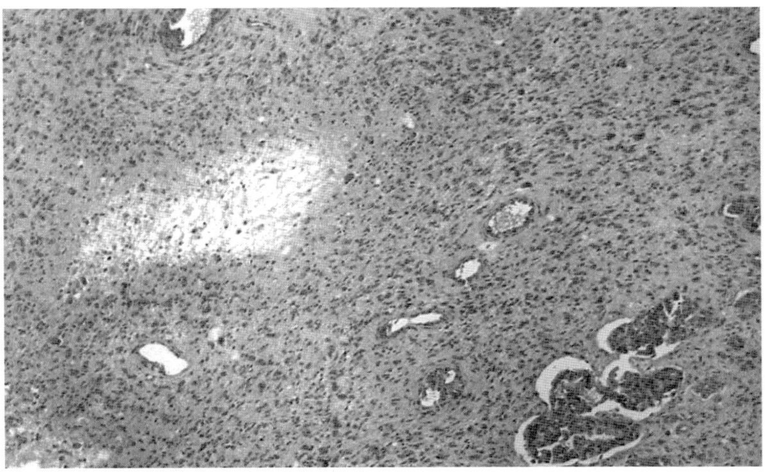

Figure 7 . Glioblastoma. Circumscribed necrosis in an area devoid of capillaries and with large glomeruloid formations, H&E, x 200

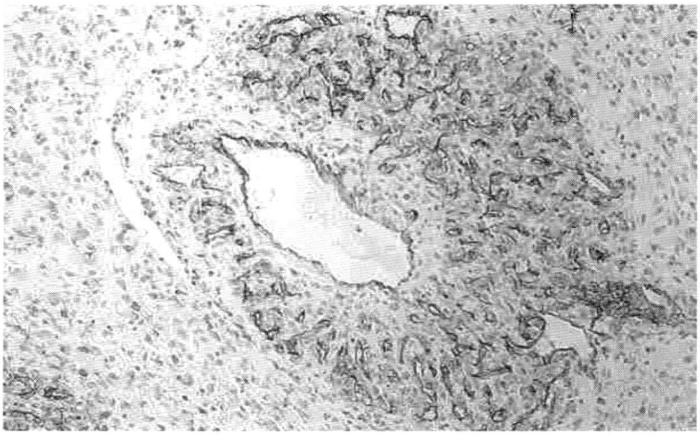

Figure 8. Glioblastoma. Large glomeruloid formation. CD31, DAB, x 400 (Schiffer et al., 1989)

More recently, a study carried out with alkaline phosphatase histochemistry on celloidin sections confirmed the description (Challa et al., 2004).The most important observations from these studies are that formation of new capillaries in areas of solid tumor may vary greatly until they produce a very dense network of small vessels; and that in the classical situations of a cortex infiltrated from the tumor coming from below, new capillary formation and endothelial buds follow and do not precede tumor cell migration, contradicting what is generally thought (Folkman, 1971). But there is agreement with other studies showing, after quantitative evaluation of many parameters such as vessel perimeter and extra-vascular nuclei, that "many intratumoral regions may not be overtly angiogenesis dependent or amenable to antiangiogenic therapy" (Wesseling et al., 1994); that neoangiogenesis may not be necessary for tumor infiltration (Wesseling et al., 1997); and that there is no correlation between MIB-1 LI and microvessel density (Wesseling et al., 1998). These conclusions are somewhat in contrast with the observations made with C6 glioma spheroids implanted in nude mice where glioma cells migrated demonstrating an affinity to perivascular spaces and pial/subpial vessels, with preference for arteriolar segments, establishing a relationship between cell invasion and angiogenesis (Vajkoczy et al., 1999). Recent observations would demonstrate that invading cells express HIF-1, because there would be a link between hypoxia and cell migration (Zagzag et al., 2000), even though on this point there are dissenting opinions (Parliament et al., 2000). These observations would favor the hypothesis that neo-vascularization follows tumor cell invasion, since neo-angiogenesis starts only when peri-vascular tumor cells regress (Holasch et al., 1999).

In a recent analysis two types of vascular patterns have been identified in glioblastoma biopsies: one with bizarre vascular formations and the other with classic angiogenesis. In tumors with classic angiogenesis there was no difference for MIB-1 LI in tumor cells and diffuse expression of HIF-1α, whereas vessel density and AI were higher in comparison with tumors with bizarre angiogensis that showed a higher VEGF. No correlation of MIB-1 LI with HIF-2α and no correlation of diffuse HIF-1α with AI, MIB-1 LI and vessel density were found. In the pattern with bizarre vascular formation cell growth seems to outpace neovascularization; whereas in the tumors with classic vascular pattern, cell growth parallels neovascularization. Survival was longer in the latter than in the former. The pattern with bizarre angiogenesis seemed to be peculiar to human glioblastomas and not present in animal models. An important deduction was that HIF-1α was not induced by hypoxia alone, but rather by oncogenic stimuli. All these observations represent a good basis for the strategy of anti-angiogenic therapies in glioblastoma (Birner et al., 2003) and algorythms for the standardized assessment of vascular patterns have been provided (Preusser et al., 2004).

Microvascular proliferations and neovascularization are obviously associated with strong VEGF expression in gliomas (Plate et al., 1994), especially glioblastomas (Pietsch et al., 1997), whereas in oligodendroglioma VEGF has not been found (Pietsch et al., 1997) or it was expressed in tumor vasculature markedly in an anaplastic variant (Christov et al., 1998).

Every tumor type has its own vascular distribution pattern which can be different even between two similar tumors (Goldbrunner et al., 1999).

The development of circumscribed necroses has been related to the imbalance between the high tumor cell proliferation rate and the low vessel cell proliferation rate. This observation has recently been confirmed by the low MIB-1 LI of the latter and by the lack of any correlation between proliferation of microvascular cells and time to recurrence (Kern et al., 2003); whereas TP53 mutations have not been found in glomeruloid vessels (Kulla et al., 2003). This casts some doubts on the possible efficacy of anti-angiogenetic therapy indiscriminately applied to glioblastomas.

In surgical samples of glioblastomas none of the angiogenetic aspects are usually present at the same time and, on the other hand, their contemporary occurrence is not required for the recognition of malignancy. It is not necessary to observe as well an increased vessel density, even though it is an ominous sign. As a matter of fact, increased vessel density has been observed to correlate negatively with TTP and survival (Leon et al., 1996), but it does not add anything to the prognostic diagnosis (Folkert, 2000). Vessel density is more uniform in low-grade astrocytomas than in high-grade tumors (Assimakopoulou et al., 1997) and astrocytomas with >7 microvessels/400 x have shorter survivals, greater risk of transformation and a greater expression of VEGF (Abdulrauf et al., 1998). Finally, tumors with no expression of the anti-angiogenetic thrombospondin-2 molecule show an increased vessel density (Kazuno et al., 1999).

In other tumors, vessel density has occasionally been investigated with no final conclusion about its correlations. The only very interesting findings have been the enhancement of MRI correlating with an increased vessel density (Tynninen et al., 1999).

3. SIGNIFICANCE OF PRODUCTIVE VESSEL CHANGES IN TUMOR BIOPSIES

In astrocytic gliomas, nuclear pleomorphism, circumscribed necroses, microvascular proliferations and mitoses derive according to an accepted rationale from the first phenotypic sign of malignant transformation, which is increased cell density

(Schiffer, 1998). They indicate a glioblastoma stage. While circumscribed necroses, even dissociated from microvascular proliferations, are a definite mark of this tumor stage, microvascular proliferations alone have long been discussed as to whether or not they are compatible with the diagnosis of anaplastic astrocytoma: recently they have been considered as incompatible, just because they express an already started angiogenesis (Kleihues et al., 2000). Microvascular proliferations are composed in their simplest form by more than one peri-luminal layer of endothelial cells. The first row of cells is given by CD31- and FVIII/Rag-positive endothelial cells, whereas the other rows are composed of α-sm-actin-positive cells, i.e. pericytes or other cell types. Endothelial cells originate from preexisting vessels or from free migrating endothelial cells and form new channels coming from buds as hyperplastic endothelial cells. Since neo-angiogenesis in tumors recapitulates normal vessel formation during embryonal life, hyperplastic endothelial cells show immature features with paucity of organelles in the cytoplasm and an increased nucleo-cytoplasmic ratio (Weller et al., 1977). Before vessels acquire multilayered walls, they remain single layered, but with hyperplastic cells. This process has been well outlined (Brem et al., 1972), but how can we recognize hyperplastic endothelial cells in the tissue? It has been established that when in transverse section a capillary shows more than 1 and ½ nuclei, and in longitudinal sections endothelial nuclei are in contact, these must be considered as indicating hyperplasia.

The recognition of hyperplastic endothelial cells in small biopsies when no other sign of angiogenesis is present, but nuclear pleomorphism of tumor cells and mitoses occur, leaves some doubts whether to call the tumor glioblastoma or to limit oneself to calling it simply malignant glioma.

The problem outlined previously concerns diffuse astrocytomas/glioblastomas and not other gliomas. Pilocytic astrocytomas, for example, may show microvascular proliferations, but with no prognostic implication, whereas in oligodendrogliomas the problem is completely different. Therefore, a diagnosis must be made before using interpretations of small vessels or endothelial cells for prognosis. In oligodendrogliomas the three patterns recognized in glioblastoma can be found as well. First of all, certain patterns of vascularization, such as "chicken wire", which roughly corresponds to the first pattern of glioblastoma, may have a diagnostic and not a prognostic relevance; secondly, microvascular proliferations do not have the same importance as astrocytic gliomas for prognosis, because the criteria used in the latter tumors for recognizing malignancy do not apply directly to oligodendrogliomas. Microvascular proliferations can be found in anaplastic oligodendrogliomas, but even though less frequently, also in classic oligodendrogliomas, so that their occurrence in a biopsy does not directly indicate malignancy, if the diagnosis is that of oligodendroglioma. Microvascular proliferations have been considered to be (Daumas-Duport et al., 1997) or not to be (Mork et al., 1986) or hardly (Schiffer et al., 1997) a prognostic factor, but never in individual cases. Endothelial hyperplasia evaluated separately from microvascular proliferations has been considered as a prognostic factor (Daumas-Duport et al., 1997), but in our experience, as mentioned before, it occurs in 70% of oligodendroglioma and is devoid of any prognostic significance, as is small vessel density (Schiffer et al., 1997).

Even though it increases with malignancy in gliomas, VEGF was not found to be an independent prognostic factor (Oehring et al., 1999).

Chapter 15

MENINGIOMAS

1. DEFINITION OF DIAGNOSTIC PROBLEMS

Histologically, meningiomas have been subdivided into different subtypes: meningothelial, fibroblastic, transitional, psammomatous, angiomatous, microcystic, secretory, lymphoplasmacellular, metaplastic, chordoid, rhabdoid, with clear cells, papillary, atypical and anaplastic (Louis et al., 2000). Setting aside the differential diagnoses towards other mesodermal and non-mesodermal neoplasias, important considerations are the diagnostic and prognostic ones related to subtypes of grades II and III, since they involve different therapeutic strategies. The verification of the prognostic meaning of a histological diagnosis is only rarely performed by survival analyses. It is made rather by the analysis of DFP or of recurrences which can be influenced by other factors than those belonging to tumor biology and mainly by location and type of surgical resection. Chordoid, with clear cells and atypical meningiomas are considered as grade II; papillary, rhabdoid and anaplastic are considered grade III. In the recognition of the different subtypes and in clinical deductions it is very important to establish how the various histological aspects are representative of the neoplasia, i.e. to give due importance to local heterogeneity (Figures 3, 4).

2. ANALYSIS OF THE SUB-TYPES IN RELATION TO DIAGNOSIS

Chordoid and clear cell meningiomas are rare variants with a supposedly more aggressive behavior. They are classified as grade II tumors. Chordoid meningioma shows trabeculae of vacuolated cells immersed in a mixoid background, similar to chordoma. Clear cell meningioma is composed of polygonal cells containing glycogen in the clear cytoplasm.

Papillary meningioma is a rare variant occurring most frequently in children. The histologic aspect is associated with other signs of malignancy. The tumors are characterized by aggressive clinical behavior with high rate of local recurrence, invasion of the brain and distant metastasis (Ludwin et al., 1975). These biological features were confirmed in 7 adults and are regularly found in all the cases described in the literature where necrosis, high mitotic rate and abundance in reticulin were the main malignant signs (Pasquier et al., 1986). They are characterized by a high MIB-1 LI and hTR (Rushing et al., 1999).

Rhabdoid meningiomas are meningiomas with malignant histological features and rhabdoid aspects: characterized by polygonal cells with a vesicular nucleus and a prominent nucleolus. The nuclei are frequently in a peripheral position and the cytoplasm is roundish and hyaline. There is a resemblance with rhabdomyoblasts. Nuclear pleomorphism, focal expression of desmin, high MI and LI for Ki-67 are also present. The rhabdoid component may be present only in recurrences. A differential diagnosis must be carried out to distinguish malignant gliomas or ependymomas, metastatic carcinomas, sarcomas, rhabdoid tumors (Klein et al., 2002). In the series of 15 cases described by Perry et al. (1998), 87% of patients had recurrence, 13% extracranial metastases and 46% died after 5.8 years. A very important question concerning these tumors is how large the rhabdoid component must be to label the tumor as rhabdoid and whether the rhabdoid component is associated with other malignant signs. Cases with focal rhabdoid aspects must be kept under control (Perry et al., 1998). Four cases were described by Kepes et al. (1998) and it was stressed that rhabdoid morphology may be acquired by meningiomas together with more aggressiveness. A case has also been described with associated rhabdoid and papillary aspects (Hojo and Abe, 2001).

Meningiomas are subdivided into classic, atypical and anaplastic with increasing aggressive behavior from grade I to III. The definition of the subgroups is uncertain and the indications given for their identification are rather vague (Louis et al., 2000). Atypical meningioma represents 5–7% of tumors and anaplastic meningiomas 1–2% of tumors (Perry et al., 1997; Louis et al., 2000). Atypical meningiomas show an increased mitotic activity with >4 (Perry et al., 1997) or 5 (Maier et al., 1997) mitoses per 10 HPF and present not less than three of the following features: increased cell density, occurrence of small cells, increased nucleus/cytoplasmic ratio, prominent nucleoli, "sheeting", circumscribed necroses (Louis et al., 2000). Anaplastic meningioma can be recognized for carcinomatous, sarcomatous and melanomatous aspects and for > 20 mitoses per 10 HPF (Ludwin et al., 1975). Only the invasion of nervous tissue is insufficient for the diagnosis (Perry et al., 1999) and it has been proposed to disregard it (Ho et al., 2002). All the observations made till now on the parameters for the recognition of the two variants are compared for their validation with the histologic aspect and mainly with the frequency of recurrence which is 7–20% for classic, 29–40% for atypical and 50–78% for anaplastic meningiomas (Louis et al., 2000).

In order to better distinguish the three variants, MIB-1 LI is very important (Figure 1). Roughly it correlates with the histological diagnosis and with other proliferation markers, such as BrdU or PCNA (Hsu et al., 1998; Abramovich and Prayson, 1999); but the correlation with the frequency of recurrence is doubtful. The MI varies in the different series (Miller et al., 1985; Maier et al., 1992; Hsu et al., 1998; Ayerbe et al., 1999; Perry et al., 1999). According to some authors it varies also with the frequency of recurrence (Ohta et al., 1994; Kolles et al., 1995; Matsuno et al., 1996; Langford et al., 1996; Madsen and Schröder, 1997; Perry et al., 1998; Abramovich and Prayson, 1998; Nakaguchi et al., 1999; Nagashima et al., 1999), but it must be taken into account that it varies also with the grade of malignancy with which recurrences increase: 0.7%–2.2% for classic, 2.1%–

9.3% for atypical and 11.0%–16.3% for anaplastic meningiomas. In every grade the range would be wide (Maier et al., 1997) with higher values in recurrent tumors, 4.0%–8.8%, than in nonrecurrent tumors, 0.98%–3.8%. There are, however, observations that this correlation does not occur (Moller and Braendsdrup, 1997; Abramovich and Prayson, 1999; Aguiar et al., 2003), especially in completely resected tumors (Kostantinidou et al., 1998). In partially resected tumors the predictive significance of MIB-1 LI can be accepted and it appeared inversely correlated with tumor doubling time of these tumors (Nakaguchi et al., 1999).

The procedures for calculating MIB-1 LI vary among laboratories. The methods for counting positive nuclei also vary: nuclei can be counted in areas selected for their highest frequency or at random or making a mean of all the areas of the tumor. With the latter modality a better prediction of the doubling time can be obtained, because the LI calculated can be attributed to the part of the tumor not removed. This does not happen when in cases with low frequency of positive nuclei there are foci of high frequency (Nakasu et al., 2001). The cut-off point of LI between recurrent and non-recurrent tumors is 2% with counts at random and 3% with counts in selected areas. It is of great importance that in recurrences MIB-1 LI has been found unchanged, increased (Matsuno et al., 1996) or decreased (Madsen and Schröder, 1996). Another important observation is that sometimes MIB-1 LI may be very high, but dissociated from MI, since MIB-1-positive cells may have a longer cell cycle or a greater mortality .

It is also worth emphasizing that MIB-1-positive nuclei in meningiomas are extremely unevenly distributed in the tumor (Siegers et al., 1989; Schiffer et al., 2003); and with a great variability of unevenness among the tumors. Morover, the areas with the highest frequency of positive nuclei do not overlap with those with circumscribed high cell density often encountered.

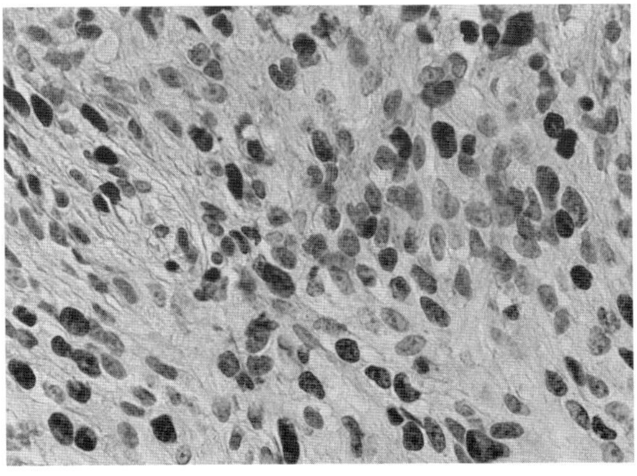

Figure 1. Malignant meningioma. High MIB.1 LI. DAB, x 400

AI has been found to correlate with MIB-1 LI (Maier et al., 1997), with the trend to recurrence, but not with histological grade or p53 expression (Kostantinidou et al., 2001). Bcl-2 shows a weak inverse correlation with apoptosis (Maier et al., 1997;

Kostantinidou et al., 2001) and it increases slightly with grade III (Abramovich and Prayson, 2000). The increase of AI in anaplastic meningiomas has been confirmed, as well as the absence of its correlation with histological sub-types, p53, Bcl-2 or c-myc (Ng and Chen, 1999). Survivin, an apoptotic inhibitory protein, has been found diffusely expressed in most tumors (Sasaki et al., 2002), but with no relationship to different parameters such as Bcl-2 and Bax (Das et al., 2003). Survivin is expressed in gliomas of every grade, but more frequently in those of high grade (Kajiwara et al., 2003). Also catepsin D, involved in the regulation of apoptosis and a negative prognostic factor in breast carcinoma, has been found to be diffusely positive, especially in the classic variant, but with no relationship to MIB-1 LI (Castilla et al., 2003).

Less used, but very important, are the survival data. Anaplastic meningiomas show a mortality rate at 5 years of 68% with a mean survival of 1.5 years against 38% and 7.5 years for the others. Multivariate analysis gives as prognostic factors the extension of surgical removal, MIB-1 LI >20 x 10 HPF and nuclear atypia. Atypical meningiomas show a behavior similar to that of classic tumors. If the invasion of the nervous tissue is used as a criterion, there is no difference between benign and atypical tumors, but the category of atypical tumors increases and that of anaplastic tumors would remain stable. Invasion of the nervous tissue has been introduced as the third criterion for defining atypical meningiomas (Perry et al., 1999). Since the recurrence frequency seems to be in relation to histological diagnosis, one wonders what happens in cases of missed recognition or overestimation of atypical meningiomas. The inclusion of this sub-type among benign meningiomas produces an increase of the recurrence rate at 5 years; whereas its inclusion among anaplastic tumors produces an increase in malignant tumors without recurrence (Ho et al., 2002). The last evaluation method, including papillary formations, endothelial proliferations and invasion of the nervous tissue, does not overlap with that previously described (Perry et al., 1999).

3. THE PROBLEM OF TUMOR PROGRESSION

An important problem is whether atypical and anaplastic meningiomas are the product of a tumor progression or they are such *ab initio*. The idea of a tumor progression in time derives from the existence of a scale of grades. However, in a series of 109 benign meningiomas, only two showed malignancy foci with increase of cell density and mitoses (Michaud et al, 1985). This would indicate that a malignant transformation rarely takes place in meningiomas (Nakasu et al, 2001). In our series of 100 recurrent tumors the histological grade at first surgery did not change in recurrences, with two exceptions only (Schiffer et al, 2005).

As already mentioned, MIB-1 LI also did not change in most cases between first surgery and recurrence (Nakasu et al., 1999; Abramovich and Prayson, 2000; Schiffer et al., 2005), with the exception of a few cases with values higher at first operation (Nakasu et al., 2001). Reports of an increase (Matsuno et al., 1996) or a decrease (Madsen and Schröder, 1997) may be within the range of variability of the counting methods.

On the contrary, molecular genetics data indicating an increase of alterations from grades I to III are interpreted as a progression. Most meningiomas

Meningioma grade I: Loss on 22q

Meningioma grade II: Gain on 1q, 9q, 12q, 15q, 17q, 20
Loss on 1p, 6q, 10, 14q, 18q

Meningioma grade III: Loss on 9 p
Ampl on 17q

Figure 2. Schema of meningioma progression, according to Weber et al., 1997

show losses on 22q and NF2 mutations (Wellenreuther et al., 1995) which must be early events. Biallelic inactivation leads to the loss of merlin (Gutmann et al., 1997) which would not be due to the activation of μ-calpain, which has been suggested as responsible, without alterations of NF2 (Kimura et al., 1998), but through genetic alterations, according to the classic "two-hit" theory of Knudson (1984) (Ueki et al., 1999), even though the hypothesis of μ-calpain has been restored (Kaneko et al., 2001). Inactivation of NF2 is found in 58% of meningiomas (Weber et al., 1997) and it is the only target of LOH of 22q (Büschges et al., 1999). In some cases, mutations of INI1 have been seen in the exon 9 on 22q which could be a second suppressor gene (Schmitz et al., 2001). Moreover, on 18p11.3 in 60% of meningiomas the loss of DAL.1 occurs, analogous to merlin, both being members of the family of proteins 4.1 and homologous to NF2 which could be involved in the tumorigenesis (Gutmann et al., 2000). There is much evidence that merlin has a role in the regulation of cell growth and motility and interacts with several other proteins, and that proteins 4.1 intervene as meningioma growth regulators (Perry et al., 1994). Very interesting is that, starting at the 4th month of life, 30% of rats in the arachnoidal cells of which a Cre-mediated excision of NF2 has been performed, develop meningiomas similar to human tumors (Kalamarides et al., 2001).

Recently, it has been demonstrated that independently of the genotype of tumors, meningiomas, even slow-growing, show numerical chromosomal instability (Van Tilborg et al, 2005).

Since in atypical meningiomas losses of 1p, 14q, 6, 9 and 18q are found, whereas in anaplastic meningiomas losses of 9p and amplification of 17q occur, a schema of clonal evolution of meningiomas by accumulation of genetic alterations is given (Weber et al., 1997) (Figure 2).

Many aberrations in meningiomas indicate 18q losses, but the supposed inactivated tumor suppressor gene remains unidentified (Büschges et al., 2001). Losses of 1p are rather frequent, even though the supposed onco-suppressor gene has not been identified. The hypothesis that it could be p18 has been discarded (Leuraud et al., 2000). Losses on chromosome 10 seem to be in relation to malignancy, but precise correlations with histological types and clinical behavior are still lacking (Mihaila et al., 2003).

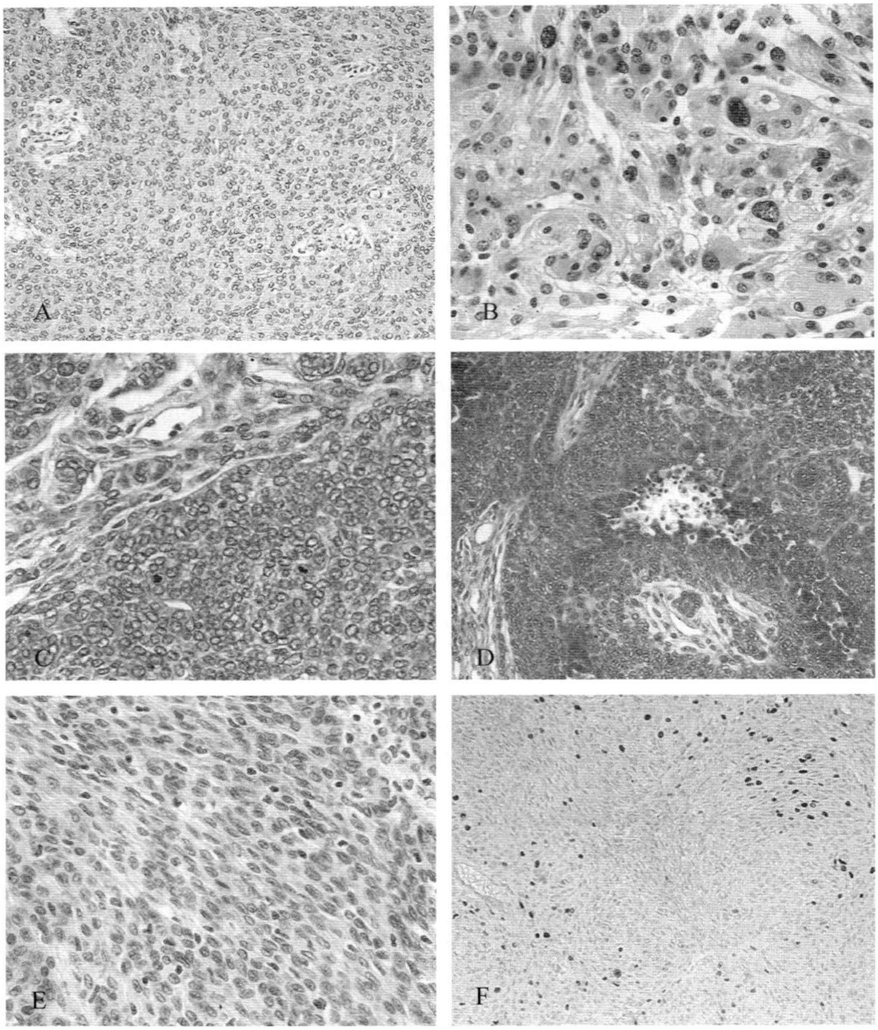

Figure 3. Meningioma A. Sheeting, H&E, x 200; B. Nuclear pleomorphism, H&E, x 200; C. Mitoses, H&E, x 400; D. Circumscribed necrosis, H&E, x 200; E. Sarcomatous aspect, H&E, x 400; F. Regional variability, MIB. 1, DAB, x 100. From Schiffer et al., 2005

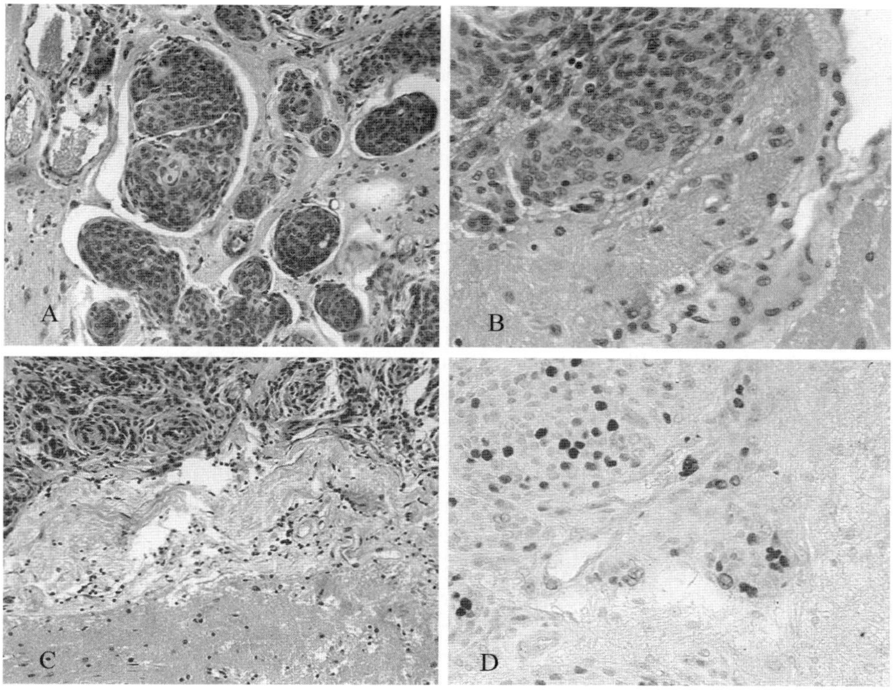

Figure 4. Meningioma A. Prongs into the nervous tissue, H&E, x 400; B. Absence of pial membrane toward the nervous tissue, H&E, x 400; C. Thickened arachnoidnal membrane, H&E, x 400; D. MIB.1 positive meningothelial cells in the CNS, MIB.1, DAB. x 400. From Schiffer et al., 2005

Homozygous deletions of CDKN2A, present in 5% of classic, 18% of atypical and 38% of anaplastic tumors are of great importance. Deletions of p14 and CDKN2B would be present in 46% of anaplastic and 3% of atypical meningiomas (Boström et al., 2000; Simon et al., 2001). In another series, deletions show similar values: 17%, 52% and 74% (Perry et al., 2002). It must be remarked that anaplastic tumors with deletions, studied by dual-color FISH, show 1 year survival against 2 years of those without deletions, whereas no difference was found between recurring and non-recurring tumors in the category of benign and atypical meningiomas. In yet another series, deletions of p16 and p15 were: 27%, 25% and 57%; and mRNA of p14 was lacking in 36% of benign and 71% of anaplastic tumors (Simon et al., 2001). In atypical and mainly anaplastic tumors telomerase activity and hTERT increase (Simon et al., 2001) and there is expression of c-myc correlated with MIB-1 (Nagashima et al., 2001).

TP53 mutations are rare in meningiomas (Perry et al, 1997; Joachim et al., 2001), mainly limited to malignant cases (Wang et al., 1995). On the contrary, in many cases there is accumulation of the protein (Ellison et al., 1995; Perry et al., 1997; Hakin-Smith et al., 2001) for other mechanisms than mutations. Sometimes associated with malignancy or recurrence, there is no correlation with histology.

Three factors drew attention to VEGF: tumor growth, neo-vascularization and edema. It is generally recognized that VEGF is up-regulated in meningiomas (Christov et al., 1999), but no correlation with histological subtypes was found. Some observations showed a relationship of VEGF mRNA and protein with the tumor vascularization (Samoto et al., 1995; Provias et al., 1997), whereas others were negative (Pietsch et al., 1997; Nishikawa et al., 1997). Also PIGf and VEGF-B have been found to be positive (Machein and Plate, 2000) as if meningiomas regulate edema and vascularization through different mechanisms. In other findings, a high level of expression of VEGF is the most useful predictor of recurrence, because its secretion from cell residues after surgery stimulate neovascularization which promotes recurrence (Yamasaki et al., 2000).

Beside EGF and PDGF, also TGFα is expressed in meningiomas, where it correlates with shorter PFS, even though it does not distinguish between classic and atypic and anaplastic tumors (Hsu et al., 2000).

From the cytogenetic point of view, different observations showed complex caryotypes in meningiomas of a higher grade and a normal caryotype or the monosomy 22 alone in most classic tumors (Cerda-Nicolas et al., 2000). Correspondingly, meningiomas with normal diploid asset and with monosomy 22 alone showed a low rate of recurrence, whereas those hypodiploid with somatic losses besides monosomy 22 and those with 1q deletion and other alterations show a high recurrence rate (Ketter et al., 2001).

Molecular genetics of meningiomas marked time in comparison with gliomas and the significance of many alterations still escapes us. A transcriptional panorama of cells can be obtained by gene expression profiling by mRNA microarrays. Not long ago, profiles associated with the three grade meningiomas have been found (Watson et al., 2002). Recently, activation of PI3K/Akt signaling has been added to the characteristics of malignant meningiomas, together with that of MAPK (Mawrinetal, 2005). In the development of a more malignant phenotype the deregulated expression of the Notch pathway seems to play a role (Cuevas et al., 2005).

4. GENESIS OF RECURRENCES

Partially resected tumors show a higher recurrence rate than totally resected tumors (Stafford et al., 1998; Ayerbe et al., 1999) and the possibility of a total resection depends on the location of the tumor. It was observed that 21% of recurrences belonged to parasagittal tumors and that 44% of these recur as sinus invasion (Christensen et al., 1983). The recurrence free rate has been calculated to be 95%, 80% and 68% at 5 years for completely resected tumors and 63%, 45% and 9% for partially resected ones. Convexity meningiomas could be totally resected in 96% of cases with a recurrence/progression rate at 5 years of only 3%, whereas for parasellar meningiomas the values were respectively 57% and 19% and for those of the cerebello-pontine angle 28% and 34% (Mirimanoff et al., 1985). Two remarks must be made: in the different series, excluding malignant and multiple tumors, it is possible that a partial resection is given as total; today with CT scan and MRI this error is much less common. For grade I tumors total resection is fundamental, because recurrence rate at 5 years is 5% against 31% of incompletely resected tumors, whereas for grade II tumors the values at 5 years are 40%, also after total removal. Malignant tumors; show a mean survival <2 years (Watson et al., 2000). The second remark is that meningiomas, contrary to previously held opinions, are not constantly capsulated and the tumor has a direct contact with the nervous tissue that might be invaded and this event has a frequency between 26% and 72% (Perry et al., 1998; Nakasu et al., 1999; Palma, 2002) (Figure 4). The idea that recurrence derived from cells left behind at the surface of the brain during surgery is not new (Crompton e Gautier-Smith, 1970). All this is convincing, but the observation that usually recurrences occur at the periphery of dural insertion must be taken into account (de Vries and Wakhloo, 1994).

Is the recurrence of a meningioma due to its malignancy ? To incomplete resection/difficult location? Or to both events? In the first case, if no tumor cell has been left behind, a promotion and tumor progression should be hypothesized for meningothelial or meningotheliomatous cells located in the area where the tumor has arisen. In the second case, the recurrence should not be the expression of a greater malignancy. As a matter of fact completely benign tumors also recur, even though more rarely and, therefore, it could be concluded that meningiomas recur for multicentricity in time or for incomplete resection or for malignancy, but in the latter case an invasion of near structures should have happened with persistence of tumor cells (Nakasu et al., 1999). In the dura in proximity of large meningiomas clusters of meningothelial cells have been found, not present in other locations, which have been considered as tumor cells. They occur between the two laminae of the dura and are not the product of a seeding or of a hematogeneous dissemination (Borovich and Doron, 1986).

Since recurrences are more frequent in tumors with greater MIB-1 LI and MI, the conclusion could be that their higher frequency in atypic and anaplastic tumors, regardless of the event responsible for the recurrence, could be due to their shorter tumor doubling time (TdT). The latter inversely correlates with MIB-1 LI in incompletely resected tumors (Nakaguchi et al., 1999). In recurrences MIB-1 LI does not increase, even though the appearance of aneuploidy has been considered a sign of great malignancy (Arai et al., 1998). Even if features of higher malignancy were found in recurrences, this does not mean that recurrence is a sign of malignancy, but only that the latter is responsible for their precociousness (du Plessis et al., 2000). The meningothelial islands found by Borovich and Doron in the dura at the basis of large meningiomas, but not in meninges without meningiomas, acquire a particular meaning as the starting point of recurrences (Wilson, 1994). The cells of the islands can be either tumor cells or normal meningothelial cells, in view of the difficulty in distinguishing them, and one wonders why they start to proliferate at a given moment. If they were tumor meningothelial cells with the TdT of a classic meningioma, that is 2 years, the time necessary for the development of a symptomatic tumor would be very long. This is not so if they are transformed cells and this would give an explanation for the greater frequency of recurrence of atypical and anaplastic tumors.

The observation that incompletely resected meningiomas recur more frequently enables us to conclude that also in completely resected tumors some cells could have been left behind and this would be in agreement with the other observation that meningiomas with "mushrooming", or of lobulated form, recur more frequently (Nakasu et al., 1999). It is not mandatory that these cells should be in the dura, because in the meningiomas of convexity they are in the thickened arachnoidal membrane between the brain and the tumor; and with the resection of the arachnoidal membrane the number of recurrences decreases (Kamitani et al., 2001). These considerations are in agreement with the conclusions of Perry et al. (1999) that the invasion of the nervous tissue around the tumor is predictive of recurrence. Another possible explanation of recurrences in completely resected tumors is that they are multiple meningiomas in time arising in the same area or, alternatively, that they derive from a first tumor, as demonstrated by their clonality, because of the same genetic alterations (Von Deimling et al., 1999; Zhu et al., 1999).

5. RADIOTHERAPY AS A PROGNOSTIC FACTOR

The use of standard fractionated radiotherapy with >60 Gy as the maximum dose is based on the rationale that the tumor proliferation rate requires a TdT lower than a certain value. Radiosurgery or stereotactic radiotherapy uses higher and different doses with specific problems related to the sparing of normal nervous tissue. In meningiomas the indication and the effects of RT are unclear. First of all, only malignant tumors are irradiated, and even then only after failure of repeated surgical resections or at least after subtotal resection. Secondly, atypical meningiomas are often included in the category of malignant meningiomas and irradiated even after complete resection. Thirdly, the treatments mentioned above have been extended to classic meningiomas when subtotally resected and even to meningiomas not candidates for surgery owing to their being in a dangerous position for life and for fundamental functions. It is difficult to deduce guidelines for individual cases from the many different observations. One of the most accepted approaches is: Benign meningiomas, total resection = no therapy, partial resection = radiosurgery or radiotherapy. Atypical meningiomas, partial resection = radiosurgery or radiotherapy.

6. CONCLUSIONS

The histological diagnosis of meningioma is very easy; caution must be taken when the differential diagnosis needs to be carried out for papillary, rhabdoid, clear cell, atypic and anaplastic variant, and when the tumor is subtotally resected or when it is lobulated or mushrooming. The examination must be carried out on the entire long axis of the surgical sample and attention must be devoted to the diffuse or regional distribution of the structures mentioned above and to their association with other signs of malignancy. Mitoses must be counted in the entire tumor section and the MI must be calculated. MIB-1 LI must be calculated after examining the areas with the highest number of positive nuclei after visual analysis. The distribution of MIB-1 positive nuclei is extremely variable in meningiomas: usually so-called proliferation centers with high cellularity are not rich either in mitoses or in MIB-1 positive nuclei. Circumscribed necroses are not important by themselves, but only when they are associated with other signs of malignancy. Of course, one should be informed when the tumor has been embolized before surgery.

Particular attention must be paid to the existence of nervous tissue adhering to the tumor, especially in convexity meningiomas. It must be carefully studied, also by proliferation markers, because it could contain tumor meningothelial cells either in the thickened arachnoidal membrane or directly in the nervous tissue adhering to the tumor without a capsule. Similar cells may occur in the adjacent resected brain tissue. If the tumor is resected with its dural implant, the latter must be studied carefully and the possible clusters of meningothelial cells or of tumor cells must be investigated for their proliferation potential.

The contribution of molecular genetics to the diagnosis may be today limited to the study of LOH of CDKN2A or to immunohistochemistry of p16 which could be absent in atypical or, especially, anaplastic cases.

One wonders whether for prediction of recurrence, for survival and for the therapeutic strategy after surgery, all the diagnostic procedures mentioned above are worth performing. We have seen that the diagnosis of atypical, anaplastic, papillary and rhabdoid meningioma as well the occurrence of infiltrated nervous tissue, of a high MIB-1 LI, high expression of VEGF and LOH of CDKN2 may influence either recurrence and survival, especially when associated with subtotal resection. Also the therapeutic strategy can be influenced, even though today post-surgical therapy is oriented toward a more diffuse employment of different kinds of irradiation, even independently of surgery and malignancy.

Once it has been accepted that chemotherapy has failed, there are two basic procedures of radiotherapy the effects of which are evalutaed in terms of local control, complications and survival. External beam radiotherapy is officially used in subtotally resected or recurrent meningiomas or following surgery for aggressive and malignant meningiomas (Wilson, 1994; Goldsmith et al., 1994). This form of therapy is not exempt from complications, the most important ones being radiation necrosis, small vessel arteriopathy and intellectual decline, not to mention the induction of secondary malignancies. Radiosurgery is applied to the same patients and also to patients with surgically inaccessible tumors or as a boost following external irradiation (Lunsford, 1994) and it shows the same complications mentioned above.

The results of external radiotherapy are variable and not encouraging (Mathiesen et al., 2003) or acceptable (Kokubo et al., 2000), whereas in benign skull base meningiomas they were similar for radiotherapy and radiosurgery with a decrease of PFS progression (Mendenhall et al., 2003). Good results have been obtained with the radiosurgery of benign unresected tumors (Kondziolka et al., 2003; Flickinger et al., 2003) with a PFS comparable to that of Simpson grade 1 tumor (Pollock et al., 2003), of subtotally resected aggressive meningiomas (Harris et al., 2003) and with fractionated stereotactic radiotherapy and gamma-knife of both resected and unresected and of aggressive tumors (Ojemann et al., 2000; Kobayashi et al., 2001; Milker-Zabel et al., 2005). But also disappointing results in dealing with aggressive meningiomas have been obtained (Stafford et al., 2001). New techniques have been introduced using proton and photon beams (Noel et al., 2002; Bolsi et al., 2003), high precision focused irradiation with fractionated stereotactic conformal radiotherapy (Jalali et al., 2002), high dose, proton 3D-conformal radiation therapy. With the latter treatment good local control and improved survival were obtained for atypical and malignant meningiomas (Hug et al., 2000). Also stereotactic implantation of iodine-125 has been succerssfully used for primary and recurrent tumors (Obasi et al., 2002).

7. MENINGIOMAS IN CHILDREN

Meningiomas in children are very rare and they have roughly the same problems as in adults. The most important one is the differential diagnosis of meningeal-based masses mimicking meningioma. The most common category is represented by sarcomas, followed by hemangiopericytomas, fibrosarcomas, malignant fibrous histiocytomas, chondrosarcomas and also benign mesenchymal tumors and histiocytic disorders (Perry and Dehner, 2003). Clear cell and papillary meningiomas are more frequent in children as well as the "sclerosing variant" of meningioma (Hope et al., 1983).

REFERENCES

Abdel-Rahman, A., Rao, M.S., and Shetty, A.K., 2004, Nestin expression in hippocampal astrocytes after injury depends on the age of the hippocampus, *Glia* 47: 299-313.

Abdulrauf, S. I., Edvardsen, K., Ho, K.L., Yang, X.Y., Rock, J.P., and Rosenblum, M.L., 1998, Vascular endothelial growth factor expression and vascular density as prognostic markers of survival in patients with low-grade astrocytoma, *J. Neurosurg.* 88: 513-520.

Abramovich, C.M., and Prayson R.A., 1999, Histopathologic features and MIB-1 labelling indices in recurrent and nonrecurrent meningiomas, *Arch. Pathol. Lab. Med.* 123: 793-800.

Abramovich, C.M., and Prayson, R.A., 2000, Apoptotic activity and Bcl-2 immunoreactivity in meningiomas. Association with grade and outcome, *Am. J. Clin. Pathol.* 114: 84-92.

Actor, B., Cobbers, J.M.J.L, Büschgess, R., Wolter, M., Knobbe, C.B., Lichter, P., Reifenberger, G., and Weber, R.G., 2002, Comprehensive analysis of genomic alteration in gliosarcoma and its two tissue components, *Genes, Chromosome & Cancer* 34: 416-427.

Adams, J., 2002, Proteasome inhibitors as a new anticancer drugs, *Curr. Opin. Oncol.*, 14: 628-634.

Adams, S.A., and Hilton, D.A., 2002, Recurrent haemangioblastoma with glial differentiation, *Neuropathol. Appl. Neurobiol.* 28:142-146.

Aguiar. P.H., Tsanaclis, A.M., Tella, O.I., Plese, J.P., 2003, Proliferation rate of intracranial meningiomas as defined by the monoclonal antibody MIB-1. *Neurosurg. Rev.* 26: 221-228.

Aguirre-Cruz. L., Mokhtari. K., Joang-Xuan, K., Marie, Y., Criniere, E., Taillibert, S., Lops, M., Delattre, J.Y., and Sanson, M., 2004, Analysis of the bHLH transcription factor Olig 1 and lig 2 in brain tumors, *J Neuro-Oncol.* 67: 265-271.

Akasaki, Y., Liu, G., Chung, N.H., Ehtesham, M., Black, K.L., and Yu, J.S., 2004, Induction of a CD4+ T regulatory type 1 response by cyclooxygenase-2-overexpressing gliomas, *J. Immunol.* 173: 4352-4359.

Aldape, K.D., Ballman, K., Furth, A., Buckner, J.C., Giannini, C., Burger, P.C., Scheithauer, B.W., Jenkins, R.B., and James, C.D., 2004, Immunoihistochemical detection of EGFRvIII in high malignancy grade astrocytomas and evaluation of prognostic significance, *J. Neuropathol. Exp. Neurol.* 63 700-707.

Alkalay., I., Yaron., A., Hatzubai, A., Orian, A., Ciechanover, A., and Ben-Neriah, Y., 1995, Stimulation-dependent I kappa B alpha phosphorylation marks the NF-kappa B inhibitor for degradation via the ubiquitin-proteasome pathway, *P.N.A.S. U.S.A.* 92:10599-10603.

Almqvist., M., Mah, R., Lendahl, U., Jacobsson, B., and Herndson, G., 2002, Immunohistochemical detection of nestin in pediatric brain tumors, *J. Histo-cytochem.* 50: 147-158.

Altieri, D.C., 2003, Validating surviving as a cancer therapeutic target, *Nat. Rev. Cancer* 3: 46-54.

Alvarez-Buylla. A., Garcia-Verdugo, J.M., and Tramontin, A.D., 2001, A unified hypothesis on the lineage of neural stem cells, *Nat. Rev. Neurosci.* 2: 287-293.

Amirlak., B., and Couldwell W.T., 2003, Apoptosis in glioma cells: review and analysis of techniques used for study with focus on the laser scanning cytometer, *J. Neuro-Oncol.* 63: 129-145.

An, J., Sun, Y.P., Adams, J., Fischer, M., Belldegrun, A., and Rettig, M.B. 2003, Drug interactions between proteasome inhibitor bortezomib and cytotoxic chemotherapy, tumor necrosis factor (TNF) alpha, and TNF-related apoptosis-inducing ligand in prostate cancer, *Clin. Cancer Res.* 9: 4537-4545.

Anda, T., Shabani, H.K., Tsunoda, K., Tokunaga, Y., Kaminogo M., Shibata, S., Hayashi, T., and Iseki, M., 2003, Relationship between expression of O6-methylguanine-DNA methyltransferase, glutathione-S-transferase pi in glioblastoma and the survival of the patients treated with nimustine hydrochloride: an immunohistochemical analysis, *Neurol. Res.* 25: 241-248.

Ang, L.C., Perry, J.R., Bilbao, J.M., Ozane, W., Peschke, E., Young, B., and Nelson, N., 1996, Postirradiated and nonirradiated gliosarcoma: immunophenotypical profile, *Can. J. Neurol. Sci.* 23: 251-256.

Arai, H., Beppu, T., Wada, T., Yoshida, Y., Kubo, Y., Suzuki, M., and Ogawa, A., 1998, Pathological analyses of early recurrence and malignant transformation in meningiomas, *Brain Tumor Pathol.* 15: 37-40.

Arends, M.J., Mc Gregor, A.H., Wyllie, A.H., 1994, Apoptosis is inversely related to necrosis and determines net growth in tumours bearing constitutively expressed myc, ras and HPV oncogenese, *Am. J. Pathol.* 144: 1045-57.

Ashbury A.K., Ojemann, R.J., Nielsen, S.L., and Sweet, W.H., 1972, Neuropathologic study of fourteen cases of manlignant brain tumor treated by Boron -10 slow neutron capture radiation, *J. Neuropathol. Exp. Neurol.* 31: 278-303.

Ashkenazy Am Dixit, V.M., 1998, Death receptors: signaling and modulation, *Science* 281: 1305-1308.

Assimakopoulou, M., Sotiropoulou-Bonikou, G., Maraziotis, T., Papadakis, N., and Varakis, I., 1997, Microvessel density in brain tumors, *Anticancer Res.* 17: 4747-4754.

Audero, E., Cascone, I., Zanon, I., Previtali, S.C., Piva, R., Schiffer, D., and Bussolino, F., 2001, Expression of Angiopoietin-1 in human glioblastomas regulates tumor-induced angiogenesis in vio and in vitro studies, *Arterioscler. Thromb. Vasc. Biol.* 21: 536-541.

Ayerbe, J., Lobato, R.D., de la Cruz, J., Alday, R., Rivas , J.J., Gomez, P.A., and Cabrera, A., 1999, Risk factors predicting recurrence in patients operated on for intracranial meningioma: a multivariate analysis, *Acta Neurochir.* 141: 824-932.

Azzarelli, B., Miravalle, L., Vidal, R., 2003, Immunolocalization of the oligodendrocyte transcription factor 1 (Olig 1) in brain tumors, *J. Neuropathol. Exp. Neurol.* 63: 170-179.

Baeza, N., Weller, M., Yonekawa, Y., Kleihues, P., and Ohgaki, H., 2003, PTEN methylation and expression in glioblastomas, Acta Neuropathol. 106: 479-485.

Balko, M.G., Blisard, K.S., Samaha, F.J., 1998, Oligodendroglial gliomatosis cerebri, *Hum. Pathol.* 23: 706-707.

Bäcklund, L.M., Nilsson, B.R., Goike, H.M., Schmidt, E.E., Liu, L., Ichimura, K., and Collins, V.P., 2003, Short postoperative survival for glioblastoma patients with a dysfunctional Bb1 pathway in combination with no wild-type PTEN, *Clin. Cancer Res.* 9: 4151-4158.

Baehring, J., Ogle, E., Sze, G., Duncan, C., and Bannykh, S., 2005, Ganglio neurocytomas in the spinal cord, *J. Neuro-Oncol.* 71: 149-150.

Bailey, P., and Cushing, H., 1926, A classification of the tumors of the glioma group on a histogenetic basis with a correlation study of prognosis, Lippincott, Philadelphia.

Baisden, B.L., Brat, D.J., Melhem, E.R., Rosenblum, M.C., King, A.P., and Burger, P.C., 2001, Dysembryoplastic neuroepithelial tumor-like neoplasm of the septum pellucidum. A lesion often misdiagnosed as glioma. Report of 10 cases, *Am. J. Surg. Pathol.* 25: 494-499.

Barker, F.G. II, Chang, S.M., Huhn, S.L., Davis, R.L., Gutin, P.H., McDermott, M.W., Wilson, C.B., Prados, M.D.,1997, Age and the risk of of anaplasia in magnetic resonance –nonenancing supratentorial cerebral tumors, *Cancer* 80: 936-941.

Bechtel, J.T., Patton, J.M., and Takey, I., 1978, Mixed mesenchymal and neuroectodermal tumor of the cerebellum, *Acta Neuropathol.* 41: 261-263.

Becker, I., Paulus, W., and Roggendorf, W., 1989, Histogenesis of stromal cells in cerebellar hemangioblastomas. An immunohistochemical study, *Am. J. Pathol.* 134: 271-275.

Behin, A., Hoang-Xuan, K., Carpenter, A.F., Delattre, J.Y., 2003, Primary brain tumors in adults, *Lancet* 361: 323-331.

Bell, H.S., Whittle, I.R., Walker, M., Leaver, H.A., and Wharton, S.B., 2001, The development of necrosis and apoptosis in glioma: experimental findings using spheroid culture systems, *Neuropathol. Appl. Neurobiol.* 27: 291-302.

Berger, F., Gay, E., Pelletier, L., Tropel, P., and Wion, D., 2004, Development of gliomas: potential role of asymmetrical cell division on neural stem cells, *Lancet Oncol.* 5: 511-514.

Bergmann, A., Tugentman, M., Shilo, B.Z., Steller, H., 2002, Regulation of cell number by MAPK-dependent control of apoptosis: a mechanism for trophic survival signaling, *Dev. Cell* 2: 159-170.

Bertolone, S.J., Yates, A.J., Boyett, J.M., Wallace, D., and Finlay, J.L., 2003, Combined modality therapy for poorly differentiated gliomas of the posterior fossa in children: a children's cancer group report, *J. Neuro-Oncol.* 63: 49-54.

Bian, J., and Sun, Y., 1997,Transcriptional activation by p53 of the human type IV collagenase (gelatinase A or matrix metalloproteinase 2) promoter, *Mol. Cell Biol.* 17: 6330-6338.

Biernat, W., Aguzzi, A., Sure, U., Grant, J.W., Kleihues, P., and Hegi, M.E., 1995, Identical mutations of the p53 tumor suppressor gene in the gliomatous and the sarcomatous components of gliosarcomas suggest a common origin from glia cells, *J. Neuropathol. Exp. Neurol.* 54: 651-656.

Biernat, W., Debiec-Rychter, M., and Liberski, P.P., 1998, Mutations of TP53, amplification of EGFR, MDM2 and CDK4, and deletions of CDKN2A in malignant astrocytomas, *Pol. J. Pathol.* 49, 267- 271.

Biernat, W., Zakrzewski, K., and Liberski, P.P., 2000, 12-year-old boy with recent onset seizures, Brain Pathol. 10: 313-314.

Biernat, W., Huang, H., Yokoo, I., Kleihues, P., and Ohgaki, H., 2004, Predominant expression of mutant EGFR (EGFRvIII) is rare in primary glioblastoma, *Brain Pathol.* 14: 131-136.

Bigner, D.D., Brown, M., Friedman, A.H., Coleman, R.E., Akabani, G., Friedman, H.S., Thorstad, W.L., Mcendon, R.E., Bigner, S.H., Zhao, X.C., Pegram, C.N., Wikstrand, C.J., Herndon, J.E. 2nd, Vick, N.A.,Paleologos, N., Cokgor, I., Provenzale, J.M. and Zalutsky, M.R., 1996, Iodine-131-labeled antitenascin monoclonal antibody 81CC6 treatment of patients with recurrent malignant gliomas: phase I trial results, *J. Neuro-Oncol.* 16: 2202-2212.

Bigner, S.H., Bullard, D.E., Pegram, C.N., Wikstrand, C.I., Bigner, D.D., 1981, Relationship of *in vitro* morphologic and growth characteristics of established human glioma-derived cell lines to their tumorigenicity in athymic nude mice, *J. Neuropathol. Exp. Neurol.* 40: 390-409.

Bigner, S.H., McLendon, R.E., Fuchs, H., McKeever, P.E., and Friedman, H.S., 1997, Chromosomal characteristics of childhood brain tumors, *Cancer Genet. Cytogen.* 97: 125-134.

Bigner, S.H., Rasheed, B.K., Wiltshire, R., and McLendon, R.E., 1999, Morphologic and molecular genetic aspects of oligodendroglial neoplasms, *Neuro-Oncology* 1: 52-60.

Bigner, S.H., Matthews, M.R., Rasheed, B.K.A., Wiltshire, R.N., Friedman, H.S., Friedman, A.H., Stenzel, T.T., Dawes, D.N., McLendon, R.E., and Bigner, D.D., 1999, Molecular genetic aspects of oligodendrogliomas including analysis by comparative genomic hybridization, *Amer. J. Pathol.* 55: 375-386.

Birner, P., Gatterbauer, B., Oberhuber, G., Schindl, M., Rossler, K., Prodinger, A., Budka, H., and Hainfellner, J.A., 2001, Expression of hypoxia-inducible factor 1 alpha in oligodendrogliomas: its impact on prognosis and on neoangiogenesis, *Cancer* 92: 165-171.

Birner, P., Piribauer, M., Fischer, I., Gatterbauer, B., Marosi, C., Ambros, P.F., Ambros, I.M., Bredel, M., Oberhuber, G., Rössler, K., Budka, H., Harris, A.L., and Hainfellner, J.A., 2003, Vascular patterns in glioblastoma influence clinical outcome and associate with variable expression of angiogenic proteins: evidence for distinct angiogenic subtypes, *Brain Pathol.* 13: 133-143.

Blümke, I., Beck, H., Lie, A.A., and Wiestler, O.D., 1999, Molecular neuropathology of human mesial temporal lobe epilepsy, *Epilepsy Res.* 36: 205-223.

Blümke, I., Becker, A.J., Normann, S., Hans, V., Riederer, B.M., Krajewski, S., Wiestler, O.D., and Reifenberger, G., 2001, Distinct expression pattern of microtubule-associated protein-2 in human oligodendrogliomas and glial precursor cells, *J. Neuropathol. Exp. Neurol.* 60: 984-993.

Blümke, I., and Wiestler, O.D., 2002, Gangliogliomas: an intriguing tumor entity associated with focal epilepsies, *J. Neuropathol. Exp. Neurol.* 61: 575-584.

Boerman, R.H., Anderl, K., Herath, J., Borell, T., Johnson, N., Schaeffer-Klein, J., Kirchhof, A., Raap, A.K., Scheithauer, B., and Jenkins, R.B., 1996, The glial and mesenchymal elements of gliosarcomas share similar genetic alterations, *J. Neuropathol. Exp. Neurol.* 55: 973-981.

Bolsi, A., Fogliata, A., and Cozzi, L., 2003, Radiotherapy of small intracranial tumours with different advanced techniques using photon and proton beams: a treatment planning study, *Radioth. Oncol.* 68: 1-14.

Bolteus, A.J., Berens, M.E., and Pilkington, G.J., 2001, Migration and invasion in brain neoplasms, *Curr. Neurol. Nerosci. Rep.* 1: 225-232.

Bonni, A., Sun, Y., Nadal-Vicens, M., Bhatt , A., Frank, D.A., Rozovsky, I., Stahl, N., and Yankopoulos, M.E., (1997), Regulation of gliogenesis in the central nervous system by the Jak-Stat signaling pathway, *Science* 278: 477-483. Bonnin, J.M., Pena, C.E., and Rubinstein, L.J., 1983, Mixed capillary hemangioblastoma and glioma. A redefinition of the “angioglioma”, *J. Neuropathol. Exp. Neurol.* 42: 504-516.

Bonnin JM, Rubinstein LJ. Astroblastomas: a pathological study of 23 tumors, with a postoperative follow-up in 13 patients. *Neurosurgery* 1989,25: 6-13

Borovich B, Doron Y. Recurrence of intracranial meningiomas: the role played by regional multicentricity. *J Neurosurg* 1986, 64: 58-63

Bortolotto S, Chiadò-Piat L, Cavalla P, Bosone I, Chiò A, Mauro A, Schiffer D. CDKN2A/p16 inactivation in the prognosis of oligodendrogliomas. *Int J Cancer* 2000, 88: 554-557

Bortolotto S, Chiadò-Piat L, Cavalla P, Bosone I, Mauro A, Schiffer D. CDKN2A/p16 in ependymomas. *J Neuro-oncol* 2001, 54: 9-13

Bosch DA. Short and long term effects of methyl- and ethylnitrosourea (MNU & & ENU) on the developing nervous system of the rat. II hort term: concluding remarks on chemical neuro-oncogenesis. *Acta Neurol Scand* 1977, 55: 106-122

Bosone I, Cavalla P, Chiadò-Piat L, Di Vito N, Schiffer D. Cyclin D1 expression in normal oligodendroglia and microglia cells: Its use in the differential diagnosis of oligodendrogliomas. *Neuropathology* 2001, 21: 155-161

Boström J, Meyer-Puttlitx B, Wolter M, Blaschke B, Weber RG, Lichter P, Ichimura K, Collins VP, Reifenberger G. Alterations of the tumor suppressor genes *CDKN2A* $(p16^{INK4a})$, $p14^{ARF}$, *CDKN2B* $(p15^{INK4b})$, and *CDKN2C* $(p18^{INK4c})$ in atypical and anaplastic meningiomas. *Am J Pathol* 2001, 159: 661-669

Bouvier, C., Bartoli, C., Aguirre-Cruz, L, Virard, I., Colin, C., Fernandez, C., Gouvernet, J., Figarella-Branger, D., 2003, Shared oligodendrocyte lineage gene expression in gliomas and oligodendrocyte progenitor cells, *J. Neurosurg.* 99: 344-350.

Brandes, A.A., Palmisano, V., and Monfardini, S., 1999, Medulloblastoma in adults: clinical characteristics and treatment, *Cancer Treat Rev.* 25: 3-12.

Brandes, A,A,, Amista, P., Gardiman, M., Volpin, L., Danieli, D., Guglielmi, B., Carollo, C., Pinna, G., Turazzi, S., and Monfardini, S., 2000, Chemotherapy in patients with recurrent and progressive central neurocytoma, *Cancer* 88: 169: 174.

Brat, D.J., Scheithauer, B.W., Staugautis, S.M., Cortez, S.C., Brecher, K., and Burger, P.C., 1998, Third ventricular chordoid glioma: a distinct clinicopathologiucal entity, *J. Neuropathol. Exp. Neurol.* 57: 283-290.

Brat, D.J., Hirose, Y., Cohen, K.J., Feuerstein, B.G., and Burger, P.C., 2000, Astroblastoma: clinicopathologic features and chromosomal abnormalities defined by comparative genomic hybridization, *Brain Pathol.* 10: 342-352.

Brat. D.J., Scheithauer, B.W., Eberhart, C.G., and Burger, P.C., 2001, Extraventricular neurocytomas: pathologic features and clinical outcome, *Am. J. Surg. Pathol.* 25: 1252-1260.

Bredel, M., Piribauer, M., Marosi, C., Birner, P., Gatterbauer, B., Fischer I., Ströbel, T., Rössler, K., Budka, H., Hainfellner, J.A., 2002, High expression of DNA topoisomerase IIα and Ki-67 antigen is associated with prolonged survival in glioblastoma patients, *Europ. J. Cancer* 38: 1343-1347.

Brem. S.S., Cotran, R.S., and Folkman, M.J., 1972, Tumor angiogenesis: a quantitative method for histologic grading, *J. Natl. Cancer Inst.* 48: 347-356.

Brown, H.G., Kepner, J.L., Perlman, E.J., Freidman, H.S., Strother, D.R., Duffner, P.K., Kun, L.E., Goldthwaite, P.T., and Burger, P.C., 2000, "Large cell/anaplastic" medulloblastomas: a pediatric oncology group study, *J. Neuropathol. Exp. Neurol.* 59: 857-865.

Buccoliero, A.M., Caldarella, A., Ammannati, F., Mennonna, P., Taddei, A., and Taddei, G.L., 2002, Extraventricular neurocytoma: morphological and immunohistochemical considerations of differential diagnosis, *Pathol. Res. Pract.* 198:627-633.

Buhl, R., Huang, H., Hugo, H.H., and Mehdorn, H.M., 2004, Ganglioneurocytoma of the third ventricle, *J. Neuro-Oncol.* 66: 341-344.

Burger, P.C., Shibata, T., and Kleihues, P., 1986, The use of monoclonal antibody Ki-67 in the identification of proliferating cells: application to surgical neuropathology, *Am. J. Surg. Pathol.* 10: 611-617.

Burger, P.C., Heinz, E.R., Shibata, T., and Kleihues, P., 1988, Topographic anatomy and CT correlations in the untreated glioblastoma multiforme, *J. Neurosurg.* 68: 698-704.

Burger, P.C., and Kleihues, P., 1989, Cytologic composition of the untreated glioblastoma with implications for evaluation of needle biopsies, *Cancer* 63: 2014-2023.

Burger, P.C., and Scheithauer, B.W., 1994, Tumors of neuroglia and choroid plexus epithelium: tumors of the central nervous system, AFIP, Washington DC, 25-161.

Burger, P.C., Scheithauer, B.W., Lee, R.R., and O'Neill, B.P., 1997, An interdiscipliunary approach to avoid overtreatment of patients with central nervosu system lesions, *Cancer* 80: 2040-2046.

Burger, PC. Comments on Kawano N. et al., *Surg Neurol* 1999, 51: 281-288.

Burger, P.C., Scheithauer, B.W., Paulus, W., Szymas, J., Giannini, C., and Kleihues, P., 2000, Pilocytic astrocytoma. In: *Pathology and genetics. Tumours of the nervous system.* IARC, Lyon, pp. 45-51.

Burger, P.C., Pearl, D.K., Aldape, K., Yates, A.J., Scheithauer, B.W., Passe, S.M., Jenkins, R.B., and James, C.D., 2001, Small cell architecture - A histological equivalent of EGFR amplification in glioblastoma multiforme ? *J. Neuropathol. Exp. Neurol.* 60: 1099-1104.

Burger, P.C., Minn, A.Y., Smith, J.S., Borell, T.J., Jedlicka, A.E., Hyntley, B.K., Goldthwaite, P.T., Jenkins, R.B., and Feuerstein, B.G., 2001, Losses of chormosomal arms 1p and 19q in the diagnosis of oligodendroglioma. A study of paraffin embedded sections, *Mod. Pathol.* 14: 842-853.

Burger, P.C., 2002, What is an oligodendroglioma? *Brain Pathol,* 12: 257-259.

Burger, P.C., Scheithauer, B.W., and Vogel, S.F., 2003, Surgical pathology of the nervous system and its coverings. Ed. 4, Churchill Livingstone, London.

Bursztajn, S., Feng, J.J., Berman, S.A., and Nanda, A., 2000, Poly(ADP-ribose) polymerase induction is an early signal of apoptosis in human neuroblastoma, *Brain Res. Mol. Brain Res.* 76: 363-376.

Buschges, R., Weber, R.G., Actor, B., Lichter, P., Collins, V.P., and Reifenberger, G., 1999, Amplification and expression of cyclin D genes (CCDND1, CCND2, CCND3) in human malignant gliomas, *Brain Pathol.* 9: 435-442.

Buschges, R., Boström, J., Wolter, M., Blaschke, B., Weber, R.G., Lichter, P., Collins, V.P., and Reifenberger, G., 2001, Analysis of human meningiomas for aberrations of the *MADH2, MADH4, APM-1* and *DCC* tumor suppressor genes on the long arm of chromosome 18, *Int. J. Cancer* 92: 551-554.

Cairncross, J.G., 1987, The biology of astrocytomas: a lesson learned from chronic myelogenous leukemia hypothesis, *J. Neuro-Oncol.* 5: 99-104.

Cairncross, J.G., and MacDonald, D.R., 1988, Successful chemotherapy for recurrent malignant oligodendroglioma, *Ann. Neurol.* 23: 360-364.

Cairncross, G., MacDonald, D., Ludwin, S., Lee, D., Cascino, T., Buckner, J., Fulton, D., Dropcho, E., Stewart, D., and Schold, C.J., 1994, Chemotherapy for anaplastic oligodendroglioma, Nat. Cancer Inst. Canada Clinical Trials Group, *J. Clin. Oncol.* 12: 2013-2021.

Cairncross, J.G., Ueki, K., Zlatescu, M.C., Lisle, D.K., Finkelstein, D.M., Hammond, R.R., Silver, J.S., Stark, P.C., Macdonald, D.R., Ino, Y., Ramsay, D.A., and Louis, D.N., 1998, Speific genetic predictors of chemotherapeutic response and survival in patients with anaplastic oligodendrogliomas, *J. Natl. Cancer Inst.* 90: 1473-1479.

Cairncross JG. Understanding low-grade gliomas: a decade of progress. *Neurology* 2000, 54: 1402-1403

Campos MG, Zentner J. Ostertun B, Wolf HBK, Schramm J. Anaplastic ganglioglòiomaa: case report and review of the literature. *Neurol Res* 1994, 16: 317-320

Caputy, A.J., McCullough, D.C., Manz, H.J., Patterson, K., and Hammock, M.K., 1987, A review of the factors influencing the prognosis of medulloblastoma. The importance of cell differentiation, *J. Neurosurg.* 66: 80-87.

Cartron, P.-F., Oliver, L., Martin, S., Moreau, C., LeCabellec, M.-T., Jezequel, P., Meflah, K., and Vallette, F.M., 2002, The expression of a new variant of the pro-apoptotic molecule Bax, Baxψ, is correlated with an increased survival of glioblastoma multiforme patients, *Hum. Mol. Genet.* 11: 675-687.

Castilla, E.A., Prayson, R.A., Abramovich, C.M., and Cohen, M.L., 2003, Immunohistochemical expression of cathepsin D in meningiomas, *Anat. Pathol.* 119: 123-128.

Cattoretti, G., Becker, M.H.G., and Key, G., 1992, Monoclonal antibodies against recombinant parts of the Ki-67 antigen (MIB-1 and MIB-3) detect proliferating cells in microwave-processed formalin fixed paraffin sections, *J. Pathol.* 168: 357-363.

Cavalla, P., Piva, R., Bortolotto, S., Grosso, R., Cancelli, I., Chiò, A., and Schiffer, D., 1998, Cyclin D1 expression in gliomas, *Acta Neuropathol.* 95: 131-135.

Cavalla, P., Piva, R., Bortolotto, S., Grosso, R., Cancelli, I., Chiò, A.,and Schiffer, D., 1999, p27/Kip.1 expression in oligodendrogliomas and its possible prognostic role, *Acta Neuropathol.* 98: 629-634.

Cenacchi, G., Giangaspero, A., Cerasoli, S., Manetto, V., and Martinelli, G.N., 1996, Ultrastructural characterization oligodendroglial-like cells in central nervous system tumors, *Ultrastruct. Pathol.* 20: 537-547.

Cenacchi, G., Roncaroli, F., Cerasoli, S., Ficarra, G., Merli, G.A., and Giangaspero, F., Chordoid glioma of the third ventricle: an ultrastructural study of three cases with a histogenetic hypothesis, *Am. J. Surg. Pathol.* 2001, 25: 401-405.

Cenacchi, G., and Giangaspero, F. 2004, Emerging tumor entities and variants of CNS neoplasms, *J. Neuropathol. Exp. Neurol.* 63: 185-192.

Cerda-Nicolas, M., and Kepes, J.J., 1993, Gliofibromas (including malignant forms), and gliosarcomas: a comparative study and review of the literature, *Acta Neuropathol.* 85: 349-361.

Cerda-Nicolas, M., Lopez-Ginés, C., Pérez-Bacete, M., Barcia-Salorio, J.L., and Llombart-Bosch, A., 2000, Histopathological and cytogenetic findings in benign atypical and anaplastic human meningiomas: a study of 60 tumors, *Clin. Neuropathol.* 19: 259-267.

Cervera-Pierot, P., Varlet, P., Chodkiewicz, J.P., and Daumas-Duport, C., 1997, Dysembryoplastic neuroepithelial tumors located in caudate nucleus. Report of four cases, *Neurosurgery* 40: 1065-1070.

Challa, V.R., Moody, D.M., Brown, W.R., and Zagzag, D., 2004, A morphologic study of the vasculature of malignant gliomas using thick celloidin sections and alkaline phosphatase stain, *Clin. Neuropathol.* 23: 167-172.

Chamberlain, M.C., 2003, Ependymomas, *Curr. Neurol. Neurosurg. Rep.* 3: 193-199.

Chandrasoma, P., and Apuzzo, M.L., 1989, Astrocytic neoplasms. *Stereotactic brain biopsy* P. Chandrasoma and M.L. Apuzzo Eds, IGAKU-SHOIN, New York, 89-109.

Chang, C.H., Housepian, E.M., and Herbert, C., Jr. An operative staging system and a megavoltage radiotherapeutic 1969, technic for cerebellar medulloblastomas, *Radiology* 93:1351-1359.

Chang, K.H., Han, M.H., Kim, D.G., Chi, J.G., Suh, D.C., Kim, S.J., Cha, S.H., and Han, M.C., 1993, MR appearance of central neurocytoma, *Acta Radiol.* 34: 520-526.

Chen, G.G., Chu, Y.S., Chak, E.C.W., Leung, B.C.S., and Poon, W.S., 2002, Induction of apoptosis in glioma cells by molecules released from activated macrophages, *J. Neuro-Oncol.* 57: 179-186.

Chen, G.G., Chak, E.C.W., Chun, Y.S., Lam, I.K.Y., Sin, F.L.F., Leung, B.C.S., Ng, H.K., Poon, W.S. (2003), Glioma apoptosis induced by macrophages involves both death receptor-dependent and independent pathways, *J. Lab. Clin. Med.* 141:190-199.

Chicoine, M.R., and Silbergeld, D.L., 1995, Assessment of brain tumor cell motility *in vivo* and *in vitro*, *J. Neurosurg.* 22: 134-143.

Chin, H.W., Maruyama, Y., Markesberry, W., and Young, A.B., 1982, Intracranial ependymoma: results of radiotherapy at the University of Kentucky, *Cancer* 49: 2276-2280.

Choe, G., Horvath, S., Cloughesy, T.F., Crosby, K., Seligson, D., Palotie, A., Inge, L., Smith, B.L., Sawyers, C.L., and Mischel, P.S., 2003, Analysis of the phosphatidylinositol 3'-kinase signaling pathway in glioblastoma patients *in vivo*, *Cancer Res.* 63: 2742-2746.

Choi, B.H., Kim, R.C., 1984, Expression of glial fibrillary acidic protein in immature oligodendroglia, *Science* 223: 407-409.

Christensen, D., Laursen, H., and Klinken, L., 1983, Prediction of recurrence in meningiomas after surgical treatment. A quantitative approach, *Acta Neuropathol.* 61: 130-134.

Christov, C., Adle-Biassette, H., Le Guerine, C., Natchev, S., and Gherardi, R.K., 1998, Immunohistochemical detection of vascular endothelial growth factor (VEGF) in the vasculature of oligodendrogliomas, *Neuropathol. Appl. Neurobiol*, 24: 29-35.

Ciechanover, A., and Brundin, P., 2003, The ubiquitin-proteasome system in neurodegenerative diseases: sometimes the chicken, sometimes the egg, *Neuron* 40: 427-446.

Ciechanover, A., and Iwai, K., 2004, The ubiquitin system: from basic mechanisms to the patient bed, *IUBMB Life* 56: 193-2001.

Clark, S.R., Shetty, A.K., Bradley, J.L., Turner, D.A. (1994), Reactive astrocytes express embryonic intermediate neurofilament nestin, *NeuroReport* 5: 1885-1888.

Coffin, C.M., Braun, J.T., Wick, M.R., and Dehner, L.P., 1990, A clinicopathologic and immunohistochemical analysis of 53 cases of medulloblastoma with emphasis on synaptophysin expression, *Mod. Pathol.* 3: 164-170.

Collins, V.P., 2002, Cellular mechanisms targeted during astrocytoma progression, *Cancer Lett.* 188: 1-7.

Collins, V.P., 2004, Brain tumours: classification and genes. *J. Neurol. Neurosurg. Psychiatry* 75(Suppl II) ii2-ii11.

Coons, S.W., and Pearl, D.K., 1988, Mitosis identification in diffuse gliomas: implication for tumor grading, *Cancer* 82: 1550-1555.

Coons, S.W., and Johnson, P.C., 1993, Regional heterogeneity in the proliferative activity of human gliomas measured by Ki-67 labeling index, *J. Neuropathol. Exp. Neurol.* 52: 609-618.

Coons, S.W., Johnson, P.C., and Pearl, D.K., 1997, The prognostic significance of Ki-67 labeling indices for oligodendrogliomas, *Neurosurg.* 41: 878-884.

Coons, S.W., Johnson, P.C., Scheithauer, B.W., Yates, A.J., and Pearl, D.K., 1999, Improving diagnostic accuracy and interobserver concordance in the classification and grading of primary gliomas, *Cancer* 79: 1381-1393.

Cox, L.B., 1933, The cytology of the glioma group; with special reference to the inclusions of cells derived from the invaded tissue, *Am. J. Pathol.* 9: 839-898.

Crompton, M.R., and Gautier-Smith, P.C., 1970, The prediction of recurrence in meningiomas, *J. Neurol. Neurosurg. Psychiatry* 33: 80-87.

Cuevas, I.C., Slocum, A.L., Costello, J.F., Bollen, A.W., Riggins, G.J., McDermott M.W., and Lal, A., 2005, Meningioma tramscript profiles reveal deregulated notch signaling pathway, *Cancer Res.* 65: 5070-5075.

Dahlstrand, J., Collins, P., and Lendhal, U., 1992, Expression of the class VI intermediate filament nestin in human central nervous system tumors, *Cancer Res.* 52: 5334-5341.

Dai, C., Celestino, J.C., Okada, Y., Louis, D.N., Fuller, G.N., and, Holland, E.C., 2001, PGDF autocrine stimulation dedifferentiates and induces oligodendrogliomas and oligoastrocytomas from neural progenitors and astrocytes in vitro, *Genes Dev.* 15: 1913-1925.

Dai, C., and Holland, E.C., 2003, Astrocyte differentiation states and glioma formation, *Cancer J.* 9: 72-81.

Day, Y., Rahmani, M., and Grant, S., 2003, Proteasome inhibitors potentiate leukemic cell apoptosis induced by the cyclin-dependent kinase inhibitor flavopiridol through a SAPK/JNK- and NF-kappaB-dependent process, *Oncogene* 22: 7108-7122.

Damert, A., Machein, M., Breier, G., Fujita, M.Q., Hanahan, D., Risau,W., and Plate, K.H., 1997, Up-regulation of vascular endothelial growth factor expression in a rat glioma is conferred by two distinct hypoxia-driven mechanisms, *Cancer* 57: 3860-3864.

Darlymple, J.S., Parisi, J.E., Roche, P.C., Ziesmer, S.C., Scheithauer, B.W., and Kelly, P.J., 1994, Changes in proliferating cell nuclear antigen expression in glioblastoma multiforme cells along a stereotactic biopsy trajectory, *Neurosurgery* 35: 1036-1045.

Das, A., Tan, W.L., and Smith, D.R., 2003, Expression of the inhibitor protein survivin in benign meningiomas, *Cancer Lett.* 193: 217-223.

Daumas-Duport, C., Scheithauer, B.W., Chodkiewicz, J.P., Laws, E.R.Jr, and Vedrenne C., (1988), Dysembryoplastic neuroepithelial tumor: a surgical curable tumor of young patients with intractable partial seizures. Report of thirty cases, *Neurosurgery* 23: 545-556.

Daumas-Duport, C., Scheithauer, B.W., O'Fallon, J., and Kelly, P., 1988, Grading of astrocytomas. A simple and reproducible method, *Cancer* 62: 2152-2165.

Daumas-Duport, C., Varlet, P., Tucker, M.L., Beuvon, F., Cervera, P., and Chodkiewicz, J.O., 1997a, Oligodendrogliomas. Part I: Patterns of growth, histological diagnosis, clinical and imaging correlations: a study of 153 cases, *J. Neuro-Oncol.* 34: 37-59.

Daumas-Duport, C., Tucker, M.L., Kolles, H., Cervera, P., Beuvon, F., Varlet, P., Udo, N., Koziak M., Chodkiewicz, J.-P., 1997b, Oligodendrogliomas. Part II: A new grading system based on morphological and imaging criteria, *J. Neuro-Oncol.* 95: 493-504.

Daumas-Duport, C., Pietsch, T., and Lantos, P.L., 2000, Dysembryoplastic neuroepithelial tumour. *Pathology and Genetics. Tumours of the nervous system.* P. Kleihues and W.K. Cavenee eds. IARC, Lyon.

Dawson, D.W., Volpert, O.V., Gillis, P., Crawford, S.E., Xu, H., Benedict, W., and Bouck, N.P., 1999, Pigmented epithelium-derived factor: inhibitor of angiogenesis. *Science* 285: 245-248.

Deck, J.H., and Rubinstein, L.J., 1981, Glial fibrillary acidic protein in stromal cells of some capillary hemangioblastomas: significance and possible implications of an immunoperoxidase study, *Acta Neuropathol.* 54: 173-181.

Deckaestecker, C., Camby, I., Gordower, L., Dewitte, O., Cras, P., Martin, J.J., Pasteels, J.-L,, Van Ham, T., Brotchi, J., Kiss, R., and Salmon, I,. 1998, Characterization of astroglial *versus* oligodendroglial phenotyps in glioblastomas by means of quantitative morphonuclear variables generated by computer-assisted microscopy, *J. Neuropathol. Exp. Neurol.* 57: 791-802.

Dehgani, F., Schachenmayr, W., Laun, A., and Korf, H.W., 1998, Prognostic implication of histopathological, immunohistochemical and clinical features of oligodendrogliomas: a study of 89 cases, *Acta Neuropathol.* 95: 493-504.

Deininger, M.H., Weller, M., Streffer, J., and Meyermann, R., 1999, Antiapoptotic Bcl-2 family protein expression increases with progression of oligodendroglioma, *Cancer* 86: 1832-1839.

Deininger, M.H., Grote, E., Wickboldt, J., and Meyermann, R., 2000, Distinct radiochemotherapy protocols differentially influence cellular proliferation and expression of p53 and Bcl-2 in glioblastoma multiforme, *J. Neuro-Oncol.* 48: 121-129.

Delgado, M.B., Anderson, J.R., Whittel, I.R., and Wharton, S.B., 1999, Expression of Bcl-2 and Bax in oligodendrogliomas and their relationship to apoptosis, *Neuropathol. Appl. Neurobiol.* 25: 400-407.

Demuth, T., and Berens, M.E., 2004, Molecular mechanism of glioma cell migration and invasion, *J. Neuro-Oncol.* 70: 217-228.

De Vries, J., and Wakhloo, A.K., 1994, Repeated multifocal recurrence of grade I, grade II, and grade III meningiomas: regional multicentricity (primary new growth) or metastases? *Surg. Neurol.* 41: 299-305.

Diedrich, U., Lucius, J., Baron, E., Behnke, J., Pabst, B., and Zoll, B., 1995, Distribution of epidermal growth factor receptor gene amplification in brain tumors and correlation to prognosis, *J. Neurol.* 242: 683-688.

Ding, H., Roncari, L., Wu, X., Shannon, P., Naggy, A., and Guha, A., Expression and hypoxic regulation of angiopoietins in human astrocytomas, *Neuro-Oncology* 2001, 3: 1-10.

Dirks, P.B., Jay, V., Becker, L.E., Drake, J.M., Humphreys, R.P., Hoffamn, H.J., and Rutka, J.T., 1994, Development of anaplastic changes in low grade astrocytomas of childhood, *Neurosurgery* 34: 68-78.

Djalilian, H.R., and Hall, W.A., 1998, Malignant gliomas of the cerebellum: an analytic review, *J. Neuro.-Oncol.* 36: 247-257.

Doetsch, F., Caille, I., Lim, Da., Garcia-Verdugo, J.M., and Alvarez-Buylla, A., 1999, Subventricular zone astrocytes are neural stem cells in the adult mammalian brain, *Cell* 97: 703-716.

Doetsch, F., Petreanu, L., Caille, I., Garcia-Verdugo, J.M., and Alvarez-Buylla, A., 2002, EGF converts transit-amplifying neurogenic precursors in the adult brain into multipotent stem cells, *Neuron* 36: 1021-1034.

Doetsch, F., 2003a, A niche for adult neural stem cells, *Curr. Opin. Gen. Dev.* 13: 543-550.

Doetsch, F., 2003b, The glia identity of neural stem cells, *Nature Neurosci.* 6: 1127-1134.

du Plessis, D.G., Yusoff, S., Roebuck, M.M., Gledhill, P.J., Rossi, M.L., and Broome, J.C., 2000, An analysis of histopathological features associated with early recurrence of meningiomas, *Brain Pathol* 10: 742.

du Plessis, D.G., Rutherfoord, G.S., Joyce, K.A., and Walker, C., 2004, Phenotypic and genotypic characterization of glioblastoma multiforme with epithelial differentiation and adenoid formations, *Clin. Neuropathol.* 23: 141-148.

Dunn, I.F., Heese, O., and Black, P.M., 2000, Growth factors in glioma angiogenesis: FGFs, PDGF, EGF and TGF, *J. Neuro-Oncol.* 50: 121-137.

Dunne, I.F., and Black, P.M., 2003, The neurosurgeon as local oncologist: cellular and molecular neurosurgery in malignant glioma therapy, *Neurosurgery* 52: 1422-1424.

Durand, B., Gao, F.B., and Raff, M., 1997, Accumulation of the cyclin-dependent kinase inhibitor p27/Kip.1 and the timing of oligodendrocyte differentiation, *EMBO J.* 16: 306-317.

Duriez, P.J., Desnoyers, S., Hoflack, J.C., Shali, G.M., Morelle, B., Bourassa, S., Poirier, G.G., and Talbot, B., 1997, Characterization of anti-peptide antibodies directed towards the automodification domain and apoptotic fragment of poly(ADP-ribose) polymerase. *Biocheim. Biophys. Acta* 1334: 65-72.

Eberhart, C.G., Kaufman, W.E., Tihan, T., and Burger, P.C., 2001, Apoptosis, neuronal maturation and neurotrophin expression within medulloblastoma nodules, *J. Neuropathol. Exp. Neurol.* 60: 462-469.

Eberhart, C.G., Kepner, J.L., Goldthwaite, P.T., Kun, L.E., Duffner, P.K., Friedman, H.S., Strother, D.R., and Burger, P.C., 2002a, Histopathologic grading of medulloblastomas. A pediatric oncology group study, *Cancer* 94: 552-560.

Eberhart, C.G., Kratz, J.E., Schuster, A., Goldthwaite, P., Cohen, K.J., Perlman, E.J., and Burger, P.C., 2002b, Comparative genomic hybridization detects an increased number of chromosomal alterations in large cell/anaplastic medulloblastomas, *Brain Pathol.* 12: 36-44.

Eberhart, C.G., and Burger, P.C., 2003c, Anaplasia and grading in medulloblastomas, *Brain Pathol.* 13: 376-385.

Eberhart, C.G., Cohen, K.J., Tihan, T., Goldthwaite, P.T., and Burger, P.C., 2003d, Medulloblastomas with systemic metastases: evaluation of tumor histopathology and clinical behavior in 23 patients, *Ped. Hematol./Oncol.* 25: 198-203.

El-Rayes BF, Norton CS, Sakr W., Maciorowski, Z., Smith, D., Pietrazkiewicz, H., Del Mar Alonso, M., Ensley J.F. (2005), Cellular DNA content parameters as prognostic indicators in human astrocytomas. *J Neuro-Oncol* 2005, 71: 85-89.

Elek, G., Slowik, F., Eross, L., Toth, S., Szabo, Z., and Balint, K., 1999, Central neurocytoma with malignant course. Neuronal and glial differentiation and craniospinal dissemination, *Pathol Oncol Res* 5: 155-159.

Ellison, D.W., Steart, P.V., Gatter, K.C., and Weller, R.O., 1995a, Apoptosis in cerebral astrocytic tumours and its relationship to expression of the Bcl-2 and p53 proteins, *Neuropathol. Appl. Neurobiol.* 21: 352-361.

Ellison, D.W., Steart, P.V., Bateman, A.C., Pickering, R.M., Palmer, J.D., and Weller, R.O., 1995b, Prognostic indicators in a range of astrocytic tumours: an immunohistochemical study with Ki-67 and p53 antibodies, *J. Neurol. Neurosurg.* Psychiatry 59: 413-419.

Ellison, D. 2002, Classifying the medulloblastoma: insights from morphology and molecular genetics, *Neuropathol. Appl. Neurobiol.* 28: 257-282.

Endo, H., Kumabe, T., Jokura, H., Shirane, R., Ariga, H., Takai, Y., and Yoshimoto, T., 2003, Leptomeningeal dissemination of cerebellar malignant astrocytomas, *J. Neuro-Oncol.* 63: 191-199.

Eng, D.Y., DeMonte, F., Ginsberg, L., Fuller, G.N., and Jackle, K., 1997, Craniospinal dissemination of central neurocytoma. Report of two cases, *J. Neurosurg.* 86: 547-552.

Evans, A.J., Fayaz, I., Cusimano, M.D., Laperriere, N., and Bilbao, J.M., 2000, Combined pleomorphic xanthoastrocytoma-ganglioglioma of the cerebellum, *Arch. Pathol. Lab. Med.* 124: 1707-1709.

Ever, L., and Gaiano, N., 2005, Radial 'glial' progenitors: neurogenesis and signaling, *Curr. Opinion Neurobiol.* 15: 29-33.

Fazeny-Dorner, B., Wenzel, C., Veitl, M., Piribauer, M., Rossler, K., Dieckmann, K., Ungersbock, K., and Marosi, C., 2003, Survival and prognostic factors of patients with unresectable glioblastoma multiforme, *Anticancer Drugs* 14: 305-312.

Feigenberg, S.J., Amdur, R.J., Morris, C.G., Mendenhall, W.M., Marcus, R.B., and Friedman, W.A., 2003, Oligodendroglioma: does deferring treatment compromise outcome ? *Am. J. Clin. Oncol.* 26: 60-66.

Feigin, I., and Gross, S.W., 1955, Sarcoma arising in glioblastoma of brain, *J. Pathol.* 31: 633-653.

Feigin, I., Allen, L.B., Lipkin, L., Gross, S.W., 1958, The endothelial hyperplasia of the cerebral blood vessels with brain tumors and its sarcomatous transformation, *Cancer* 11: 264-277.

Feldkamp, M.M., Lala, P., Nau, N., Roncart, L., and Guha, A., 1999, Expression of activated epidermal growth factor receptors, Ras-guanosine, triphosphate, and mitogen-activated protein kinase in human glioblastoma multiforme specimens, *Neurosurgery* 45: 1442-1453.

Fernandez, C., Figarella-Branger, D., Girard, N., Boouvier-Labit, C., Gouvernet, J., Paz Paredes, A., and Lena, G., 2003, Pilocytic astrocytomas in children: prognostic factors - A retrospective study of 80 cases, *Neurosurgery* 53: 544-553.

Fernandez-Terijeiro, A., Betensky, R.A., Sturla, L.M., Kim, J.Y., Tamayo, P., and Pomeroy, S.L., 2004, Combining gene expression profiles and clinical parameters for risk stratification in medulloblastomas, *J. Clin. Oncol.* 15: 994-998.

Fiano, V., Ghimenti, C., and Schiffer D. 2003, Expression od cyclins, cyclin-dependent kinases and cyclin-dependent kinase inhibitors in oligodendrogliomas in humans, *Neurosci. Lett.* 347: 111-115.

Figarella-Branger, D., Gambarelli, D., Dollo, C., Devictor, B., Perez-Castillo, A.M., Genitori, L., Lena, G., Choux, M., and Pellissier, J.F., 1991, Infratentorial ependymomas of childhood. Correlation between histological features, immunological phenotype, silver nucleoalr organizer region staining values and post-operative survival in 16 cases, *Acta Neuropathol.* 82: 208-216.

Figarella-Branger, D., Söylemezoglu, F., Kleihues, P., Hassoun, J., 2000, Central neurocytoma. *Pathology and genetics. Tumours of the nervous system*. P: Kleihues P and W.K. Cavenee Eds, IARC, Lyon.

Finkelstein, S.D., Mohan, D., Hamilton, R.L., Sasatomi, E., Swalsky, P., and Lieberman, F.S., 2004, Microdissection-based genotyping assists discrimination of reactive gliosis from glioma, *Am. J. Clin. Pathol.* 121: 671-678.

Fiveash, J.B., and Spencer, S.A., 2003, Role of radiation therapy and radiosurgery in glioblastoma multiforme, *Cancer J.* 9: 222-229.

Flickinger, J.C., Kondziolka, D., Maitz, A.H., and Lunsford, L.D., 2003, Gamma knife radiosurgery of imagimg-diagnosed intracranial meningioma, *Int. J. Radiat. Oncol. Biol. Phys.* 56: 801-806.

Folkert, R.D., 2000, Descriptive analysis and quantification of angiogenesis in human brain tumors, *J. Neuro-Oncol.* 50: 165-172.

Folkman, J., 1971, Tumor angiogenesis: Therapeutic implications, *N. Eng. J. Med.* 285: 1182-1186.

Fortin, D., MacDonald, D.R., Stitt, L., and Cairncross, J.G., 2001, PCV for oligodendroglial tumors: in search of prognostic factors for response and survival, *Can. J. Neurol. Sci.* 28: 215-223.

Fouladi, M., Jenkins, J., Burger, P., Langston, J., Merchant, T., Heideman, R., Thompson, S., Sanford, A., Kun, L., and Gajjar, A., 2001, Pleomorphic xanthoastrocytoma: Favorable outcome after complete surgical resection, *Neuro-Oncology* 3: 184-192.

Fouladi, M., Helton, K., Dalton, J., Gilger, E., Gajjar, A., Merchant, T., Kun, L., Newsham, I., Burger, P., and Fuller, C., 2003, Clear cell ependymoma: a clinicopathologic and radiographic analysis of 10 patients, *Cancer* 98: 2232-2244.

Foulds, L., 1965, Multiple etiologic factors in neoplastic development, *Cancer Res.* 25: 1339-1347.

Franco, A.V., Zhang, X.D., Van Berkel, E., Sanders, J.E., Zhang, X.Y., Thomas, W.D., Nguyen, T., and Hersey, P., (2001), The role of NF-kappa B in TNF-related apoptosis-inducing ligand (TRAIL)-induced apoptosis of melanoma cells, *J. Immunol.* 166: 5337-5345.

Frank, T.S., Trojanowski, J.Q., Roberts, S.A., and Brooks, J.J., 1989, A detailed immunohistochemical analysis of cerebellar hemangioblastoma: an undifferentiated mesenchymal tumor, *Mod. Pathol.* 2: 638-651.

Frank, A.J., Hernan, R., Hollander, A., Lindsey, J.C., Lusher, M.E., Fuller, C.E., Clifford, S.C., and Gilbertson, R.J., 2004, The TP53-ARF tumor suppressor pathway is frequently disrupted in large cell/anaplastic medulloblastoma, *Brain Res. Mol. Brain Res.* 121: 137-140.

Frankel, B., Longo, S.L., Kyle, M., Canute, G.W., and Ryken, T.C., 2001, Tumor Fas (APO-1/CD95) up-regulation results in increased apoptosis and survival times for rats with intracranial malignant gliomas. *Neurosurgery* 49: 168-175.

Frankfurt, O.S., Robb, J.A., Sugarbaker, E.V., and Villa, L., 1996, Monoclonal antibody to single-straned DNA is a specific amd sensitive cellular marker of apoptosis, *Exp. Cell Res.* 226: 387-397.

Fruwald, M.C., O'Dorisio, M.S., Rush, L.J., Reiter, J.L., Smiraglia, D.J., Wenger, G., Costello, J.F., White, P.S., Krahe, R., Brodeur, G.M., and Plass, C., 2000, Gene amplification in PET/medulloblastomas: mapping of a novel amplified gene within the MYCN amplikon, *J. Med. Genet.* 37: 501-509.

Fujimaki, T., Matsuno, A., Sasaki, T., Toyoda, T., Matsuura, R., Ogai, M., Kitanaka, C., Asai, A., Matsutani, M., and Kirino, T., 1997, Proliferative activity of central neurocytoma: measurement of tumor volume doubling time, MIB-1 staining index and bromodeoxyuridine labeling index, *J. Neuro-Oncol.* 32: 103-109.

Fujimoto, Y., Izumoto, S., Suzuki, T., Kinoshita, M., Kagawa, N., Wada, K., Hashimoto, N., Maruno, M., Nakatsuji, Y, and Yoshimine, T., 2005, Ganglioside GM3 inhibits proliferation and invasion on glioma, *J. Neuro-Oncol.* 71:99-106.

Fujisawa, H., Marukawa, K., Hasegawa, M., Tohma, Y., Hayashi, Y., Uchiyama, N., Tachibana, O., and Yalashita, J., 2002, Genetic differences between neurocytoma and dysembyoplastic neuroepithelial tumors and oligodendroglial tumors, *J. Neurosurg.* 97: 1350-1355.

Fulda, S., Meyer, E., and Debatin, K.-M., 2002, Inhibition of TRAIL-induced apoptosis by Bcl-2 overexpression, *Oncogene* 21: 2283-2294.

Fuller, G.N., Hess, K.R., Rhee, C.H., Yung, W.K., Sawaya, R.A., Bruner, J.M., and Zhang, W., 2002, Molecular classification of human diffuse gliomas by multidimensional scaling analysis of gene expression profiles parallels morphology-based classification, correlates with survival and reveals clinically-relevant novel glioma subsets, *Brain Pathol.* 12: 108-116.

Fuller, C.E., Schmidt, R.E., Roth, K.A., Burger, P.C., Scheithaer, B.W., Baneriee, R., Trinkaus, K., Lytle, R., and Perry, A., 2003, Clinical utility of fluorescence in situ hybridization (FISH) in morphologically ambiguous gliomas with hybrid oligodendroglial/astrocytic features, *J. Neuropathol. Exp. Neurol.* 62: 1118-1128.

Fung, K.-M., Rorke, L.B., Giasson, B., Lee, V.M.-Y., and Trojanowski, J.Q., 2003, Expression of α-, β-, and γ-synuclein in glial tumors and medulloblastomas, *Acta Neuropathol.* 106: 167-175.

Funato, H., Inoshita, N., Okeda, R., Yamamoto, S., and Aoyagi, M., 1997, Cystic ganglioneurocytoma outside the ventricular region, *Acta Neuropathol.* 94: 95-98.

Gage, F.H., 2000, Mammalian neural stem cells, *Science* 287: 1433-1438.

Gage, F.H. 2002, Neurogenesis in the adult brain, *J. Neurosci.* 22, 612-613.

Galli, R., Binda, E., Orfanelli, U., Cipelletti, B., Gritti, A., De Vitis, S., Fiocco, R., Foroni, C., Dimeco, F., and Vescovi, A., 2004, Isolation and characterizxation of tumorigenic, stem-like neural precursors from human glioblastoma. *Cancer Res,* 64: 7011-7021.

Gajjar, A., Hernan, R., Kocak, M., Fuller, C., Lee, Y., McKinnon, P.J., Wallace, D., Lau, C., Chintagumpala, V., Ashley, D.M., Kellie, S.J., Kun, L., and Gilbertson, R.J., 2004, Clinical, histopathological, and molecular markers of prognosis: toward a new disease risk stratification system for medulloblastoma, *J. Clin. Oncol.* 22: 984-993.

Gazit, R., Krizhanovsky, V., and Ben-Arie, N., 2004, Math1 controls cerebellar granule cell differentiation by regulating multiple componentns of the Notch signaling pathway, *Development* 131: 903-913.

Gelpi, E., Ambros, I.M., Birner, P., Luegmayr, A., Drlicek, M., Fischer, I., Kleinert, R., Maier, H., Huemer, M., Gatterbauer, B., Anton, J., Rossler, K, Budka, H., Ambros, P.F., and Hainfellner, J.A., 2003, Fluorescent in situ hybridization on isolated tumor cell nuclei: a sensitive method for 1p and 19q deletion analysis in paraffin-embedded oligodendroglial tumor specimens, *Mod. Pathol.* 16: 708-715.

Gerdes, J., Schwab, U., and Lemke, H., 1983, Production of a mouse monoclonal antibody reactive with human nuclear antigen associated with cell proliferation, *Int. J. Cancer* 31: 13-20.

Ghimenti, C., Fiano, V., Chiadò-Piat, L., Chiò, A., Cavalla, P., and Schiffer, D., 2003, Deregulation of the $p14^{ARF}$Mdm2/p53 pathway and G1/S transition in two glioblastoma sets, *J. Neuro-Oncol.* 61: 95-102.

Giangaspero, F., Rigobello, L., Badiali, M., Loda, M., Andreini, L., Basso, G., Zorzi, F., and Montaldi, A., Large cell medulloblastomas. A distinct variant with highly aggressive behavior. *Am. J. Surg. Pathol.* 1992, 16: 687-693.

Giangaspero, F., Cenacchi, G., Losi, L., Cerasoli, S., Bisceglia, M., and Burger, P.C., 1997, Extraventricular neoplasms with neurocytoma features. A clinicopathological study of 11 cases, *Am. J. Surg. Pathol.* 21: 206-212.

Giangaspero, F., Bigner, S.H., Kleihues, P., Pietsch, T., and Trojanowski, J.Q., 2000, Medulloblastoma. *Pathology and genetics. Tumours of the nervous system*, P. Kleihues and W.K. Cavenee Eds, Iarc, Lyon, pp. 129-137.

Giannini, C., Scheithauer, B.W., Burger, P.C., Brat, D.J., Woltan, P.C., Lach, B., and O'Neill, B.P., 1999a, Pleomorphic xanthoastrocytoma. What do we really know about it. *Cancer* 85: 2033-2045.

Giannini, C., Scheithauer, B.W., Burger, P.C., Christensen, M.R., Wollan, P.C., Sebp, T.J., Forsyth, P.A., and Hayostek, C.J., 1999b, Cellular proliferation in pilocytic and diffuse astrocytomas, *J. Neuropathol. Exp. Neurol.* 58: 46-53.

Giannini, C., Hebrink, D., Scheithauer, B.W., Dei Tos, A.P., and James, C.D., 2001, Analysis of p53 mutation and expression in pleomorphic xanthoastrocytoma, *Neurogenetics* 3: 159-162.

Giannini, C., Scheithauer, B.W., Weaver, A.L., Burger, P.C., Kros, J.M., Mork, S., Graeber, M.B., Bauserman, S., Buckner, J.C., Burton, J., Riepe, R., Tazelaar, H.D., Nascimento, A.G., Crotty, T., Keeney, G.L., Pernicone, P., and Altermatt, H., 2001, Oligodendrogliomas: reproducibility and prognostic value of histologic diagnosis and grading, *J. Neuropathol. Exp. Neurol.* 60: 248-262.

Giannini, C., Scheithauer, B.W., Lopes, M.B.S., Hirose, T., Kros. J.M., and VandenBerg, S.R., 2002, Immunophenotype of pleomorphic xanthoastrocytoma, *Am. J. Surg. Pathol.* 26: 479-485.

Gilbertson, R.J., Jaros, E., Perry, R.H., Kelly, P.J., Lunec, J., and Pearsin, A.D., 1997, Mitotic percentage index: a new prognostic factor for childhood medulloblastoma, *Eur. J. Cancer* 33, 609-615.

Gilhuis, H.J., Anderl, K.L., Boerman, R.H., Jeuken, J.M., James, C.D., Raffel, C., Scheithauer, B.W., and Jenkins, R.B., 2001, Comparative genomic hydridization of medulloblastoma and clinical relevance: eleven new cases and a review of the literature, *Clin. Neurol. Neurosurg.* 102: 203-209.

Gilhuis, H.J., van der Laak, J., Wesseling, P., Boerman, R.H., Beute, G., Teepen, L.M.J., Grotenhuis, J.A., and Kappelle, A.C., 2003, Inverse correlation between genetic aberrations and malignancy grade in ependymal tumors: a paradox? *J. Neuro-Oncol.* 66: 111-116.

Ginzinger, D.G., Godfrey, T.E., Nigro, J.M., Moore, G.H. 2nd, Suzuki, S., Pallavicini, M.G., Gray J.W., and Jensen, R.H., 2000, Measurement of DNA copy number at microsatellite loci using quantitatiove PCR analysis, *Cancer Res.* 60: 5405-5409.

Giordana, M.T., Cavalla, P., Chiò, A., Marino, S., Soffietti, R., Vigliani, M.C., and Schiffer, D., 1995, Prognostic factors in adult nedulloblastoma. A clinico-pathologic study, *Tumori* 81: 338-346.

Giordana, M.T., Cavalla, P., Dutto, A., Borsotti, L., Chiò, A., and Schiffer D., Is medulloblastoma the same tumor in children and adults? *J. Neuro-Oncol.* 1997, 35: 169-176.

Giordana, M.T., Duò, D., Gasverde, S., Trevisan, E., Boghi, A., Morra, I., Pradotto, L., Mauro, A., and Chiò, A., 2002, MDM2 overexpression is associated with short survival in adults with medulloblastoma, *Neuro-Oncology* 4: 115-122.

Glantz, M.J., Burger, P.C., Herndon, J.E., 2nd, Friedman, A.H., Cairncros, J.G., Vick, N.A., and Schold, S.C. Jr, 1991, Influence of the type of surgery on the histologic diagnosis in patients with anaplastic gliomas, *Neurology* 41: 1741-1744.

Glass, J., Hochberg, F.H., Gruber, M.L., Louis, D.N., Smith, D., and Rattner, B., 1992, The treatment of oligodendrogliomas and mixed oligodendroglioma-astrocytomas with PCV chemotherapy, *J. Neurosurg.* 76: 741-745.

Glioma meta-analysis trialists (GMT) group. 2002, *Cochraine Database Syst Rev* CD003913.

Glioma meta-analysis trialists (GMT) group. Chemotherapy for high-grade glioma. 2002, *Cochraine Database Syst Rev* CD003913.

Goetz, C., Riva, P., Poepperl, G., Gildehaus, F.J., Hischa, A., Tatsch, K., and Reulen, H.J., 2003, Locoregional radioimmunotherapy in selected patients with malignant glioma: experiences, side effects and survival times. *J. Neuro-Oncol.* 62: 321-328.

Gold, R., Schmied, M., Rothe, G., Zischler, H., Breightschopf, H., Wekerle, H., and Lassmann, H., 1993, Detection of DNA fragmentation in apoptosis: application of in situ nick translation to cell culture system and tissue sections, *J. Histochem. Cytochem.* 41: 1023-1030.

Gold, R., Schmied, M., Gegerich, G., Breightschopf, H., Hartung, P., Toyka, K.V., and Lassmann, H., 1994, Differentiation between cellular apoptosis and necrosis by the combined use of in situ tailing and nick translation techniques, *Lab. Invest.* 71: 219-225.

Goldbrunner, R.H., Pietsch, T., Vince, G.H., Bernstein, J.J., Wagner, S., Hageman, H., Selby, D.M., Krauss, J., Soerensen, N., and Tonn, J.C., 1999a, Different vascular patterns of medulloblastoma and supratentorial primitive neuroectodermal tumors, *Int. J. Dev. Neurosci.* 17: 593-599.

Goldbrunner, R.H., Bernstein, J.J., Plate, K.H., Vince, G.H., Roosen, K., and Tonn, J.C., 1999b, Vascularization of human glioma spheroids implanted into rat cortex is conferred by two distinct mechanisms, *J. Neurosci. Res.* 55: 486-495.

Goldman, J.E., 2000, Glial differentiation and lineages, *J. Neurosci. Res.* 59: 410-412.

Goldsmith, B.J., Wara, W.M., Wilson, C.B., and Larson, D.A., 1994, Postoperative irradiation for subtotally resected meningiomas. A retrospective analysis of 140 patients treated from 1967 to 1990, *J. Neurosurg.* 94: 195-201.

Gorczyca, W., Gong, J., and Darzynkiewicz, Z., 1993, Detection of DNA strand breaks in individual apoptotic cells by the in situ terminal deoxynuelotidyl transferase and nick translation assays, *Cancer Res.* 53; 1945-1951.

Goth, R., and Rajewsky, M., 1974, Persistence of O6-ethylguanine in rat brain DNA. Correlation with nervous system specific carcinogenesis by ethylnitrosourea, *PNAS USA* 71: 639-653.

Grant, J.W., Steart, P.V., Aguzzi, A., Jones, D.B., and Gallagher, P.J., 1989, Gliosarcoma: an immunohistochemical study, *Acta Neuropathol.* 79: 305-309.

Green, D.R., and Reed, J,C., 1998, Mitochondria and apoptosis, *Science* 281: 1309-1312

Grigelioniené, G., Blennow, M., Törrök, K., Fried, G., Dahlin, I., Lendahl, U., and Lagercrantz, H., 1996, Cerebrospinal fluid of newborn infants contains a deglycosylated form of the intermediate filament nestin, *Pediatr. Res.* 40: 809-881.

Gross, A., Yin, X.M., Wang, K., Wei, M.C., Jockel, J., Milliman, C., Erdjument-Bromage, H., Tempst, P., and Korsmeyer, S.J., 1999, Caspase cleaved BID targets mitochondria and is required for cytochrome c release, whicle BCL-XL prevents this release but not tumor necrosis factor R1-/Fas death, *J. Biol. Chem.* 274: 1156-1163.

Grossman, S.A., 2003, Arguments against the routine use of currently available adjuvant chemotherapy in high-grade gliomas, *Sem. Oncol.* 20: 19-22.

Grotzer, M.A., Janss, A.J., Fung, K., Biegel, J.A., Sutton, L.N., Rorke, L.B., Zhao, H., Cnaan, A., Phillips, P.C., Lee, V.M., and Trojanowski, J.Q., 2000, Trk expression predicts good clinical outcome in primitive neuroectodermal brain tumors, *J. Clin. Oncol.* 18: 1027-1035.

Grotzer, M.A., Janss, A.J., Fung, K.M., Sutton, L.N., Zhao, H., Trojanowski, J.Q., Rorke, L.B., and Phillips, P.C., 2001, Abundance of apoptotic neoplastic cells in diagnostic biopsy samples is not a prognostic factor in childhood primitive neuroectodermal tumors of the central nervous system, *J. Pediatr. Hematol. Oncol.* 23: 25-29.

Grzybicki, D.M., Liu, Y., Moore, S.A., Brown, H.G., Silverman, J.F., D'Amico, F., and Raab, S.S., 2001, Interobserver variability associated with the MIB-1 labeling index. High levels suggest limited prognostic usefulness for patients with primary brain tumors, *Cancer* 92: 2720-2726.

Guarda, L.G., Ordonez, L.G., Smith, J.L., and Hanssen, G., 1982, Immunopweroxidase localization of Factor VIII in angiosarcomas, *Arch. Pathol. Lab. Med.* 106: 515-516.

Gultekin, S.H., Dalmau, J., Graus, Y., Posner, J.B., and Rosenblum, M.K., 1998, Anti-Hu immunolabeling as an index of neuronal differentiation in human brain tumors: a study of 112 central neuroepithelial neoplasms, *Am. J. Surg. Pathol.* 22: 195-200.

Gutmann, D.H., Donahoe, J., Perry, A., Lemke, N., Gorse, K., Kittiniyom, K., Rempel, S.A., Gutierrez, J.A., and Newsham, I.F., 2000, Loss of DAL-1, a protein 4.1-related tumor suppressor, is an important event in the pathogenesis of meningiomas, *Hum. Mol. Genet.* 9: 1495-1500.

Ha, H.C., and Snyder, S.H., 1999, Poly(ADP-ribose) polymerase is a mediator of necrotic cell death by ATP depletion, *PNAS USA* 96: 13978-13982.

Haapasalo, H., Sallinen, S.-L., Sallinen, P., Helén, P., Jääscheläinen, J., Salmi, T.T., Paetau, A., Paljärvi, L., Visakorpi, T., and Kalimo, H., 1999, Clinicopathological correlation of cell proliferation, apoptosis and p53 in cerebellar pilocytic astrocytomas, *Neuropathol. Appl. Neurobiol.* 25: 134-142.

Haddad, S.F., Moore, S.A., Schelper, R.L., and Goeken, J., 1991, Smooth muscle cells can comprise the sarcomatous component of gliosarcoma, *J. Neuropathol. Exp. Neurol.* 51: 493-498.

Hakin-Smith, V., Battersby, R.D., Maltby, E.L., Timperley, W.R., and Royds, J.A., 2001, Elevated p53 expression in benbign meningiomas potects against recurrence and may be indicative of senescence, *Neuropathol. Appl. Neurobiol.* 27: 40-49.

Hamilton, R.L., and Pollack, I.F., 1997, The molecular biology of ependymomas. *Brain Pathol.* 7: 807-822.

Hanzély, Z., Polgar, C., Fodor, J., Brucher, J.-M., Vitanovics, D., Mangel, L.C., and Afra, D., 2003, Role of early radiotherapy in the treatment of supratentorial WHO grade II astrocytomas: long-term results of 97 patients, *J. Neuro-Oncol.* 63: 305-312.

Hao, C., Parney, I.F., Roa, W.H., Turner, J., Petruk, K.C., and Ramsay, D.A., 2002, Cytokyne and cytokine receptor mRNA expression in human gliobastomas: evidence of Th1, Th2 and Th3 cytokine dysregulation, *Acta Neuropathol.* 103: 171-178.

Hara, M., Aoyagi, M., Yamamotoro, M., Maehara, T., Takada, Y., Nojiri, T., and Ohno, K., 2003, Rapid shrinkage of remnant central neurocytoma after gamma knife radiosurgery: a case report, *J. Neuro-Oncol.* 62: 269-273.

Harris, A.E., Lee, J.Y., Omalu, B., Flickinger, J.C., Kondziolka, D., and Lunsford, L.D., 2003, The effect of radiosurgery during management of aggressive meningiomas, *Surg. Neurol.* 60: 298-305.

Hartmann, C., Mueller, W., Lass, U:, Kamel-Reid, S., and von Deimling, A., 2005, Molecular genetic analysis of oligodendroglial tumors, *J. Neuropathol. Exp. Neurol.* 64: 10-14.

Haslam, R.H., Lamborn, K.R., Becker, I.E., and Israel, M.A., 1998, Tumor cell apoptosis present at diagnosis may predict treatment outcome for patients with medulloblastoma, *J. Pediatr. Hematol. Oncol.* 20: 520-527.

Hartmann, C., Johnk, L., Kitange, G., Wu, Y., Ahworth, L.K., Jenkins, R.B., and Louis, D.N., 2002, Transcript map of the 3.7-Mb D195S112-D195S246 candidate tumor suppressor region on the long arm of chromosome 19, *Cancer* 62: 4100-4108.

Hasselblatt, M., and Paulus, W., 2003, Sensitivity and specificity of epithelial membrane antigen staining patterns in ependymomas, *Acta Neuropathol.* 106: 385-388.

Hassoun, J., Gambarelli, D., Grisoli, F., Pellet, W., Salamon, G., Pellissier, J.F., and Toga, M., 1982, Central neurocytoma. An electron microscopi study of two cases, *Acta Neuropathol.* 56: 151-156.

Hassoun, J., Söylemezoglu, F., Gambarelli, D., Figarella, B., von Ammon, K., and Kleihues, P., 1993, Central neurocytoma: a synopsis of clinical and histological features, *Brain Pathol.* 3: 297-306.

Hatanpaa, K.J., Burger, P.C., Eshekman, J.R., Murphy, K.M., and Berg, K.D., 2003, Molecular diagnosis of oligodendroglioma in paraffin sections, *Lab. Invest.* 83: 419-428.

Hatva, E., Bohling, T., Jaaskelainen, J., Persico, M.G., Haltia, M., Alitalo, K., 1996, Vascular growth factors and receptors in capillary hemangioblastomas and hemangiopericytomas, *Am. J. Pathol.* 14: 763-775.

Hayashi, S., Yamamoto, M., Ueno, Y., Ikeda, K., Ohshima, K., Soma, G., and Fukushima, T., 2001, Expression of nuclear factor-kappa B, tumor necrosis factor receptor type 1, and c-Myc in human astrocytomas, *Neurol. Med. Chir. (Tokyo)* 41: 187-195.

He, J., Mokhtari, K., Sanson, M., Marie, Y., Kujas, M., Huguet, S,, Leuraud, P, Capelle, L., Delattre, J.Y., Poirier, J., and Hoang-Xuan, K., 2001, Glioblastomas with an oligodendroglial component: a pathological and molecular study, *J. Neuropathol. Exp. Neurol.* 60: 863-871.

Heesters, A.A.M., Koudstaal, J., Gwann, G.K., and Molenaar, W.M., 1999, Analysis of proliferation and apoptosis in brain. gliomas: prognosis and clinical value, *J. Neuro-Oncol*. 44: 255-266.

Heimberger, A.B., Hlatky, R, Suki, D, Yang, D., Weinberg, J., Gilbert, M., Sawaya, R., and Aldape, K., 2005, Prognostic effect of epidermal growth factor and EGFRvIII in glioblastoma multiforme patients. *Clin. Cancer Res*. 11: 1462-1466.

Hemmati, H.D., Nakano, I., Lazareff, J.A., Masterman-Smith, M., Geschwind, D.H., Bonner-Fraser, M., and Kornblum, H.I., 2003, Cancerous stem cells can arise from pediatric brain tumors, *PNAS USA* 2003, 100: 15178-15183.

Hengartner, M.O., 2002, Biochemistry of apoptosis, *Nature* 2000, 407: 770-776.

Herbst, R.S. and Shin, D.M., 2002, Monoclonal antibodies to target epidermal growth factor receptor-positive tumors: a new paradigm for cancer therapy. *Cancer,* 94: 1593-1611.

Herms, J., Neidt, I., Luscher, B., Sommer, A., Schurmann, P., Schroder, T., Bergamnn, M., Wilken, B., Probst-Cousin, S., Hernaiz-Driever, P., Behnke, J., Fanefeld, F., Pitsch, T., and Kretschmar, H.A., 2000, C-MYC expression in medulloblastoma and its prognostic value, *Int. J. Cancer* 89: 395-402.

Hernan, R., Fasheh, R., Calabrese, C., Frank, A.J., Maclean, K.H., Allard, D., Barraclough, R., and Gilbertsoin, R.J., 2003, ERBB2 up-regulates *S1004* and several other prometastatic genes in medulloblastoma, *Cancer Res*. 63: 140-148.

Herpers, M.J.H.M., Budka, H., 1984, Glial fibrillary acidic protein (GFAP) in oligodedroglial tumors: gliofibrillary oligodendroglioma and transitional oligoastrocytoma as subtypes of oligodendroglioma, *Acta Neuropathol*., 64: 265-272.

Herrlinger, U., Felsberg, J., Küker, W., Bornemann, A., Plasswilm, L., Knobbe, C.B., Strik, H., Wick, W., Mayermann, R., Dichgans, J., Bamberg, M., Reifenberger, G., and Weller, M. 2002, Gliomatosis cerebri: molecular pathology and clinical course, *Ann. Neurol*. 52: 390-399.

Hirose, T., Scheithauer, B.W., Lopes, M.B., Gerber, H.A., VandenBerg, S.R., 1994, Dysembryoplastic neuroepithelial tumor (DNT): an immunohistochemical and ultrastructural study. *J. Neuropathol. Exp. Neurol*. 53: 184-195.

Hirose, T., Scheithauer, B.W., Lopes, M.B.S., Gerber, H.A., Altermatt HJ, snd VandenBerg, S.R., 1997, Ganglioglioma: an ultrastructural and immunohistochemical study, *Cancer* 79: 989-1003.

Hirose, T., and Scheithauer, B.W. 1998, Mixed dysembryoplastic neuroepithelial tumor and ganglioglioma, *Acta Neuropathol*. 36: 649-654.

Ho, D.M.-T., Hsu, C.-Y., Ting, L.-T., and Chiang, H., 2002, Histopathology and MIB-1 labeling index predicted recurrence of meningiomas, *Cancer* 94: 1538-1547.

Ho, D.M., Hsu, C.Y., Ting, L.T., and Chiang, H. 2003, MIB-1 and DNA topoisomerase I.I alpha could be helpful for predicting long-term survival of patients with glioblastoma, *Amer. J. Clin. Pathol*. 119: 715-722

Hoang-Xuan, K., He, J., Huguet, S., Mokhtari, K., Marie, Y., Kujas, M., Leuraud, P., Capelle, L., Delattre, J.Y., Poirier, J., Broet, P., and Sanson, M. 2001, Molecular heretogeneity of oligodendrogliomas suggests alternative pathways in tumor progression, *Neurology* 57: 1278-1281.

Hockfield, S., and McKay, R.D.G., 1985, Identification of major cell classes in the development of mammalian nervous syste, *J. Neurosci*. 5: 3310-3328.

Hojo, H., and Abe, M., 2001, Rhabdoid papillary meningioma, *Am. J. Surg. Pathol.* 25: 964-969.

Holash, J., Maisonpierre, P.C., Compton, D., Boland, P., Alexander, C.R., Zagzag, D., Yankopoulos, G.D., and Wiegand, S.J., 1999, Vessel cooption, regression, and growth in tumors mediated by angiopoietin and VEGF, *Science* 284: 1994-1998.

Holland, E.C., Celestino, C., Dai, C., Schaefer, L., Sawaya, R.E., and Fuller, G.N., 2000, Combined activation of Ras and Akt in neural progenitors induces glioblastoma formation in mice, *Nat. Genet.* 25: 55-57.

Holland, EC. 2001, Gliomagenesis: genetic alterations and mouse models. *Nature Rev Genet* 2: 120-129.

Holland, E.C., 2001, Progenitor cells and glioma formation, *Curr. Opinion in Neurol.* 14: 683-688.

Hope, J.K., Armstrong, D.A., Babyn, P.S., Humphreys, R.R., Harwood-Nash, D.C., Chuang, S.H., and Marks, P.V., 1992, Primary meningeal tumors in children: Correlation of clinical and CT findings with histologic type and prognosis, *Am. J. Neuroradiol.* 13: 1353-1364.

Hopman, A.H.N., and Ramaekers, F.C.S., 1998, Processing and staining of cell and tissue material for interphase cytogenetics. *Current protocols in cytometry,.* Robinson P Ed, John Wiley and Sons, New York.

Hsu, D.W., Efird, J.T., and Hedley-White, E.T., 1998, MIB-1 (Ki-67) index and transforming growth factor-alpha (TGFα) immunoreactivity are significant prognostic predictors for meningiomas, *Neuropathol. Appl. Neurobiol.* 24: 441-452.

Huang, H., Mahler-Araujo, B.M., Sankila, A., Chimelli, L., Yomekawa, Y., Kleihues, P., and Ohgaki, H., 2000, APC mutations in sporadic medulloblastomas, *Am. J. Pathol.* 156: 433-437.

Huang, H., Colella, S., Kurrer, M. et al. 2000, Gene expression profiling of low-grade diffuse astrocytomas by cDNA arrays, *Cancer Res.* 60: 6868-6874.

Huang, B., Starostik, P., Kühl, J., Tonn, J.C., and Roggendorf, W., 2002, Loss of heterozygosity on chromosome 22 in human ependymomas, *Acta Neuropathol.* 103: 415-420.

Huang, B., Starostik, P., Schraut, A., Krauss, J., Sörensen, N., and Roggendorf, W., 2003, Human ependymomas reveal frequent deletions on chromosome 6 and 9, *Acta Neuropathol.* 106: 357-362.

Hug, E.B., Devries, A., Thornton, A.F., Munzunride, J.E., Pardo, F.S., Hedley-White, E.T., Bussiere, M.R., and Ojemann, R., 2002, Management of atypical and malignant meningiomas: role of high 3D-conformal radiation therapy, *J. Neuro-Oncol.* 48: 151-160.

Huncharek, M., and Kupelnick, B., 2000, Epidermal growth factor receptor gene amplification as a prognostic marker in glioblastoma multiforme: results of a metanalysis, *Oncol. Res.* 12: 107-112.

Husemann, K., Wolter, M., Büschgess, R., Boström, J., Sabel, M., and Reifenberger, G., 1999, Identification of two distinct deleted regions on the short arm of chromosome 1 and rare mutation of the CDKN2C gene at 1p32 in oligodendroglial tumors, *J. Neuropathol. Exp. Neurol.* 58: 1041-1050.

Ilgren, E.B., Stiller, C.A., Hughes, J.T., Silberman, D., Steckel, N., and Kaye, A., 1984, Ependymomas: A clinical and a pthological study. II Survival features. *Clin. Neuropathol.* 3: 122-127.

Im, S.-H., Paek, S.H., Choi, Y.-L., and Kim, D.G., 2003, Clinicopathological study of seven cases of sympomatic supratentorial subependymoma, *J. Neuro-Oncol.* 61: 57-67.

Im, S.-H., Chung, C.K., Kim, S.-K., Cho, B.-K., Kim, M.-K., and Chi, J.G., 2003, Pleomorphic xanthoastrocytoma: a developmental glioneuronal tumor with prominent glioproliferative changes, *J. Neuro-Oncol.* 66: 17-27.

Ino, Y., Betensky, R.A., Zlatescu, M.C., Sasaki, H., MacDonald, D.R., Stemmer-Rachamimov, A.O., Ramsay, D.A., Cairncross, J.G., and Louis, D.N., 2001, Molecular subtypes of anaplastic oligodendroglioma; implication for patient management at diagnosis, *Clin. Cancer Res.* 7: 839-845.

Inoue, R., Isono, M., Abe, M, Abe, T., Kobayashi, H., A genotype of the polymorphic DNA repair gene MGMT is associated with de novo glioblastoma, *Neurol. Res.* 2003, 25: 875-879.

Ironside, J.W., Moss, T.H., Louis, D.N., Lowe, J.S., and Weller, R.O., 2002, *Diagnostic pathology of nervous system tumours*, Churchill Livingstone, London

Hirato, J., Nakazato, Y., and Ogawa, A., 1994, Expression of non-glial intermediate filament proteins in gliomas, *Clin. Neuropathol.* 13: 1-11.

Iseki, S., 1986, DNA strand breaks in rat tissues as detected by in situ nick translation. *Exp. Cell Res*, 167: 311-326.

Hosrtmann, S., Perry, A., Reifenberger, G., Giangaspero F, Huang, H., Hara, A., Masuoka, J., Rainov, N.G., Bergmann, M., Heppner, F.L., Brandner, S., Chimelli, L., Montagna, N., Jackson, T., Davis, D.G., Markesbery, W.R., Ellison, D.W., Weller, R.O., Taddei, G.L., Conti, R., Del Bigio, M.R., Gonzales-Campora, R., Radhakrishnan, V.V., Soylemezoglu, F., Uro-Coste, E., Qian, J., Kleihues, P., and Ohgaki, H., 2004, Genetic and expression profiles of cerebellar liponeurocytoma, *Brain Pathol.* 14: 281-289.

Katoh, M., Wilmotte, R., Belkouch, M.C., de Tribolet, N., Pizzolato, G., and Dietrich, P.Y., 2003, Survivin in brain tumors: an attractive target for immunotherapy, *J. Neuro-Oncol.* 64: 71-76.

Ito, S., Hoshino, T., Prados, M.D., and Edwards, M.S., 1992, Cell kinetics of medulloblastomas, *Cancer* 70: 671-678.

Iuzzolino, P., Ghimenton, G., Nicolato, A., Giorgetti, F., Fina, P., Doglioni, C., and Barbareschi, M.l., 1994, p53 protein in low grade astrocytomas: a study with long-term follow-up, *Br. J. Cancer* 69: 586-591.

Jackson, R.J., Fuller, G.N., Abi-Said, D., Lang, F.F., Gokaslan, Z.L., Shi, W.M., Wildrick, D.M., and Sawaya, R., 2001, Limitations of stereotactic biopsy in the initial management of gliomas, *Neuro-Oncology* 3: 193-200.

Jalali, R., Loughrey, C., Baumert, B., Perks, J., Warrington, A.P., Reaish, D., Ashley, S., and Brada, M., 2002, High precision focused irradiation in the form of fractionated stereotactic conformal radiotherapy (SCRT) for benign meningiomas predominantly in the skull base location, *Clin. Oncol.* 14: 103-109.

Jang, T., Litofsky, N.S., Ross, A., and Recht, L., 2001, Analysis of subventricular zone (SVZ) kinetics during ethylnitrosourea (ENU) induced neuro-carcinogensis, *J. Neuro-Oncol.* 3: 276 quoted by Recht et al., 2003.

Jiang, H., Conrad, C., Fuego, J., Gomez-Marzano, C., and Liu, T.J., 2003, Oncolytic adenoviruses for malignant glioma therapy, *Front. Biosci.* 8: 577-588.

Janss, A.J., Yachnis, A.T., Silber, J.H., Trojanowski, J.Q., Lee, V.M., Sutton, L.N., Perilongo, G., Rorke, L.B., and Phillips, P.C., 1996, Glial differentiation predicts poor clinical outcome in primitive neuroectodermal brain tumors, *Ann. Neurol.* 39: 481-489.

Jay, V., Edwards, V., Squire, J., and Rutka, J., 1993, Astroblastoma: report of a case with ultrastructure, cell kinetic, and cytogenetic analysis, *Pediatr. Pathol.* 13: 323-332.

Jay, V., Squire, J., Bayani, J., Alkhani, A.M., Rutka, J.T., and Zielenska, M., 1999, Oncogene amplification in medulloblastoma: analysis of a case by comparative genomic hybridization and fluorescence in situ hybridization, *Pathology* 31: 337-344.

Jenkins, D., Shabanah, M.A., Shail, E.A., Gray, A., Hassounah, M., Khafaga, Y., Kofide, A., Mustafa, M., and Schultz, H., 2000, Prognostic factors for medulloblastoma, *Int. J. Radiat. Oncol. Biol. Phys.* 47: 573-584.

Jeuken, J.W.M., von Deimling, A., and Wesseling, P., 2004, Molecular pathogenesis of oligodendroglial tumors, *J. Neuro-Oncol.* 70: 1161-181.

Joachim, T., Ram, Z., Rappaport, Z.H., Simon, M., Schramm, J., Wiestler, O.D., and von Deimling, A., 2001, Comparative analysis of the *NF2, TP53, PTEN, KRAS, NRAS*, and *HRAS* genes in sporadic and radiation-induced meningiomas, *Int. J. Cancer* 94: 218-221.

Johnstone, R.W., Ruefli, A.A., and Lowe, S.W., 2002, A link between cancer genetics and chemotherapy, *Cell* 108: 153-164.

Jones, T.R., Pizzo, S.V., and Bigner, D.D., 1985, Localisation of Factor VIII/von Willebrand factor and glial fibrillary acidic protein in the haemangioblastoma: implications for stromal cells histogenesis, *Acta Neuropathol.* 67: 174-176.

Jones, H., Steart, P.V., and Weller, R.O., 1991, Spindle-cell glioblastoma or gliosarcoma? *Neuropathol. Appl. Neurobiol.* 17: 177-187.

Kajiwara, Y., Yamasaki, F., Hama, S., Yahara, K., Yoshioka, H., Sugiyama, K., Arita, K., and Kurisu, K., 2003, Expression of survivin in astrocytic tumors. Correlation with malignant grade and prognosis, *Cancer* 97: 1077-1083.

Kalamarides, M., Niwa-Kawakita, M., Leblois, H., Abramowski, V., Perricaudet, M., Janin, A., Thomas, G., Gutmann, D.H., and Giovannini, M., 2002, *Nf 2* gene inactivation in arachnoidal cells is rate-limiting for meningioma development in the mouse, *Genes & Develop.* 16: 1060-1065.

Kamitani, H., Masuzawa, H., Kanazawa, I., and Kubo, T., 2001, Recurrence of convexity meningiomas: tumor cells in the arachnoidal membrane, *Surg. Neurol.* 56: 228-235.

Kaneko, T., Yamashima, T., Tohma, Y., Nomura, M., Imajoh-Ohmi, S., Saido, T.C., Nakao, M., Saya, H., Yamamoto, H., and Yamashita, J., 2001, Calpain-dependent proteolysis of merlin occurs by oxidative stress in meningiomas: a novel hypothesis of tumorigenesis, *Cancer* 92: 2662-2672.

Karim, A.B., Maat, B., Hatlevoll, R., Menten, J., Rutten, E.H., Thomas, D.G., Mascarenhas, F., Horiot, J.C., Parvinen, L.M., Van Reijn, M., Jager, J.J., Fabrini, M.G., Van Alphen, A.M., Hamers, H.P., Gaspar, L., Noordman, E., Pierart, M., and Van Glabbeke, M., 1996, A randomized trial on dose-response in radiation therapy of low-grade cerebral glioma. EORTC study 22844, *Int. J. Radiat. Oncol. Biol. Phys*. 36: 549-556.

Karim, A.B., Afra, D., Cornu, P., Bleehan, N., Schraub, S., De Witte, O., Darcel, F., Stenning, S., Pierart, M., and Van Glabbeke, M., 2002, Randomized trial on the efficacy of radiotherapy for cerebral low-grade glioma in the adult. EORTC Study 22845, *Int. J. Radiat. Oncol. Biol. Phys*. 52: 316-324.

Kaschten, B., Flandroy, P., Reznik, M., Hainaut, H., Stevenaert, A., 1995, Radiation-induced gliosarcoma. Case report and review of the literature, *J. Neurosurg*. 83: 154-162.

Katoh, M., Wilmotte, R., Belkouch, M.C., de Tribolet, N., Pizzolato, G., and Dietrich, P.Y., 2003, Survivin in brain tumors: an attractive target for immunotherapy, *J. Neuro-Oncol*. 64: 71-76.

Katsetos, C.D., and Burger, P.C., 1994, Medulloblastoma, *Semin. Diagn. Pathol*. 11: 85-97.

Katsetos, C.D., Herman, M.M., Krishna, L., Vender, J.R., Vinores, S.A., Agamanolis, D.P., Schiffer, D., Burger, P.C., and Urich, H., 1995, Calbindin-D29k in subsets of medulloblastomas and in the human medulloblastoma cell line D283 Med, *Arch. Pathol. Lab. Med*. 119: 734-743.

Katsetos, C.D., Del Valle, L., Geddes, J.E., Aldape, K., Boyd, J.C., Legido, A., Khalili, K., Perentes, E., and Mörk, S.J., 2002, Localization of the neuronal class β-tubulin in oligodendrogliomas: comparison with Ki-67 proliferative index and 1p/19q status, *J. Neuropathol. Exp. Neurol*. 61: 307-320.

Kaulich, K., Blaschke, B., Nümann, A., von Deimling, A., Wiestler, O.D., Weber, R.G., and Reifenberger, G., 2002, Genetic alterations commonly found in diffusely infiltrating cerebral gliomas are rare or absent in pleomorphic xanthoastrocytomas, *J. Neuropathol. Exp. Neurol*. 61: 1092-1099.

Kawano, N., Yada, K., and Yagishita, S., 1989, Clear cell ependymoma. A histological variant with diagnostic implications, *Virch. Arch*. 415: 467-472.

Kawano, N., Yagishita, S., Komatsu, K., Suwa, T., Oka, H., Utsuki, S., and Fuji, K., 1999, Cerebellar clear cell epednymoma mimicking hemangioblastoma. Its clinical and pathological features, *Surg. Neurol*. 51: 281-288.

Kawarabuki, K., Ohta, T., Hashimoto, N., Wada, K., Maruno, M., Yamaki, T., and Ueda, S., 2005, Cerebellar glioblastoma genetically defined as a secondary one, *Clin. Neuropath*. 24: 64-68.

Kazuno, M., Tokunaga, T., Oshika, Y., Tanaka, Y., Tsugane, R., Kijima, H., Yamazaki, H., Ueyama, Y., and Nakamura, M., 1999, Thrombospondin-2 (TSP2) expression is inversely correlated with vascularity in glioma, *Eur. J. Cancer* 35: 502-506.

Kelly, P.J., Daumas-Duport, C., Scheithaer, B.W., Kall, B.A., and Kispert, D.B., 1987, Stereotactic histologic correlation of computed tomography and magnetic resonance imaging-defined abnormalities in patients with glial neoplasms, *Mayo Clin. Proc*. 62: 450-459.

Kenney, A.M., Widlund, H.R., and Rowitch, D.H., 2004, Hedgehog and PI-3 kinase signaling converge on Nmyc1 to promote cell cycle progression in cerebellar neuronal precursors, *Development* 131: 217-228.

Kepes, J.J., 1979, "Xanthomatous" lesions of the central nervous system: definition classification and some recent observations. *Progress in Neuropathology*, Zimmermann H.M. Ed 4, Raven Press, New York, 179-213.

Kepes, J.J., Fulling, K.H., and Garcia, J.H., (1982), The clinical significance of "adenoid" formations of neoplastic astrocytes, imitating metastatic carcinoma, in gliosarcomas. A review of five cases, *Clin. Neuropathol.* 1: 139-150.

Kepes. J.J., Rengachary, S.S., and Lee, S.H., 1979, Astrocytes in hemangioblastomas of the central nervous system and their relationship to stromal cells, *Acta Neuropathol.* 47: 99-104.

Kepes, J.J., Moral, L.A., Wilkinson, S.B., Abdullah, A., and Llena, J.F., 1998, Rhabdoid transformation of tumor cells in meningiomas: a histologic indication of increased proliferative activity, *Am J. Surg. Pathol.* 22: 231-238.

Kepes, J.J., Louis, D.N., Giannini, C., and Paulus, W., 2000, Pleomorphic xanthoastrocytoma. *Pathology and genetics. Tumours of the nervous system*, P. Kleihues and, W.K. Cavenee, Eds, IARC, Lyon.

Kern, M.A., Feisel, K.D., Friese, M., Ernestus, R.-I., and Schröder, R., 2003, Proliferative activity of microvascular cells in glioblastomas does not correlate with time to recurrence, *J. Neuro-Oncol.* 63: 9-13.

Kernohan, J.W., Mabon, R.F., Svien, H.J., and Adson, A.W., 1949, Simplified calssification of gliomas, *Proc. Staff Meet. Mayo Clin.* 24: 71-75.

Ketter, R., Henn, W., Niedermayer, I., Steilen-Gimbel, H., König, J., Zang, K.D., and Steudel, W.-I., 2001, Predictive value of progression-associated chromosomal aberrations for the prognosis of meningiomas: a retrospective study of 198 cases, *J. Neurosurg.* 95: 601-606.

Kim, D.G., Yang, H.J., Park, I.A., Chi, J.G., Jung, H.W., Han, D.H., Choi, K.S., and Cho, B.K., 1998, Gliomatosis cerebri: clinical features, treatment and prognosis, *Acta Neurochir. (Wien),* 140: 755-762.

Kim, S., Choi, K., Kwon, D., Benveniste, E.N., and Choi, C., (2004), Ubiquitin-proteasome pathway as a primary defender against TRAIL-mediated cell death, *Cell Mol. Life Sci.* 61: 1075-1081.

Kimura, Y., Koga, H., Araki, N., Mugita, N., Fujita, N., Takeshima, H., Nishi, T., Yamashima, T., Saido, T.C., Yamasaki, T., Moritake, K., Saya, H., and Nakao. M., 1998, The involvement of calpain-dependent proteolysis of the tumor suppressor NF (merlin) in schwannomas and meningiomas. Nat. Med. 4: 915-922. Kirby, S., Macdonald, D., Fisher, B., Gaspar, L., and Cairncross, G., 1996, Pre-radiation chemotherapy for malignant glioma in adults, *Can. J. Neurol.* Sci. 23: 123-127.

Kleihues, P., Rajewsky, M.F., 1984, Clinical neuro-oncogenesis: role of structural DNA modifications, DNA repair and neural target cell population, *Progr. Exp. Tum. Res.* 27: 1-16.

Kleihues, P., and Ohgaki, H., 1999, Primary and secondary glioblastomaa, from concept to clinical diagnosis, *Brain Pathol.* 1: 44-51.

Kleihues, P., and Cavenee, W.K., *Pathology and genetics. Tumours of the nervous system.* 2000, IARC, Lyon.

Kleihues, P., Davis, R.L., Coons, S.W., Burger, P.C., 2000, Anaplastic astrocytoma. *Pathology and genetics. Tumours of the nervous system*. P. Kleihues ans W.K. Cavenee Eds, IARC, Lyon.

Kleihues, P., Chimelli, L., and Giangaspero, F., 2000, Cerebellar liponeurocytoma. *Pathology and genetics. Tumours of the nervous system,* P. Kleihues and W. K. Cavenee Eds, IARC, Lyon.

Klein, R., Bendszus, M., Perez, J., and Roggendorf, W., 2002 Rhabdoides Meningeom. Ein neuer maligner Subtyp. Der Pathologe 23:297-302.

Knobbe, C.B., Merlo, A., and Reifenberger, G., 2002, PTEN signaling in gliomas. *Neuro-Oncology* 4: 196-211.

Knobbe, C.B., and Reifenberger, G. 2003, Genetic alterations and aberrant expression of genes related to the phosphatidyl-inositol-3'-kinase/protein kinase B (Akt) signal transduction pathway in glioblastomas, *Brain Pathol.* 13: 507-518.

Knobbe, C.B., Reifenberger, J., Blaschke, B., and Reifenberger, G., 2004, Hypermethylation and transcriptional downregulation of the carboxyl-terminal modulator protein gene in glioblastomas, *J. Natl. Cancer Inst.* 96:483-486.

Kobayashi, D., Kobayashi, M., Matsumoto, K., Ogura, T., Nakafuku, M., and Shimamura, K. ,2001, Early subdivisions in the neural plate define distinct competence for inductive signals. *Development,* 129: 83-93.

Kobayashi, T., Kida, Y., and Mori, Y., 2001, Long-term results of stereotactic gamma radiosurgery of meningiomas, *Surg. Neurol* 55: 325-331.

Kochi, M., 1997, Cell proliferation and apoptosis in human glioma, *Kyoto Minami Hospital* 16: 11-20.

Kochi, N., and Budka, H., 1987, Contribution of histiocytic cells to sarcomatous development of the gliosarcoma, *Acta Neuropathol* 73: 124-130.

Kokubo, M., Shibamoto, Y., Takahashi, J.A., Sasai, K., Oya, N., Hashimoto, N., and Hiraoka, M., 2000, Efficacy of conventional radiotherapy, for recurrent meningioma. *J. Neuro-Oncol.*, 48: 51-55.

Kolles H, Niedermayer I, Scmitt C, Henn W, Feld R, Strudel WI, Zang KD, Feiden W., 1995, Triple approach for diagnosis and grading of meningiomas: histology, morphometry of Ki-67/Feulgen staining, and cytogenetics. *Acta Neurochir (Wien)*, 137: 174-181.

Komine C, Watanabe T, Katayama Y, Yoshino A, Yokoyama T, Fukushima T., 2003, Promoter hypermethylation of the DNA repair gene *O^6-methylguanine-DNA methyltransferase* is an independent predictor of shortened progression free survival in patients with low-grade diffuse astrocytomas. *Brain Pathol* 13: 176-184.

Komori, T., Scheithauer, B.W., Anthony, D.C., Rosenblum, M.K., McLendon, R.E., Scott, R.M., Okazaki, H, and Kobayashi, M., 1998, Papillary glioneural tumor: a new variant of mixed neuronal-glial neoplasm, *Am. J. Surg. Pathol.*, 22: 1171-1183.

Komori, T., Scheithauer, B.W., Hirose, T., 2002, A rosette-forming glioneuronal tumor of the fourth ventricle. *Am. J. Surg. Pathol.*, 26: 582-591.

Kondziolka, D., Nathoo, N., Flickinger, J.C., Nirajan, A., Maits, A.H., and Lunsford, L.D., 2003, Long-term results after radiosurgery for benign intracranial tumors, *Neurosurgery* 53: 821-822.

Kondziolka, D., Lunsford, L.D., and Martinez, A.J., 1993, Unreliability of contemporary neurodiagnostic imaging in evaluating suspected adult supratentorial (low-grade) astrocytoma, *J. Neurosurg.* 79: 533-536.

Kopelson, G., 1982, Cerebellar glioblastoma, *Cancer,* 50: 308-311.

Kordek, R., Biernat, W., Sapiaja, W., Alwasiak, J., and Liberski, P.P., 1995, Pleomorphic xanthoastrocytoma with a gangliogliomatous component: an immunohistochemical and ultrastructural study, *Acta Neuropathol.* 89: 194-197.

Kordek, R., Hironishi, M., Liberski, P.P., Yanagihara, R., and Gajdusek, D.C., 1996, Apoptosis in glial tumors as determined by in situ nonradioactive labeling of DNA breaks, *Acta Neuropathol.* 91: 112-116.

Korkopoulou, P.A., Kostantinidou, A.E., Patsouris, E.S., Christodoulou, P.N., Thomas-Tsagli, E.A., and Davaris, P.S., 2001, Detection of apoptotic cells in archival tissue from diffuse astrocytomas using a monoclonal antibody to single-stranded DNA, *J. Pathol.* 193: 377-82.

Korkolopoulou, P., Patsouris, E., Kostantinidou, A.E., Pavlopoulos, P.M., Kavantzas, N., Boviatsis, E., Thymara, I., Perdiki, M., Thomas-Tsagli, E., Angelidakis, D., Rologis, D., and Sakkas, D., 2004, Hypoxia-inducible factor 1alpha/vascular endothelial growth factor axis in astrocytomas. Associations with microvessel morphometry, proliferation and prognosis. *Neuropathol. Appl. Neurobiol.* 30: 267-278.

Korshunov, A., Golanov, A., Sicheva, R., and Ronin, I., 1999, Prognostic value of tumour associated antigen immunoreactivity and apoptosis in cerebral glioblastomas: an analysis of 168 cases, *J. Clin. Pathol.* 52: 574-80.

Korshunov, A., and Golanov, A., 2001, The prognostic significance of DNA topoisomerase II-alpha (Ki-Si), p21/Cip-1, and p27/Kip-1 protein immunoexpression in oligodendroglioma, *Arch. Pathol. Lab. Med.* 125:892-898.

Korshunov, A., Savostikova, M., and Ozerov, S., 2002, Immunohistochemical markers for prognosis of average-risk pediatric medulloblastomas. The effect of apoptotic index, TrkC, and c.myc expression, *J. Neuro-Oncol.* 58: 271-279.

Kortmann, R.D., 2003, Radiotherapy for low-grade gliomas, *Proc. Semin. Oncol.* 30:29-33.

Kortmann, R.D., Jeremic, B., Weller, M., Olasswilm, L., and Bamberg, M., 2003, Radiochemotherapy of malignant glioma in adults. *Clinical experiences,* 79: 219-232.

Kostantinidou, A., Korkolopoulou, P., Patsouris, E., Mahera, H., and Hranioti S. 2001, Apoptosis detected with monoclonal antibody to single-stranded DNA is a predictor of recurrence in intracranial meningiomas, *J. Neuro-Oncol.* 55: 1-9.

Kostantinidou, A. E., Korkolopoulou, P., and Patsouris, E., 2005, Apoptotic markers for primary brain tumor prognosis, *J. Neuro-Oncol.* 72: 151-156.

Krajewski, S., Krajewska, M., Ehrmann, J., Sikorska, M., Lach, B., Chatten, J., and Reed, J.C., 1997, Immunohistochemical analysis of Bcl-2, Bcl-X, Mcl-1, and Bax in tumors of central and peripheral nervous system origin, *Am. J. Pathol.* 150: 805-814.

Kraus, J.A., Koopmann, J., Kaskel, P., Maintz, D., Brandner, S., Schramm, J., Louis, D.N., Wiestler, O.D., and von Deimling, A., 1995, Shared allelic losses on chromosome 1p and 19q suggest a common origin of oligodendroglioma and oligoastrocytoma, *J. Neuropathol. Exp. Neurol.* 54: 91-95.

Kraus, J.A., Wenghoefer, M., Schmidt, M.C., von Deimling, A., Berweiler, U., Roggendorf, W., Diete, S., Dietzmann, K., Muller, B., Heuser, K., Reifenberger, G., and Schlegel, U., 2000, Long-term survival of glioblastoma multiforme: importance of histopathological reevaluation, *J. Neurol.* 247: 455-460.

Kraus, J.A., Wenghoefer, M., Glesmann, N., Mohr, S., Beck, M., Schmidt, M.C., Schroder, R., Berweiler, U., Roggendorf, W., Diete, S., Dietzmannn, K., Heuser, K., Muller, B., Fimmers, R., von Deimling, A., and Schlegel, U., 2001, TP53 gene mutations, nuclear p53 accumulation, expression of Waf/p21, Bcl-2, and CD95 (APO-1/Fas) proteins are not factors un de novo glioblastoma multiforme, *J. Neuro-Oncol.* 52: 263-272.

Kros, J.M., Troost, D., Van Eden, C.G., van der Werf, A.J., and Uylings, H.B., 1988, Oligodendroglioma. A comparison of two grading systems, *Cancer* 61: 2251-2259.

Krouwer, H.G., van Duinen, S.G., Kamphorst, W., van der Valk, P., and Algra, A., 1997, Oligoastrocytomas: a clinicopathological study of 52 cases, *J. Neuro-Oncol.* 3: 223-238.

Krum, J.M., and Rosenstein, J.M., 1999, Transient coexpression of nestin, GFAP, and vascular endothelial growth factor in mature reactive astroglia following neural or brain wounds, *Exp. Neurol.* 160: 348-360.

Kuan, C.T., Reist, C.J., Foulon, C.F., Lorimer, I.A., Archer, G., Pegram, C.N., Pastan, I., Zalutscky, M.R., and Bigner, D.D., 1999, 1251-labeled anti-epidermal growth factor receptor-vIII single-chain Fv exhibits specific and high-level targeting of glioma xenografts, *Clin. Cancer Res.* 5: 1539-1549.

Kuchelmeister, K., Demirel, T., Schlorer, E., Bergmann, M., and Gullotta F., 1995, Dysembryoplastic neuroepithelial tumour of the cerebellum, *Acta Neuropathol.* 89: 385-390.

Kuchiki, H., Kayama, T., Sakurada, K., Saino, M., Kawakami, K., and Sato, S., 2002, Two cases of atypical central neurocytomas, *Brain Tumor Pathol.* 19: 105-110.

Kulla, A., Burkhardt, K., Meyer-Puttlitz, B., Teesalu T, Asser, T., Wiestler, O.D., and Becker, A.J., 2003, Analysis of the TP53 gene in laser-microdissected glioblastoma vasculature, *Acta Neuropathol.* 105: 328-332.

Kuriyama, H., Lamborn, K.R., O'Fallon, J.R., Iturria, N., Sebo, T., Schaefer, P.L., Scheithauer, B.W., Buckner, J.C., Kuriyama, N., Jenkins, R. B., and Israel, M.A., 2002, Prognostic significance of apoptotic index and apoptosis/proliferation ratio for patients with high-grade astrocytomas, *Neuro-Oncology* 4: 179-186.

Kyritsis, A.P., Yung, W.K., Bruner, J., Gleason, M.J., and Levin, V.A., (1993), The treatment of anaplastic oligodendrogliomas and mixed gliomas, *Neurosurgery* 32: 365-370.

Lach, B., Duggal, N., DaSilva, V.E., and Benoit, B.G., 1996, Association of pleomorphic xanthoastrocytoma with cortical dysplasia and neuronal tumors, *Cancer* 78: 2551-2563.

Lach, B., Gregor, A., Rippstein, P., and Omulecka, A., (1999), Angiogenic histogenesis of stromal cells in hemangioblastoma: ultrastructural and immuniohistochemical study, *Ultrastruct. Pathol.* 23: 299-310.

Lamszus, K., Laterra, J., Westphal, M., and Rosen, E.M., 1999, Scatter factor/hepatocyte growth factor (SF/HGF) content and function in human gliomas, *Int. J. Dev. Neurosci.*, 17: 517-530.

Langford, I.A., Cooksley, C.S., and DeMonte, F., 1996, Comparison of MIB-1 (Ki-67) antigen and bromodeoxyuridine proliferation indices in meningiomas, *Hum. Pathol.* 27: 350-354.

Lantos, P.L., and Rosenblum, M.K., 2000, Astroblastoma. *Pathology and genetics. Tumors of the nervous system*, P. Kleihues and K.W. Cavenee Eds, IARC, Lyon.

Lantos, P.L., and Bruner, J.M., 2000, Gliomatosis cerebi. *Pathology and genetics. Tumours of the nervous system*, P. Kleihues and K. W. Cavenee Eds., IARC, Lyon.

Lawinger, P., Venugopal, R., Guo, Z.S., Immaneni, A., Sengupta, D., Lu, W., Rastelli, L., Marin Dias Carneiro, A., Levin, V., Fuller, G.N., Echelard, Y., and Majumder, S., 2000, The neuronal repressor REST/NRSF is an essential regulator in medullooblastoma cells, *Nature Med.* 6: 826-831.

Laywell, E.D., Rakic, P., Kukekov, V.G., Holland, E.C., and Steindler, D.A., 2000, Identification of a multipotent astrocytic stem cell in the immature and adult mouse brain, *PNAS USA*, 97: 13883-13888.

Lee, Y., Miller, H.L., Jensen, P., Hernan, R., Connelly, M., Wetmore, C., Zindy, F., Roussel, M.F., Curran, T., Gilbertson, R.J., and McKinnon, P.J., 2003, A molecular fingerprint for medulloblastoma, *Cancer Res.* 63: 5428-5437.

Lendahl, U., Zimmerman, L.B., and McKay, R.D.G., 1990, CNS stem cells express a new class of intermediate filament protein, *Cell* 60: 585-595.

Leon, S.P., Folkert, R.D., and Black, O.M., 1996, Microvessel density is a prognostic indicator for patients with astroglial brain tumors, *Cancer* 77: 362-372.

Leonard, J.R., D'Sa, C., Klocke, B.J., Roth, K.A., 2001, Neural precursor cell apoptosis and glial tumorigenesis following transplacental ethyl-nitrosourea exposure, *Oncogene* 20: 8281-8286.

Leonard, J.R., Cai, D.X., Rivet, D.J., Kaufman, B.A., Park, T.S., Levy, B.K., and Perry, A., 2001 , Large cell/anaplastic medulloblastomas and medullomyoblastoma: clinicopathological and genetic features, *J. Neurosurg.* 95: 82-88.

Leung, S.Y., Gwi, E., Ng, H.K., Fung, C.F., and Yam, K.Y., 1994, Dysembryoplastic neuroepithelial tumor: A tumor with small neuronal cells resembling oligodendroglioma, *Am. J. Surg. Pathol.* 18: 604-614.

Leung, C., Lingbeek, M., Shakhova, O., Liu, J., Tanger, E., Saremasiani, P., van Lohuizen, M., and Marino, S., 2004, Bmi1 is essential for cerebellar development and is overexpressed in human medulloblastomas, *Nature* 428: 337-341.

Leuraud, P., Marie, Y., Robin, E., Huguet, S., He, J., Moktari, K., Cornu, P., Hoang-Xuan, K., and Sanson, M., 2000, Frequent loss of 1p32 region but no mutation of the p18 tumor suppressor gene in meningiomas, *J. Neuro-Oncol.* 50: 207-213.

Levison, S.W., and Goldman, J.E., 1997, Multipotential and lineage restricted precursors coexist in the mammalian perinatal subventricular zone, *J. Neurosci. Res.* 48: 83-94.

Li, H., Zhu, H., Xu, C.J., and Yuan, Y., 1998, Cleavage of BID by caspase 8 mediates the mitochondrial damage in the Fas pathway of apoptosis, *Cell* 94: 491-501.

Ligon, K.L., Alberta, J.A., Kho, A.T., Weiss, J., Kwaan, M.R., Nutt, C.L., Louis, D.N., Stiles, C.D., and Rowitch, C.H., 2004, The oligodendroglial lineage marker OLIG2 is universally expressed in diffuse gliomas, *J. Neuropathol. Exp. Neurol.* 63: 499-509.

Lin, G., Finger, E., and Gutierrez-Ramos, J.C., 1995, Expression of CD34 in endothelial cells, hematopoietic progenitors and nervous cells in fetal and adult mouse tissues, *Eur. J. Immunol.* 25: 1508-1516.

Lin, R.C., Matesic DF, Marvin M., McKay, R.D., and Brustle, O., 1995, Re-expression of the intermediate filament nestin in reactive astrocytes, *Neurobiol. Dis.* 2: 79-85.

Liotta, L.A., and Stetler-Stevenson, W.G., 1991, Tumor invasion and metastasis: an imbalance of positive and negative regulation, *Cancer Res.* 57 (Suppl): 5054-5059.

Litofsky, N.S., Mix, T.C., Baker, S.P., Recht, L.D., and Smith, T.W., 1998, Ki-67 (clone MIB.1) proliferation index in recurrent glial neoplasms: no prognostic significance, *Surg. Neurol.* 50: 579-585.

Lopes, M.B.S., 2003, Angiogenesis in brain tumors, *Micr. Res. Techn.* 60: 225-230.

Lou, Q.R., Park, J.K., Noll, E., Chan, J.A., Alberta, J., Yuk, D., Garcia Altamora, M., Louis, D.N., Stlies, C.D., Rowitch, D.H., and Black, P.M., 2001, Oligodendrocyte lineage genes (OLIG) as molecular markers for human brain tumors, *PNAS* 28: 10851-10856.

Louis, D.N., 1997, A molecular genetic model of astrocytoma histopathology, *Brain Pathol.* 7: 755-764.

Louis, D.N., Scheithauer, B.W., Budka, H., von Deimling, A., and Kepes, J.J., 2000, Meningiomas. *Pathology and genetics. Tumours of the nervous system*, P. Kleihues and W.K. Cavenee Eds. IARC, Lyon.

Louis, D.N., Holland, E.C., and Cairncross, J.G., 2001, Glioma classification. A molecular reappraisal, *Am. J. Pathol.* 159: 779-786.

Louis, D.N., and Cavenee, W.K., 2001, *Molecular biology of central nervous system tumors. Cancer: Principles and practice of oncology*, Lippincot-Raven, Ed 6, Philadelphia.

Lu, Q.R., Park, J.K., Noll, E., Chan, J.A., Alberta, J., Yuk, D., Alzamora, M.G., Louis, D.N., Stiles, C.D., Rowitch, D.H., Black, P.M., 2001, Oligodendrocyte lineage genes (OLIG) as molecular markers for human glial brain tumors, *PNAS USA* 98: 10851-10856.

Lu, Q.R., Sun, T., Zhu, Z., Ma, N., Garcia, M., Stiles, C.D., Rowitch, D.H., 2002, Common developmental requirement for Oligo function indicates a motor neuron/oligodendrocyte connection, *Cell* 109: 75-86.

Ludwin, S.K., Rubinstein, L.J., and Russell, D.S., 1975, Papillary meningioma: a malignant variant of meningioma, *Cancer*, 36: 1363-1373.

McGirt, M.J., Villavicenioi, A.T., Bulsara, K.L., and Friedman, A.H., 2003, MRI-guided stereotactic biopsy in the diagnosis of glioma: comparison of biopsy and surgical resection specimen, *Surg. Neurol.* 59: 277-282.

McManamy, C.S., Lamont, B.J.M., Taylor, R.E., Cole, M., Pearson, A.D.J., Clifford, S.C., and Ellison, D.W., 2003, Morphophenotypic variation predicts clinical behavior in childhood non-desmoplastic medulloblastomas, *J. Neuropathol. Exp. Neurol.* 62: 627-632.

Macaulay, R.J.B., Hoffman, H.J., and Becker, L.E., 1993, Increased mitotic activity as a negative prognostic indicator in pleomorphic xanthoastrocytoma. Case report, *J. Neurosurg.* 79: 761-768.

Macdonald, D.R., Gaspar, L.E., and Cairncross, J.G., 1990, Successful chemotherapy for newly diagnosed aggressive oligodendrogliomas, *Ann. Neurol.* 27: 573-574.

Macdonald, D.R., 2003, New frontiers in the treatment of malignant glioma, *Semin. Oncol.* 30: 72-76.

MacDonald, T.J., Brown, K.M., LaFleur, B., Peterson, K., Lawlor, C., Chen, Y., Packer, R.J., Cogen, P., and Stephan, D.A., 2001, Expression profiling of medulloblastoma: PDGFRA and the RAS/MAPK pathway as therapeutic target for metastatic disease, *J. Clin. Oncol.* 29: 143-152.

Machein, M.R., and Plate, K.H., 2000, VEGF in brain tumors, *J. Neuro-Oncol.* 50: 100-120.

Mackenzie, I.R., 1999, Central neurocytoma: histologic atypia, proliferation potential, and clinical outcome, *Cancer* 85:1606-1610.

Machein, M.R., Renninger, S., de Lima-Hahn, E., and Plate, K.H., 2003, Minor contribution of bone marrow-derived endothelial progenitors to the vascularization of murine gliomas, *Brainpathol.* 13: 582-597.

Madsen, C., Schröder, H.D., 1997, Ki-67 immunoreactivity in meningioma-determination of the proliferative potential of meningiomas using the monoclonal antibody Ki-67, *Clin. Neuropathol.* 16: 137-142.

Magavi, S.S., Leavitt, B.R., and Macklis, J.D., 2000, Induction of neurogenesis in the neocortex of adult mice, *Nature* 405: 951-955.

Maher, E., Furnari, F.B., Bachoo, R.M., Rowitch, D.H., Mouis, D.N., Cavenee, W.K., and De Pinho, R.A., 2001, Malignant gliomas: genetic and biology of a grave matter. *Genes & Develop.* 15: 1311-133.

Maier, H., Wanschitz, J., Sedivy, R., Rössier, K., Öffner, D., and Budka, H., 1997, Proliferation and DNA fragmentation in meningioma subtypes, *Neuropathol. Appl. Neurobiol.* 23: 496-506.

Maier, D., Jones, G., Li, X., Schonthal, A.H., Gratzl, O., Van Meir, E.G., and Merlo, A., 1999, The PTEN lipid phosphatase domain is not required to inhibit invasion of glioma cells, *Cancer Res.* 59: 5479-5482.

Maintz, D., Fiedler, K., Koopamnn, J., Rollbrocker, B., Nechev, S., Lenartz, D., Stangl, A.P., Louis, D.N., Schramm, J., Wiestler, O.D., and von Deimling, A., (1997), Molecular genetic evidence for subtypes of oligodendrogliomas. *J. Neuropathol. Exp. Neurol.* 56: 1098-1104.

Marie, Y., Sanson, M., Mokhtari, K., Leuraud, P., Kujas, M., Delattre, J.-Y., Poirie, J., Zaic, 0.B., and Hoang-Xuan, K., 2001, OLIG2 as a specific marker of oligodendroglial tumour cells, *The Lancet* 358: 298-300.

Martin, S., Toquet, C., Oliver, L., Cartron, P.F., Perrin, P., Meflah, K., Cuillere, P., and Vallette, F.M., 2001, Expression of bcl-2, bax and bcl-xl in human gliomas: a re-appraisal, *J. Neuro-Oncol.* 52: 129-139.

Martin, J.M., Katati, M., Loipez, E., Bullejos, J.A., Arregui, G., Busquier, H., Minguez, A., Olivares, G., Hernandez, V., and Arjona, V., 2003, Linear accelerator radiosurgery in treatment of central neurocytomas, *Acta Neurochir. (Wien)* 145: 749-754.

Mandenhall, W.M., Morris, C.G., Amddur, R.J., Foote, K.D., and Friedman, W.A., 2003, Radiotherapy alone or after subtotal resection for benign skull base meningiomas, *Cancer* 98: 1473-1482.

Martinez-Diaz, H., Kleinschmidt-DeMasters, B.K., Powell, S.Z., and Yachnis, A.T., 2003, Giant cell glioblastoma and pleomorphic xanthoastrocytoma show different immunohistochemical profiles for neuronal antigens and p53 but share reactivity for class III β-tubulin, *Arch. Pathol. Lab. Med.* 127: 1187-1191.

Mason, W.P., Krol, G.S., and DeAngelis, I.M., 1996, Low-grade oligodendroglioma responds to chemotherapy, *Neurology* 46: 203-207.

Mathiesen, T., Kihlström, L., Karlsson, B., and Lindquist, C., 2003, Potential complications following radiotherapy for meningiomas, *Surg. Neurol.* 60: 193-200.

Matsuno, A., Fujimaki, T., Sasaki, T., Nagashima, T., Ide, T., Asai, A., Matsuura, R., and Utsunomiya, H., 1996, Clinical and histopathological analysis of proliferative potentials of recurrent and non recurrent meningiomas, *Acta Neuropathol.* 91: 504-510.

Mawrin, C., Kirches, E., Schneider-Stock, R., Scherlach, C., Vorwerk, C., von Deimling, A., van Landeghem, F., Meyermann, R., Bornemann, A., Müller, A., Romeike, B., Stoltenburg-Didinger, G., Wickboldt, J., Pilz, P., and Dietzmann, K., 2003, Analysis of TP53 and PTEN in gliomatosis cerebri, *Acta Neuropathol.* 105: 529-536.

Mawrin, C., Sasse, T., Kirches, E., Schneider, T., Grimm, C., Pambor, C., Vorwerk, CK., Firsching, R., Lendeckel, U., and Dietzmann, K., 2005, Different activation of mitogen-activated protein kinase and Akt signaling is associated with aggressive phenotype of human meningiomas. *Clin. Cancer Res.* 11: 4074-4082.

Mawrin, C., Kirches, E., Schneider-Stock, R., Boltze, C., Vorwerk, C.K., von Deimling, A., Stoltenburg-Didinger, G., Bornemann, A., Romeike, B., Sellhaus, B., and Dietzmann K., 2005, Alterations of cell cycle regulators in gliomatosis cerebri. *J. Neuro-Oncol.* 72: 115-122.

McGirt, M.J., Villavicencio, A.T., Bulsara, K.R., and Friedman, A.H., 2003, MTI-guided stereotactic biopsy in the diagnosis of glioma: Comparison of biopsy and surgical resection specimen, *Surg. Neurol.* 59: 277-282.

McLendon, R.E., and Provenzale, J., 2002, Glioneural tumors of the central nervous system, *Brain Tumor. Pathol.* 19: 51-58.

McLendon, R.E., and Halperin, E.C., 2003, Is the long-term survival with intracranial glioblastoma multiforme overstated ? *Cancer* 98: 1745-1748.

McNeil, D,E., Cote, T.R., Clegg, L., and Rorke, L.B., 2002, Incidence and trends in pediatric malignancies medulloblastoma/primitive neuroectodermal tumor: a SEER update. Surveillance, Epidemiology and End Results. *Med. Pediatr. Oncol.* 39: 190-194.

Mendenhall, W.M., Morris, C.G., Amdur, R.J., Foote, K.D., and Friedman, W.A., 2003, Radiotherapy alone or after subtotal resection for benign skull base meningiomas, *Cancer* 98: 73-82.

Merchant, T.E., Jenkins, J.J., Burger, P.C., Sanford, R.A., Sherwood, S.H., Jones-Wallace, D., Heideman, R.L., Thompson, S.J., Helton, K.J., and Kun, L.E., 2002, Influence of tumor grade on time to progression after irradiation for localized ependymoma in children, *Int. J. Radiat. Oncol. Biol. Phys*. 53: 52.57.

Merzak, A., Koochekpour, S., Dkissi, F., Raynal, S., Lawrence, D., and Pilkington, G.J., 1995, Synergism between growth factors in the control of glioma cell proliferation, migration and invasion *in vitro, Int. J. Oncol*. 6: 1079-1085.

Mezey, E., Chandross, K.J., Harta, G., Maki, R.A., and McKercher, S.R., 2000, Turning blood into brain; cells bearing neuronal antigens generated in vivo from bone marrow, *Science* 290: 1779-1782.

Michaud, J., Barrés, D., Privé, M., and Duht, M.-A., 1985, Focal atypical transformation in benign meningiomas, *J. Neuropathol. Exp. Neurol*. 44: 314.

Miettinen, H.E., Kononen, J., Sallinen, H., Alho, H., Helen, P., Helin, H., Halimo, H., Paliarvi, L., Isola, J., and Haapasalo, H., 1999, CDKN2A/p16 predicts survival in oligodendrogliomas: comparison with astrocytomas, *J. Neuro-Oncol*. 41: 205-211.

Miettinen, H.E., Järvinen, T.A.H., Kellner, U., Kauraniemi, P., Parwaresch, R., Rantala, I., Kalimo, H., Paliärvi, L., Isola, J., and Haapasalo, H., 2000, High topoisomerase IIα expression associates with high proliferation rate and poor prognosis in oligodendrogliomas, *Neuropathol. Appl. Neurobiol*. 26: 504-512.

Miettinen, H.E., Paunu, N., Rantala, I., Kalimo, H., Paljarvi, L., Helin, H., and Haapasalo, H., 2001, Cell cycle regulators (p21, p53, pRb) in oligodendrocytic tumors: a study by novel tumor microarray technique, *J. Neuro-Oncol*. 55: 29-37.

Migheli, A., Cavalla, P., Marino, S., and Schiffer, D., 1994, A study of apoptosis in normal and pathologic nervous tissue after in situ end-labeling of DNA strand breaks, *J. Neuropathol. Exp. Neurol*. 53: 606-616.

Mihaila, D., Jankowski, M., Gutiérrrez, A., Rosenblum, M.L., Newsham, I.F., Bögler, O., and Rempel, S.A., 2003, Meningiomas: loss of heterozigosity on chromosome 10 and marker-specific correlations with grade, recurrence, and survival, *Clin. Cancer Res*. 9: 4443-4451.

Milker-Zabel, S., Zabel, A., Schulz-Ertner, D., Schlegel, W., Wannenmacher, M., and Debus, J., 2003, Fractionated stereotactic radiotherapy in patients with benign or atypical intracranial meningioma: long-term experience and prognostic factors, *Int. J. Radiation. Oncol. Biol. Phys*. 61: 809-816.

Miller, D.C., Ojemann, R.G., Proppe, KH., McGinnis, B.D., and Grillo, H.C., 1985, Benign metastasizing meningioma. Case report, *J. Neurosurg*. 62: 763-766.

Min, K.W., and Scheithauer, B.W. 1997, Clear cell ependymona: a mimic of oligodendroglioma: clinicopathologic and ultrastructural considerations, *Am. J. Surg. Pathol*. 21: 820-826.

Mirimanoff, R.O., Dosoretz, D.E., Linggood, R.M., Ojemann, R.G., and Martuza, R.I., 1985, Meningioma: analysis of recurrence and progression following neurosurgical resection, *J. Neurosurg*. 62: 18-24.

Mirimanoff, R.O., and Stupp, R., 2003, Radiotherapy in low-grade gliomas: Cons, *Semin. Oncol*. 30: 34-38.

Mischel, P.S., and Cloughesy, T.F., 2003, Targeted molecular therapy of GBM, *Brain Pathol*. 13: 52-61.

Mishima, K., Johns, T.G., Luwor, R.B., Scott, A.M., Stockert, E., Jungbluth, A.A., Ji, X.D., Suvarna, P., Voland, J.R., Old, L.J., Huang, H.J., and Cavenee, W .K., 2001, Growth suppression of intracraniual xenografted glioblastomas over-expressing mutant epidermal growth factor receptors by systemic administration of monoclonal antibody (mAb) 806, a novel monoclonal antibody directed to the receptor, *Cancer Res.* 61: 5349-5354.

Mizoguchi, M., Inamura, T., Shono, T., Ikezaki, K., Inoha, S., Ohgami, S., and Fukui, M. A., 2000, Comparative study of apoptosis and proliferation in germinoma and glioblastoma, *Neuro-Oncology* 2: 96-102.

Mizoguchi, M., Nutt, C.L., Mohapatra, G., and Louis, D.N., 2004, Genetic alterations of phosphoinositide 3-kinase subunit genes in human glioblastomas, *Brain Pathol.* 14: 372-377.

Mollemann, M., Wolter, M., Felsberg, J., Collins, V.P., and Reifenberger, G., 2005, Frequent promoter hypermethylation and low expression of the MGMT gene in oligodendroglial tumors. *Int. J. Cancer* 113: 379-385.

Mollenhauer, J., Wiemann, S., Scheurlen, W., Korn, B., Hayashi, Y., Wilgenbus, K.K., von Deimling, A., and Poustka, A., 1997, DMBT1, a new member of the SRCR superfamily, on chromosome 10q25.3-26 is deleted in malignant brain tumors, *Nat. Genet.* 17: 32-39.

Moller, M.L., and Braendsdrup, O., 1997, No prediction of recurrence of meningiomas by PCNA and Ki-67 immunohisytochemistry, *J. Neuro-Oncol.* 34: 241-246.

Montine, T.J., Vandersteenhoven, J.J., Aguzzi, A., Boyko, O.B., Dodge, R.K., Kerns, B.J., and Burger, P.C., 1994, Prognostic significance of Ki-67 proliferation index in supratentorial fibrillary astrocytic neoplasms, *Neurosurgery*, 34: 674-678.

Mørk, S.J., and Loken, A.C., 1977, Ependymoma: A follow-up study of 101 cases, *Cancer*, 40: 907-915.

Mørk, S.J., Halvorsen, T.B., Lindegaard, K.F., and Eide, G.E., 1986, Oligodendroglioma. Histologic evaluation and prognosis, *J. Neuropathol. Exp. Neurol.* 45: 65-78.

Morshead, C.M., Reynolds, B.A., Craig, C.G., McBurney, M.W., Staines, W.A., Morassutti, D., Weiss, S., and Van der Kooy, D., 1994, Neural stem cells in the adult mammalian forebrain: a relatively quiescent subpopulation of subependymal cells, *Neuron* 13: 1071-1082.

Mrak, R.E., Yasargil, G., Mohapatra, G., Earel, J.Jr, and Louis, D.N., 2004, Atypical extraventricular neurocytoma with oligodendroglioma-like spread and an unusual pattern of chromosome 1p and 19q loss,. *Human Pathol.* 35: 1156-1159.

Mujtaba, T., Mayer-Pröschel, M., and Rao, M.S., 1998, A common neural progenitor for the CNS and PNS, *Dev. Biol.* 200: 1-15.

Müller, W., Lass, U., Herms, J., Kuchelmeister, K., Bergmann, M., and von Deimling, A., 2001, Clonal analysis in glioblastoma with epithelial differentiation, *Brain Pathol.* 11: 39-43.

Mukasa, A., Ueki, K., Matsumoto, S., Tatsumi S., Nishikawa, R., Fujimaki T., Asai, A., Kirino, T., and Aburatani, H., 2002, Distinction in gene expression profiles of oligodendrogliomas with and withoput allelic loss of 1p, *Oncogene* 21: 3961-3968.

Mukasa, A., Ueki, K., Ge, X., Ishikawa, S., Ide, T., Fujimaki, T., Nishikawa, R., Asai, A., Kirino, T., and Aburatani, H., 2004 , Selective expression of a subset of neuronal genes in oligodendroglioma with chromosome 1p loss, *Brain Pathol.* 14: 34-42.

Mundle, S.B., and Raza, S., 1995, The two in situ techniques do not differentiate between apoptosis and necrosis but rather reveal distinct patterns of DNA fragmentation in apoptosis, *Lab. Invest.* 72: 615-618.

Munoz, E.L., Eberhard, D.A., Lopes, M.B.S., Schneider, B.F., Gonzalez, F., and VandenBerg, S.R., 1996, Proliferative activity and p53 mutation as prognostic indicators in pleomorphic xanthoastrocytoma, *J. Neuropathol. Exp. Neurol.* 55: 606 (abstract)

Murphy, M., Loosemore, A., Ferrer, I., Wesseling, P., Wilkins, P.R., and Bell, B.A., 2002, Neuropathological diagnostic accuracy, *Br. J. Neurosurg.* 16: 461-464.

Nafe, R., Franz, K., Schlote, W., and Schneider, B., 2005, Morphology of tumor cell nuclei is significantly related with survival time of patients with glioblastoma. *Clin. Cancer Res.* 11: 2141-2148.

Nagane, M., Leviztki, A., Gazit, A., Cavenee, W.K., and Huang, H.J., 1998, Drug resistance of human glioblastoma cells conferred by a tumor-specific mutant epidermal growth factor receptor through modulation of Bcl_XL and caspase-3-like proteases, *PNAS USA* 95: 5724-5729.

Nagashima, G., Aoyagi, M., Yamamoto, M., Yamamoto, S., Wakimoto, H., Ohno, K., Yamamoto, K., Jirakawa, K., 1999, P53 overexpression and proliferative potential in malignant meningiomas, *Acta Neurochir. (Wien)* 141: 53-61.

Nakagawa, S., Shiraishi, T., Kihara, S., and Tabuchi, K., 1995, Detection of DNA strand breaks associated with apoptosis in human brain tumors, *Virchows Arch.* 427: 175-179.

Nakaguchi, H., Fujimaki, T., Matsuno, A., Matsuura, R., Asai, A., Suzuki, T., and Kirino, T., 1999, Postoperative residual tumor growth of meningioma can be predicted by MIB-1 immunohistohcemistry, *Cancer* 85: 2249-2254.

Nakamura, M., Rieger, J., Weller, M., Kim, J., Kleihues, P., and Ohgaki, H., 2000, APO2L/TRAIL expression in human brain tumors, *Acta Neuropathol.* 99: 1-6.

Nakamura, M., Watanabe, T., Klangby, U., Asker, C., Wiman, K., Yonekaza, Y., Kleihues, P., and Ohgaki, H., 2001, $p14^{ARF}$ deletion and methylation in genetic pathways to glioblastomas, *Brain Pathol.* 11: 159-168.

Nakamura, T., Xi, G., Hua, Y., Hoff, J.T., and Keep R.F., 2003 Nestin expression after experimental intracerebral hemorrhage, *Brain Res* 981: 108-117.

Nakasu, S., Nakasu, Y., Nioka, H., Nakajima, M., and Handa, J., 1994, Bcl-2 protein expression in tumors of the central nervous system, *Acta Neuropathol.* 88: 520-526.

Nakasu, S., Nakasu, Y., Nakajima, M., Matsuda, M., and Handa, J., 1999, Preoperative identification of meningiomas that are highly likely to recur, *J. Neurosurg.* 90: 455-462.

Nakasu, S., Li, D.H., Okabe, H., Nakajima, M., and Matsuda, M., 2001, Significance of MIB-1 staining indices in meningiomas. Comparison of two counting methods, *Am. J. Surg. Pathol.* 25: 472-478.

Narita, Y., Nagane, M., Mishima, K., Huang, H.J., Furnari, F.B., and Cavenee, W.K., 2002, Mutant epidermal growth factor receptor signaling down-regulates p27 through activation of the phosphatidylinositol 3-kinase/Akt pathway in glioblastomas, *Cancer Res*. 62: 6764-6769.

Neben, K., Korschunov, A., Benner, A., Wrobel, G., Hahn, M., Kokocinski, F., Golanov, A., Joos, S., and Lichter, P., 2004, Microarray-based screening for molecular markers in medulloblastoma revealed *STK15* as independent predictor for survival, *Cancer Res*. 64: 3103-3111.

Neder, L., Marie SK, Carlotti, Jr C.G., Gabbai, A.A., Rosemberg, S., Malheiros, S.M., Siquiera, R.P., Oba-Shinjo, S.M., Uno, M., Aguiar, P.H., Miura, F., Chammas, R., Colli, B.O., Silva, W.A.Jr, and Zago, M.A., 2004, Galectin-3 as an immunohistochemical tool to distinguish pilocytic astrocytomas from diffuse astrocytomas, and glioblastomas from anaplastic oligodendrogliomas, *Brain Pathol.* 14: 399-405.

Nevin, S., 1938, Gliomatosis cerebri, *Brain* 61: 170-191.

Ng, H.-K., and Chen, L., 1998, Apoptosis is associated with atypical or malignant change in meningiomas. An in situ labeling and immunohistochemical study, *Histopathology* 33: 64-70.

Nigro, J.M., Takahashi, M.A., Ginzinger, D.G., Law, M., Passe, S., Jenkins, R.B., and Aldape, K., 2001, Detection of 1p and 19q loss in oligodendroglioma by quantitative microsatellite analysis, a real-time quantitative polymerase chain reaction assay, *Amer. J. Pathol.* 158: 1253-1262.

Nishikawa, R., Cheng, S.Y., Nagashima, R., Huang, H.J., Cavenee, W.K., and Matsutani, M., 1998, Expression of vascular endothelial growth factor in human brain tumors, *Acta Neuropathol*. 96: 453-462.

Nishio, S., Korosue, K., Tateishi, J., Fukui, M., and Kitamura, K., 1982, Ventricular and subarachnoidal seeding of intracranial tumors of neuroectodermal origin. A study of 26 consecutive autopsy cases with reference to focal ependymal defect, *Clin. Neuropathol*. 1: 83-91.

Noble, M., 1994, The O-2A lineage: from rats to humans, *Rec. Res. Cancer Res.* 135: 67-75.

Noble, M., Gutowski, N., Bevan, K., Engel, U., Linskey, M., Urenjak, L., Bhakoo, K., and Williams, S., 1995, From rodent glial precursor cell to human glial neoplasia in the oligodendrocyte-type 2 astrocytic lneage, *Glia* 15: 222-230.

Noble, M., and Mayer-Pröschel, M., 1997, Growth factors, glia and gliomas, *J. Neuro-Oncol*. 35: 193-209.

Noel, G., Habrand, J.L., Mammar, H., Haie-Meder, C., Pontvert, D., Dederke, S., Ferrand, R., Beaudre, A., Gaboriaud, G., Boisserie, G., and Mazeron, J.J., 2002, Highly conformal therapy using proton component in the management of meningiomas. Preliminary experience of the centrer de protontherapie d'Orsay. *Strahlenther Onkol*. 178: 480-485.

Nutt, C.L., Noble, M., Chambers, A.F., and Cairncross, J.G., 2000, Differential expression of drug resistance genes and chemosensitivity in glial cell lineages correlate with differential response of oligodendrogliomas and astrocytomas to chemotherapy, *Cancer Res*. 60: 4812-4818.

Nutt, C.L., Mani, D.R., Betensky, R.A., Tamayo, P., Cairncross, J.G., Ladd, C., Pohl, U., Hartmann, C., McLaughlin, M.E., Batchelor, T.T., Black, P.M., von Deimling, A., Pomeroy, S.L., Golub, T.R., and Louis, D.N., 2003, Gene expression-based classification of malignant gliomas correlates better with survival than histological classification, *Cancer Res*. 63: 1602-1607.

Obasi, P.C., Barnett, G.H., and Suh, J.H., 2002, Brachytherapy for intracranial meningioma using a permanently implanted iodine-125 seed, *Stereotact. Funct. Neurosurg.* 79: 33-43.

Oehring, R.D., Miletic, M., Valter, M.M., Pietsch, T., Neumann, J., Fimmers, R., und Schlegel, U., 1999, Vascular endothelial growth factor (VEGF) in astrocytic gliomas - a prognostic factor ? *J. Neuro-Oncol.* 45: 117-125.

Ohta, M., Iwaki, T., Kitamoto, T., Takeshita, I., Takeishi, J., and Fukui, M., 1994, MIB-1 staining index and scoring of histologic features in meningioma, *Cancer* 74: 3176-3188.

Ojemann, S.G., Sneed, P.K., Larson, D.A., Gutin, P.H., Berger, M.S., Verhey, L., Smith, V., Petti, P., Wara, W., Park, E., and McDermott, M.W., 2000, Radiosurgery for malignant meningioma: results in 22 patients, *J. Neurosurg.* 93: 622-627.

Ohnishi, A., Sawa, H., Tsuda, M., Sawamura, Y., Itoh, T., Iwasaki, Y., and Nagashima, K., 2003, Expression of the oligodendroglial lineage-associated makers Olig 1 and Olig 2 in different types of human gliomas, *J. Neuropathol. Exp. Neurol.* 62: 1052-1059.

Oliver, F.J., de la Rubia, G., Rolli, V., Ruiz-Ruiz, M.C., de Murcia, G., and Murcia, J.M., 1998, Importance of poly(ADP-ribose) polymerase and its cleavage in apoptosis. Lesson from an uncleavable mutant, *J. Biol. Chem.* 273: 33533-33539.

Olson, J.D., Riedel, E., and DeAngelis, L.M., 2000, Long-term outcome of low-grade oligodendroglioma and mixed glioma, *Neurology* 54: 1442-1448.

Packer RJ., 1999, Childhood medulloblastoma: progress and future. *Brain Dev.* 21: 75-81

Packer, R.J., Sutton, L.N., Rorke, L.B., Littman, P.A., Sposto, R., Rosenstock, J.G., Bruce, D.A., and Schut, L., 1984, Prognostic importance of cellular differentiation in medulloblastoma of childhood, *J. Neurosurg.* 61: 296-301.

Paglin, S., Hollister, T., Delohery, T., Hackett, N., McMahill, M., Sphicas, E., Domingo, D., Yahalom, J., 2001, A novel response of cancer cells to radiation involves autophagy and formation of acidic vesicles, *Cancer Res.* 61: 439-444.

Palma, L., 2003, Why do meningiomas recur? *J. Neurosurg. Sci.* 47: 65-68.

Papahill, P.A., Ramsay, D.A., and Del Maestro, R.F., 1996, Pleomorphic xanthoastrocytoma: case report and analysis of the literature concerning the effect of resection and the significance of necrosis, *Neurosurgery* 38: 822-828.

Parliament, M.B., Allalunis-Turner, M.J., Franko, A.J., Olive, P.L., Mandyam, R., Santos, C., and Wolokoff, B., 2000, Vascular endothelial growth factor expression is independent of hypoxia in human malignant glioma spheroids and tumours. *Br. J. Cancer*, 82: 635-641.

Park, K.I., Ourednik, L., Ourednik, V., Taylor, R.M., Aboody, K.S., Auguste, K.I., Lachyankar, M.B., Redmond, D.E., and Snyder, E.Y., 2002, Global gene and cell replacement strategies via stem cells, *Gene Ther.* 9: 613-624.

Pasquier, B., Gasnier, F., Pasquier, D., Keddari, E., Morens, A., and Coudere, P., 1986, Papillary meningioma: Clinicopathologic study of seven cases and review of the literature, *Cancer* 58: 299-305.

Patt, S., Gries, H., Girakìldo, M., Cervos-Navarro, J., Martin, H., Janisch, W., and Brockmoller, J., 1996, p53 gene mutations in human astrocytic brain tumors including pilocytic astrocytomas, *Hum. Pathol.* 27:586-589.

Patan, S., 2000, Vasculogenesis and angiogenesis as mechanisms of vascular network formation, growth and remodeling, *J. Neuro-Oncol.* 50: 1-15.

Patsouris, E., Davachi, P., Kapranos, N., Davaris, P., and Papageorgiou, K., 1996, A study of apoptosis in brain tumors by in situ end-labeling method, *Clin. Neuropathol.* 15: 337-341.

Patt, S., Schmidt, H., Labrakakis, C., Weydt, P., Fritsch, M., Cervos-Navarro, J., and Kettenmann, H., 1996, Human central neurocytoma cells show neuronal physiological properties in vitro, *Acta Neuropathol.* 91: 209-214.

Paulus, W., and Jellinger, K., 1992, Desmoplastic spindle-cell glioblastoma or gliosarcoma ? *Neuropathol. Appl. Neurobiol.* 18: 207-208.

Paulus, W., Lisle, D.K., Tonn, J.C., Wolf, H.K., Roggendorf, W., Reeves, S.A., and Louis, D.N., 1996, Molecular genetic alterations in pleomorphic xantho-astrocytoma, *Acta Neuropathol.* 91: 293-297.

Pearl, G.S., Takei, Y., Stefanis, G.S., and Hoffman, J.C., 1981, Intraventricular neuroblastoma in a patient with von Hippel-Lindau's disease. Light and electron microscopic study, *Acta Neuropathol.* 53: 253-256.

Pegg, A.E., and Byers, T.L., 1992, Repair of DNA containing O^6–alkylguanine, *FASEB J.* 6: 2302-2310.

Pereira, D.S., Dorrell, C., Ito, C.Y., Gan, O.I., Murdoch, B., Rao, V.N., Zou, J.P., Reddy, E.S., and Dick, J.E., 1998, Retroviral transduction of TLS-ERG initiates a leukemogenic program in normal human hematopèoietic cells, *PNAS USA* 95: 8239-8244.

Perry, A., 2002, Medulloblastomas with favorable versus unfavorable histology: how many small blue cell tumor types are there in the brain? *Adv. Anat. Pathol.* 9: 945-950.

Perry, A., Tonk, V., McIntire, D.D., and White, C.L., 1997, Interphase cytogenetic analysis of astrocytomas using archival, formalin-fixed, paraffin-embedded tissue and non-fluorescent light microscopy, *Am. J. Clin. Pathol.* 108: 166-174.

Perry, A., Stafford, S.L., Scheithauer, B.W., Suman, V.J., and Lohse, C.M., 1997, Meningioma grading. An analysis of histologic parameters, *Am. J. Surg. Pathol.* 21: 1455-1465.

Perry, A., Giannini, C., Scheithauer, B.W., Ajiani, A.M., Yachnis, A.T., Seo, I.S., Johnson, P.C., Kho, J., and Shapiro, S., 1997, Composite pleomorphic xanthoastrocytoma and ganglioglioma: report of four cases and review of the literature, *Am. J. Surg. Pathol.* 21: 763-771.

Perry, A., Stafford, S.L., Scheithauer, B.W., Suman, V.J., and Lohse, C.M., 1998, The prognostic significance of MIB-1, p53, and DNA flow cytometry in completely resected primary meningiomas, *Cancer* 82: 2262-2269.

Perry, A., Scheithauer, B.W., Stafford, S.L., Abell-Aleff, P.C., and Meyer, F.B., 1998, "Rhabdoid" meningioma: an aggressive variant, *Am. J. Surg. Pathol.* 22: 1482-1490.

Perry, A., Scheithauer, B.W., Stafford, S.L., Lohse, C.M., and Wollan, P.C., 1999, "Malignancy" in meningiomas. A clinicopathologic study of 116 patients, with grading implications, *Cancer* 85: 2046-2056.

Perry, A., Scheithauer, B.W., Macaulay, R.J., Raffel, C., Roth, K.A., and Kros, J.M., 2002, Oligodendrogliomas with neurocytic differentiation. A report of 4 cases with diagnostic and histogenetic implications, *J. Neuropathol. Exp. Neurol.* 61: 947-955.

Perry, A., Banerjee, R., Lohse, C.M., Kleinschmidt-DeMasters, K., and Scheithauer, B.W., 2002, A role for chromosome 9p21 deletions in the malignant progression of meningiomas and the prognosis of anaplastic meningiomas, *Brain Pathol.* 12: 183-190.

Perry, A., and Dehner, L.P., 2003, Meningeal tumors of childhood and infancy. An update and literature review, *Brain Pathol.* 13: 386-408.

Perry, A., Gutmann, D.H., and Reifenberger, G., 2004, Molecular pathogenesis of meningiomas, *J. Neuro-Oncol.* 70: 183-202.

Perry, A., Aldape, K.D., George, D.H., and Burger PC., 2004, Small cell astrocytoma: an aggressive variant that is clinicopathologically and genetically distinct from anaplastic oligodendroglioma, *Cancer* 101: 2318-2326.

Perry, J.R., Ang, L.C., Bilbao, J.M., and Muller, P.J., 1995, Clinicopathologic features of primary and postirradiation gliosarcoma, *Cancer* 75: 2910-2918.

Peterson, K., Paleologos, N., Forsyth, P., MacDonald, D.R., and Cairncross, J.G., 1996, Salvage chemotherapy for oligodendroglioma, *J. Neurosurg.* 85: 597-601.

Piccinini, M., Tazartes, O., Mostert, M., Musso, A., DeMarchi, M., and Rinaudo, M.T. 2000, Structural and functional characterization of 20S and 26S proteasomes in bovine brain, *Res. Mol. Brain Res.* 76: 103-114.

Piccinini, M., Mostert, M., Croce, S., Baldovino, S., Papotti, M., and Rinaudo, M.T., 2003, Interferon-gamma-inducible subunits are incorporated in human brain 20S proteasome, *J. Neuroimmunol.* 135: 135-14.

Piccinini, M., Rinaudo, M.T., Anselmino, A., Ramondetti, C., Buccinà, B., Fiano, V., Ghimenti, C., and Schiffer, D., 2005, Characterization of the 205 proteasome in human glioblastomas, *Anticancer Res.* 25:3203-3210.

Piepmeyer, J.M., 1987, Observations on the current treatment of low-grade astrocytic tumors of the cerebral hemispheres, *J. Neurosurg.* 67: 177-181.

Pietsch, T., Waha, A., Koch, A., Kraus J., Albrecht, S., Tonn, J., Sorensen, N., Berthold, F., Henk, B., Schmandt, N., Wolf, H.K., von Deimling, A., Wainwrigt, B., Chenevix-Trench, G., Wiestler, O.D., and Wicking, C., 1997, Medulloblastomas of the desmoplastic variant carry mutations of the human homologue *Drosophila patched*. *Cancer Res*. 1997, 57: 2085-2088.

Pietsch, T., Valter, M.M., Wolf, H.K., von Deimling, A., Huang, H.J., Cavenee, W.K., and Wiestler, O.D., 1997, Expression and distribution of vascular endothelial growth factor protein in human brain tumors. *Acta Neuropathol.* 93: 109-117.

Pilkington, G.J., and Lantos, P.L., 1979, The development of experimental brain tumors. A sequential light and electron microscope study of subependymal plate. II. Microtumors, *Acta Neuropathol.* 45: 177-185.

Pilkington, G.J., 1992, Glioma heterogeneity *in vitro*: the significance of growth factors and gangliosides, *Neuropathol. Appl. Neurobiol.* 18: 434-442.

Pilkington, G.J., 1994, Tumor cell migration in the central nervous system. *Brain Pathol.* 4: 157-166.

Piva, R., Cancelli, I., Cavalla, P., Bortolotto, S., Dominguez, J., Draetta, G., and Schiffer, D., 1999, Proteasome-dependent degradation of p27/Kip1 in gliomas, *J. Neuropathol. Exp. Neurol.* 58: 691-696.

Plate, K.H., Breier, G., Farrell, C.L., and Risau, W., 1992, Platelet-derived growth factor receptor-beta is induced during tumor progression in endothelial cells in human gliomas, *Lab. Invest.* 67: 529-534.

Plate, K.H., Breier, G., Weich, H.A., Mennel, H.D., and Risau, W., 1994, Vascular endothelial growth factor and glioma angiogenesis: coordinate induction of VEGF receptors, distribution of VEGF protein and possible in vitro regulatory mechanisms, *Int. J. Cancer* 59: 520-529.

Plate, K.H., 1999, Mechanisms of angiogenesis in the brain, *J. Neuropathol. Exp. Neurol.* 58: 313-320.

Plopper, G.E., McNamee, H.P., Dike, L.E., Bojanowski, K., and Ingber, D.E., 1995, Convergence of integrin and growth factor receptor signaling pathways within the focal adhesion complex, *Mol. Biol. Cell* 6: 1349-1365.

Pollock, B.E., Stafford, S.L., Utter, A., Giannini, C., and Schreiner, S.A., 2003, Stereotactic radiosurgery provides equivalent tumor control to Simpson Grade 1 resection for patients with small- to medium-size meningiomas *Int. J. Radiat. Oncol. Biol. Phys.* 55: 1000-1005.

Pomeroy, S.L., Tamayo, P., Gaasenbeek, M., Sturla, L.M., Angelo, M., McLaughlin, M.E., Kim, J.Y., Goumnerova, L.C., Black, P.M., Lau, C., Allen, J.C., Zazag, D., Olson, J.M., Curran, T., Wetmore, C., Biegel, J.A., Poggio, T., Mukherjee, S., Rifkin, R., Califano, A., Stolovitsky, G., Louis, DN., Mesirov, J.P., Lander, E.S., and Golub, T.R., 2003, Prediction of central nervous system embryonal tumour outcome based on gene expression, *Nature* 2002, 415: 436-442.

Pomeroy, S.L., and Sturla, L.-M., 2003, Molecular biology of medulloblastoma therapy, *Pediatr. Neurosurg.* 39: 299-304.

Popko, B., Pearl, D.K., Walker, D.M., Comas, T.C., Baerwald, K.D., Burger, P.C., Scheithauer, B.W., and Yates, A.J., 2002, Molecular markers that identify human astrocytomas and oligodendrogliomas, *J. Neuropathol. Exp. Neurol.* 61: 329-338.

Powell, S.Z., Yachnis, A.T., Rorke, L.B., Rojani, A.M., and Eskin, T.A., 1996, Divergent differentiation in pleomorphic xanthoastrocytoma. Evidence for a neuronal element and possible relationship to ganglion cell tumors, *Am. J. Pathol.* 20: 80-85.

Prayson, R.A., 1999, Clinicopathologic study of 61 patients with ependymoma including MIB-1 immunohistochemistry, *Ann. Diagn. Pathol.* 3: 11-18.

Prayson, R.A., 2000, Papillary glioneural tumor, *Arch. Pathol. Lab. Med.* 124: 1820-1823.

Prayson, R.A., 2002, Cell proliferation and tumors of the central nervous system. Part I: Evaluation of mitotic activity, *J. Neuropathol. Exp. Neurol.* 61: 501-509.

Prayson, R.A., and Suh, J.H., 1999, Subependymomas: clinicopathological study of 14 tumors, including comparative MIB-1 immunohistochemical analysis with other ependymal neoplasms, *Arch. Pathol. Lab. Med.* 123: 306-309.

Preusser, M., Dietrich, W., Czech, T., Prayer, D., Budka, H., and Hainfellner, J.A., 2003, Rosette-forming glioneural tumor of the fourth ventricle, *Acta Neuropathol.* 106: 506-508.

Preusser, M., Birner, P., and Heinfellner, J.A., 2004, Algorythm for the standardized assessment of vascular patterns in glioblastoma specimens, *Clin. Neuropathol.* 23: 238-240.

Provias, J., Claffey, K., DelAguila, L., Nelson, L., Feldkamp, M., and Uha, A., 1997, Meningiomas: Role of vascular endothelial growth factor/vascular permeablity factor in angiogenesis and peritumoral edema, *Neurosurgery*. 40: 1016-1026.

Puchner, M.J., Hermann, H.D., Berger, J., and Cristante, L., 2000, Surgery, tamoxifen, carboplatin, and radiotherapy in the treatment of newly diagnosed glioblastoma patients, *J. Neuro-Oncol.* 49: 147-155.

Puduvalli, V.K., Hashmi, M., McAllister, L.D., Levin, V.A., Hess, K.R., Prados, M., Jaeckle, K.A., Yung, W.K., Buys, S.S., Bruner, J.M., Townsend, J.J., Davis, R., Sawaya, R., and Kyritsis, A.P., 2003, Anaplastic oligodendrogliomas: prognostic factors for tumor recurrence and survival, *Oncology* 65: 259-266.

Purow, B.W., Haque, R.M., Noel, M.W., Su, Q., Burdick, M.J., Lee, J, Sundaresan, T., Pastorion, S., Park, J.K., Mikolaenko, I, Marie, D., Eberhart, C.G., and Fine, H.A., 2005, Expression of Notch-1 and its ligands, Delta-like-1 and Jagged-1, is critical for glioma cell survival and proliferation, *Cancer Res.* 65: 2353-2363.

Quan, A.L., Barnett, G.H, Lee, S.Y., Vogelbaum, M.A., Toms, S.A., Staugaitis, S.M., Prayson, R.A., Peereboom, D.M., Stevens, G.H., Cohen, B.H., and Sub, J.H., 2005, Epidermal growth factor receptor amplification does not have prognostic significance in patients with glioblastoma multiforme, *Int. J. Radiat. Oncol. Biol. Phys.* 2005 (Publ. ahead of print).

Quinn, B., 1998, Synaptophysin staining in normal brain: importance for diagnosis of ganglioglioma, *Am. J. Surg. Pathol.* 22: 550-556.

Rades, D., Fehlauer, F., Schild, S., Lamszus, K., and Alberti, W., 2003, Treatment for central neurocytoma. A meta-analysis based on data of 358 patients, *Strahlk. Enther. Onkol.* 179: 213-218.

Raff, M.C., 1989, Glia cell diversification in the rat optic nerve, *Science,* 243: 1450-1455.

Raghavan, R., Balani, J., Perry, A., Margraf, L., Vono, M.B., Cai, D.X., Wyatt, R.E., Rushing, E.J., Bowers, D.C., Hynan, L.S., and White, III C.L., 2003, Pediatric oligodendrogliomas: a study of molecular alterations on 1p and 19q using fluorescence in situ hybridization, *J. Neuropahol. Exp. Neurol.* 62: 530-537.

Ralte, A.M., Sharma, M.C., Karak, A.K., Mehta, V.S., and Sarkar, C., 2001, Clinicopathologial features, MIB-1 labeling index and apoptotic index in recurrent astrocytic tumors, *Pathol. Oncol. Res.* 7: 267-278.

Rao, G., Pedone, C.A., Coffin, C.M., Holland, E.C., and Fult, D.W., 2003, C-Myc enhances sonic hedgehog-induced medulloblastoma formation from nestin-expressing neural progenitors in mice, *Neoplasia* 5: 198-204.

Rao, G., Pedone, C.A., Valle, L., Reiss, K., Holland, E,C,, and Fults, D.W., 2004, Sonic hedgehog and insulin-like growth factor signaling sinergize to induce medulloblastoma formation from nestin-expressing neural progenitors in mice, *Neoplasia*, 12: 6156-6162.

Rasheed, B.K., McLendon, R.E., Herndon, J.E., Friedman, H.S., Friedman, A.H., Bigner, D.D., and Bigner, S.H.. 1994, Alterations of TP53 gene in human gliomas, *Cancer Res*. 54: 324-330.

Rawlings, C.E., Giangaspero, F., Burger, P.C., and Bullard, D.E., 1988, Ependymomas: a clinico-pathologic study, *Surg. Neurol*. 29: 271-281.

Ray, S.K., Patel, S.J., Welsh, C.T., Wilford, G.G., Hogan, E.L., and Banik, N.L., 2002, Molecular evidcence of apoptotic death in malignant brain tumor including glioblastoma multiforme: upregulation of calpain and caspase-3, *J. Neurosci. Res*. 15: 197-206.

Raza, S.M., Lang, F.F., Aggarwal, B.B., Fuller, G.N., Wildrick, D.M., and Awaya, R., 2002, Necrosis and glioblastoma: a friend or a foe? A review and a hypothesis, *Neurosurgery* 51: 2-13.

Reavey-Cantwell, J.F., Haroun, R.I., Zahurak, M., Clatterbuck, R.E., Parker, R.J., Mehta, R., Fruehauf, J.P., and Brem, H., 2001, The prognostic value of tumor markers in patients with glioblastoma multiforme: analysis of 32 patients and review of the literature, *J. Neuro-Oncol*. 55: 195-204.

Recht, L., Jang, T., and Litofsky, N.S., 2003, Neural stem cells and neuro-oncology: Quo vadis? *J. Cell Biochem*. 88: 11-19.

Reed, J., 1997, Double identity for proteins of the Bcl-2 family, *Nature* 387: 773-776.

Rees, J., 2003, Advances in magnetic resonance imaging of brain tumors, *Curr. Opin. Neurol*. 16: 643-650.

Reifenberger, J., Reifenberger, G., Liu, L., James, C.D., Wechsler, W., and Collins, V.P., 1994, Molecular genetic analysis of oligodendroglial tumors shows preferential allelic deletions on 19q and 1p, *Am. J. Pathol*. 145: 1175-1190.

Reifenberger, G., Kros, J.M., Burger, P.C., Louis, D.N., and Collins, V.P., 2000, Oligodendroglioma. *Pathology and genetics. Tumours of the nervous system.* IARC, Lyon, pp 56-64.

Reifenberger, G., and Louis, D.N., 2003, Oligodendroglioma: toward molecular definitions in diagnostic neuro-oncology, *J. Neuropathol. Exp. Neurol*. 62: 111-126.

Reifenberger, G., Kaulich, K., Wiestler, O.D., and Blumkee, I., 2003, Expression of the CD34 antigen in pleomorphic xanthoastrocytomas, *Acta Neuropathol.* 105: 358-364.

Reis, R.M., Könü-Lebleblicioglu, D., Lopes, J.M., Kleihues, P,, and Ohgaki, H., 2000, Genetic profile of gliosarcomas, *Am. J. Pathol*. 156: 425-432.

Reya, T., Morrison, S.J., Clarke, M.F., and Weissman, I.L., 2001, Stem cells, cancer, and cancer stem cells, *Nature* 414: 105-111.

Reynolds, B.A., and Weiss, S., 1992, Generation of neurons and astrocytes from isolate cells of the adult mammalian nervous system, *Science* 255: 1707-1710.

Rhodes, R.H., 1998, Biological evaluation of biopsies from adult cerebral astrocytomas: cell-growth/cell-suicide ratios and their relationship to patient survival, *J. Neuropathol. Exp. Neurol*. 57: 746-756.

Rice, W.G., Hillyer, C.D., Harten, B., Schaeffer, C,A,, Dorminy, M., Lackey, D.A. 3rd, Kirsten, E., Meneleyev, J., Buki, K.G., Hakam, A. et al. 1992, Induction of endonuclease-mediated apoptosis in tumor cells by C-nitroso-substituted ligands of poly(ADP-ribose) Polymerase, *PNAS USA* 89: 7703-7707.

Rich, T., Allen, R.L., and Wyllie, A.H., 2000, Defying death after DNA damage, *Nature* 407: 777-782.

Riemenschneider, M., Koy, T.H., and Reifenberger, G., 2004, Expression of oligodendrocyte lineage genes in oligodendroglial and astrocytic gliomas, *Acta Neuropathol.* 107: 277-282.

Robbins, P., Segal, A., Narula, S., Stokes, B., Lee, M., Thomas, W., Caterina, P., Sinclair, I., and Spagnolo, D., 1995, Central neurocytoma. A clinicopathological, immunohistochemical and ultrastructural study of 7 cases, *Pathol. Res. Pract.* 191: 100-111.

Roessler, K., Bertalanffy, A., Jezan, H., Ba-Ssalamah, A., Slavc, I., Czech, T., and Budka, H., 2002, Proliferative activity as measured by MIB-1 labeling index and long-term outcome of cerebellar juvenile pilocytic astrocytoma, *J. Neuro-Oncol.* 58: 141-146.

Rorke, L.B., 1983, The cerebellar medulloblastoma and its relationship to primitive neuroectodermal tumors, *J. Neuropathol. Exp. Neurol.* 42: 1-15,

Rosenblum, M.L., 1995, Factors influencing tumor cell traffic in the central nervous system, *Surg. Neurol.* 43:595.

Ross, G.W., and Rubinstein, L.J., 1989, Lack of histopathological correlation of malignant ependymomas with postoperative survival, *J. Neurosurg.* 70: 31-36.

Rosso, S.M., Van Dekken, H., Krisnadath, K.K., Alers, J.C., and Kros, J.M., 1997, Detection of chromosomal changes by interphase cytogenetics in recurrent astrocytomas and oligodendrogliomas, *J. Neuropathol. Exp. Neurol.* 56: 1125-1131.

Roth, W., and Weller, M., 1999, Chemotherapy and immunotherapy of malignant glioma: molecular mechanisms and clinical perspectives, *Cell Mol. Life Sci.* 56: 481-506.

Rousseau, E., Ruchoux, M.-M., Scaravilli, F., Chapon, F., Vinchon, M., De Smet, C., Godfraind, C., and Vikkula, M., 2003, CDKN2A, CDKN2B and $p14^{ARF}$ are frequently and differentially methylated in ependymal tumours, *Neuropathol. Appl. Neurobiol.* 29: 574-583.

Rubinstein, L.J., 1985, Embryonal central neuroepithelial tumors and their differentiating potential: a cytogenetic view of a complex neuro-oncological problem, *J. Neurosurg.* 62: 795-805.

Rubinstein, L.J., 1987, The correlation of neoplastic vulnerability with central neuroepithelial cytogeny and glioma differentiation, *J. Neuro-Oncol.* 5: 11-27.

Rubinstein, L.J, and Herman, M.M., 1989, The astroblastoma and its possible cytogenetic relationship to the tanycyte: an electron microscopic, immunohistochemical, tissue and organ-culture study, *Acta Neuropathol.* 78: 472-483.

Rushing, E.J., Yashima, K., Brown, D.F., White, C.L., Shay, J.W., Risser, R.C., and Grazdar, A.F., 1997, Expression of telomerase RNA component correlates with the MIB-1 proliferation index in ependymomas, *J. Neuropathol. Exp. Neurol.* 56: 1142-1146.

Rushing, E.J., Colvin, S.M., Gazdar, A., Miura, N., White, C.L. 3rd, Coimbra, C., and Burns, D.K., 1999, Prognostic value of proliferation index and expression of the RNA component of human telomerase (hTR) in papillary meningiomas, *J. Neuro-Oncol.* 45: 199-207.

Russell, D.S., and Rubinstein, L.J., 1989, *Pathology of tumours of the nervous system.*, 5th Ed, Arnold, London.

Rutka, J.T., Ivanchuk, S., Mondal, S., Taylor, M., Sakai, K., Dirks, P., Jun, P., Becker, L.E., and Ackerley, C., 1999, Co-expression of nestin and vimentin intermediate filaments in invasive human astrocytoma cells, *Int. J. Dev. Neuroscience* 17: 503-515.

Saito, R., Mizuno, M., Kumabe, T., Yoshimoto, T., Tanuma, S., and Yoshida, J., 2003, Apoptotic DNA endonuclease (DNAase-γ) gene transfer induces cell death accompanying DNA fragmentation in human glioma cells, *J. Neuro-Oncol.* 63: 25-31.

Salazar, O.M., 1981, Primary malignant cerebellar astrocytomas in children: a signal for postoperative craniospinal irradiation, *Int. J. Radiat. Oncol. Biol. Phys.* 7: 1661-1665.

Salsano, E., Pollo, B., Eoli, M., Giordana, M.T., Finocchiaro, G., 2004, Expression of *MATH1*, a marker of cerebellar granule cell progenitors, identifies different medulloblastomaa sub-types, *Neurosci. Lett.* 370: 180-185.

Samoto, K., Kyonobu, I., Mayumi, O., Tadhisa, S., Kimitoshi, K., Kuwano, M., and Fukui, M., 1995, Expression of vascular endothelial growth factor and its possible relation with neovascularization in brain tumors, *Cancer Res.* 55: 1189-1193.

Sanoudou, D., Tingby, O., Ferguson-Smith, M.A., Collins, V.P., and Coleman, N., 2000, Analysis of pilocytic astrocytoma by comparative genomic hybridization, *Brit. J. Cancer* 82: 1218-1222.

Sampson, J.H., Crotty, L.E., Lee, S., Archer, G.E., Ahsley, D.M., Wikstrand, C.J., Hale, L.P., Small, C., Dranoff, G., Friedman, A.H., Friedman, H.S., and Bigner, D.D., 2000, Unarmed, tumor-specific monoclonal antibody effectively treats brain tumors, *PNAS USA* 97: 7503-7508.

Sarkar, C., Pramanik, P., Karak, A.K., Mukhopadhyay, P., Sharma, M.C., Singh, V.P., and Mehta, V.S., 2002, Are childhood and adult medulloblastoma different? A comparative study of clinicopathological features, proliferation index and apoptotic index, *J. Neuro-Oncol.* 59: 49-61.

Sasaki, H., Zlatescu, M.C., Betensky, R.A., Ino, Y., Cairncross, J.R., and Louis, D.N., 2001, PTEN is a target of chromosome 10q loss in anaplastic oligodendrogliomas and PTEN alterations are associated with poor prognosis, *Amer. J. Pathol.* 159: 359-367.

Sasaki, H., Zlatescu, M.C., Betensky, R.A., Johnk, L.B., Cutone, A.N., Cairncross, J.G., and Louis, D.N., 2002, Histopathological-molecular genetic correlation in referral pathologist-diagnosed low-grade "oligodendroglioma", *J. Neuropathol. Exp. Neurol.* 61: 58-63.

Sasaki, T., Lopes, N.B.S., Hankins, G.R., and Helm, G.A., 2002, Expression of survivin, an inhibitor of apoptosis protein, in tumors of the nervous system, *Acta Neuropathol.* 104: 105-109.

Sato, K., Schauble, B., Kleihues, P., and Ohgaki, H., 1996, Infrequent alterations of the p15, p16, CDK4 and cyclin D1 genes in non-astrocytic human brain tumors, *Intern. Cancer* 66: 305-308.

Sato, K., Kubota, T., Ishida, M., Yoshida, K., Takeuchi, H., and Handa, Y., 2003, Immunohistochemical and ultrastructural study of chordoid glioma of the third ventricle: its tanycytic differentiation, *Acta Neuropathol.* 106:176-180.

Scheil, S., Brüderlein, S., Eicker, M., Herms, J., Herold-Mende, C., Steiner, H.H., Barth, T.F., and Möller, P., 2001, Low frequency of chromosomal imbalances in anaplastic ependymomas as detected by comparative genomic hybridization, *Brain Pathol.* 11: 133-143.

Schiebe, M., Ohneseit, P., Hoffmann, W., Meyermann, R., Rodemann, H.P., and Bamberg, M., 2000, Analysis of mdm2 and p53 gene alteration in glioblastoma and its correlation with clinical factors, *J Neuro-Oncol.* 49: 197-203.

Schiffer, D., Fabiani, A., Monticone, G.F., and Cognazzo, A., 1966, On the nature of the lymphocyte-like cells of medulloblastoma, *Acta Neuropathol.* 6: 290-297.

Schiffer, D., Fabiani, A., Grossi-Paoletti, E., and Paoletti, P., 1970, Experimental brain tumors induced in rats by nitrosourea dertivatives. Part I. Morphological aspects of methylnitrosourea tumours, *J. Neurol. Sci.* 11: 559-572.

Schiffer, D., Fabiani, A., 1971, Contribution of histochemistry to knowledge of brain tumor metabolism. *The experimental biology of brain tumors*. W.M. Kirsch, E. Grossi-Paoletti and P. Paoletti, Eds, Thomas, Sprinfield Ill.

Schiffer, D., Giordana, M.T., Pezzotta, S., Lechner, C., Paletti, P., 1978, Cerebral tumors induced by transplacental ENU: study of the different tumoral stages, particularly of early proliferations. *Acta Neuropathol.* 41: 27-31.

Schiffer, D., Giordana, M.T., Mauro, A., Racagni, G., Bruno, F., Pezzotta, S., Paoletti, P., 1980, Experimental brain tumors by transplacental ENU. Multifactorial study of the latency poeriod. *Acta Neuropathol.* 49: 117-122.

Schiffer, D., Giordana, M.T., Mauro, A., and Migheli, A., 1983, Glial fibrillary acidic protein (GFAP) in human cerebral tumors. An immunoihistochemical study, *Tumori* 69: 95-104.

Schiffer, D., Giordana, M.T., Mauro, A., and Migheli, A., 1984, GFAP, FVIII/Rag, laminin and fibronectin in gliosarcomas: an immunohistochemical study, *Acta Neuropathol.* 63: 108-116.

Schiffer, D. 1986, Neuropathology and imaging: the ways in which glioma spreads and varies in its histological aspect. M.D. Walker, and D.G.T. Thomas Eds, *Biology of brain tumors,* Nijhoff, Boston.

Schiffer, D., Chiò, A., Giordana, M.T., Leone, M., Soffietti, R., 1988, Prognostic value of histologic factors in adult cerebral astrocytoma, *Cancer* 61: 1386-1393.

Schiffer, D., Chiò, A., Mauro, A., Migheli, A., and Vigliani, M.C., 1989, The vascular response to tumor infiltration in malignant gliomas, *Acta Neuropathol.* 77: 369-378.

Schiffer, D., Chiò, A., Giordana, M.T., Mauro, A., Migheli, A., Soffietti, R., and Vigliani, M.C., 1990, Vascular response to irradiation in malignant gliomas, *J. Neuro-Oncol.* 8: 73-84.

Schiffer, D., Chiò, A., Giordana, M.T., Migheli, A., Palma, L., Pollo, B., Soffietti, R., and Tribolo, A., 1991a, Histologic prognostic factors in ependymomas, *Child's Nerv. Syst.* 7: 177-182.

Schiffer, D., Chiò, A., Cravioto, H., Giordana, M.T., Migheli, A., Soffietti, R., and Vigliani, MC., 1991b, Ependymoma: internal correlations among pathologic signs and the anaplastic variant, Neurosurgery, 29: 206-210.

Schiffer, D., 1991c, Patterns of tumor growth. *Neurobiology of brain tumors,* M. Salcman Ed, Williams and Wilkins, Baltimore, 85-135.

Schiffer, D., Giordana, M.T., Pezzotta, S., Pezzulo, T., and Vigliani, M.C., 1992, Medullomyoblastoma: report of two cases, *Child's Nerv. Syst.* 8: 268-272.

Schiffer, D., Chiò, A., Giordana, M.T., Pezzulo, T., and Vigliani, M.C., 1993, Proliferating cell nuclear antigen expression in brain tumors, and its prognostic role in ependymomas: an immunohistochemical study, *Acta Neuropathol.* 85: 495-502.

Schiffer, D., Cavalla, P., Chiò, A., Giordana, M.T., Marino, S., Mauro, A., and Migheli, A., 1994, Tumor cell proliferation and apoptosis in medulloblastoma. *Acta Neuropathol.* 87: 362-370.

Schiffer, D., Cavalla, P., Migheli, A., Chiò, A., Giordana, M.T., Marino, S., and Attanasio, A., 1995, Apoptosis and cell proliferation in human neuroepithelial tumors, *Neurosci. Lett.* 195: 81-84.

Schiffer, D., Cavalla, P., Migheli, A., Chiò, A., Giordana, M.T., and Chiadò-Piat, L., 1996, Bcl-2 distribution in neuroepithelial tumors: an immunohistochemical study, *J. Neuro-Oncol.* 27: 101-109.

Schiffer, D., Cavalla, P., Dutto, A., and Borsotti, L., 1997, Cell proliferation and invasion in malignant gliomas, *Anticancer. Res.* 17: 61-70.

Schiffer, D., Dutto, A., Cavalla, P., Bosone, I., Chiò, A., Villani, R., and Bellotti, C., 1997, Prognostic factors in oligodendroglioma, *Can. J. Neurol. Sci.* 24: 313-319.

Schiffer, D., Dutto, A., Cavalla, P., Chiò, A., Migheli, A., and Piva, R., 1997, Role of apoptosis in the prognosis of oligodendrogliomas, *Neurochem. Int.* 2: 245-250.

Schiffer, D., 1997, *Brain Tumors. Biology, Pathology and Clinical Correlates*, Springer, Heidelberg-Berlin-New York.

Schiffer, D., and Cavalla, P., 1998, Proliferation properties of malignant brain tumors. *Brain tumor invasion. Biological, clinical and therapeutic considerations,* T. Mikkelsen, R. Bjervieg, O.D. Laerum and M.L. Rosenblum Eds, John Wiley - Liss, New York.

Schiffer, D., Bosone, I., Dutto, A., Di Vito, N., and Chiò, A., 1999, The prognostic role of vessel productive changes and vessel density in oligodendroglioma, *J. Neuro-Oncol.* 44: 99-107.

Schiffer, D., Ferracini, R., and Cavalla, P., 2001, Heterogeneity of apoptotic pathways and c.Jun/JUNK expression in malignant gliomas, *Anticancer. Res.* 21: 2531 manca.

Schiffer, D., Fiano, V., Chiadò-Piat, L., Mortara, P., Richiardi, P., and Cavalla, P., 2001, Distribution of activated caspase-3 in relation with apoptosis in human malignant gliomas, *Neurosci. Lett.* 300: 37-40.

Schiffer, D., Fiano, V., and Ghimenti, C., 2002, Apoptosis and its meaning in brain tumors, *Curr. Topics in Neurochem.* 3: 2-18.

Schiffer, D., Piccinini, M., Fiano, V., Anselmino, A., and Rinando, M.T., 2004, Proteins targeted by ubiquitin and the ubiquitin-proteasome system in malignant gliomas. *Anticancer Res.* 24: 3622 (abstract).

Schiffer, D., Ghimenti, C., and Fiano, V., 2005, Absence of histological signs of tumor progression in recurrencies of completely resected meningiomas. *J. Neuro-Oncol.* 73: 125-130.

Schmechel, D.E., and Rakic, P., 1979, A Golgi study of radial glia cells in developing monkey telencephalon: Morphogenesis and transformation into astrocytes, *Anat. Embryol.* 156: 115-152.

Schmidt, M.C., Antweiler, S., Urban, N., Mueller, W., Kuklik, A., Meyer-Puttlitz, B., Wiestler, O.D., Louis, D.N., Fimmers, R., and von Deimling, A., 2002, Impact of genotype and morphology on the prognosis of glioblastoma, *J. Neuropathol. Exp. Neurol.* 61: 321-328.

Schmidt, M.H., Gottfried, O.N., von Koch, C.S., Chang, S.M., and McDermott, M.W., 2004, Central neurocytoma: a review, *J. Neuro-Oncol.* 66: 377-384.

Schmitz, U., Mueller, W., Weber, M., Sevenet, N., Delattre, O., and von Deimling, A., 2001, INI1 mutations in meningiomas at a potential hotspot in exon 9, *Br. J. Cancer* 84: 199-201.

Schroder, R., Feisel, K.D., and Ernestus, R.I., 2002, Ki-67 labeling is correlated with the time to recurrence in primary glioblastomas, *J. Neuro-Oncol.* 56: 127-132.

Schwechheimer, K., Huang, S., and Cavenee, W.K., 1995, EFGR gene amplification-rearrangement in human glioblastomas, *Int. J. Cancer* 62: 145-148.

Shankar, S.L., Mani, S., O'Guin, K.N., Kandimalla, E.R., Agrawal, S., Shafit-Zagardo, B., 2001, Survivin inhibition induces human neural tumor cell death through caspase-independent and - dependent pathways, *J. Neurochem.* 79: 426-436.

Shapiro, J.R., and Shapiro, W.R., 1984, Clonal tumor cell heterogeneity, *Progr. Exp. Tumor Res.* 27: 49-66.

Sgonc, R., and Wick, G., 1994, Methods for the detection of apoptosis, *Int. Arch. Allergy Immunol.* 105: 327-332.

Shaw, G., Daumas-Duport, C., Scheithauer, B.W., Gilbertson, D.T., O'Fallon, J.R., Earle, J.D., Laws, E.R. Jr, and Okazaki, H., 1989, Radiation therapy in the management of low-grade supratentorial astrocytomas, *J. Neurosurg.* 70: 853-861.

Shaw, E.G., Scheithauer, B.W., O'Fallon, J.R., Tazelaar, H.D., and Davis, D.H., 1992, Oligodendrogliomas. The Mayo Clinic experience, *J. Neurosurg.* 76: 428-434.

Shaw, E., Arusell, R., Scheithauer, B., O'Fallon, J., O'Neill, B., Dinapoli, R., Nelson, D., Earle, J., Jones, C., Cascino, T., Nichols, D., Ivnik, R., Hellman, R., Curran, W., and Abrams, R., 2002, Prospective randomized trial of law- versus high-dose radiation therapy in adults with supratentorial low-grade glioma: initial report of a North Central Cancer Treatment Group/Radiation Therapy Oncology Group/Eastern Cooperative Oncology Group study, *J. Clin. Oncol.* 20: 2267-2276.

Shaw, E.G., and Wisoff, J.H., 2003, Prospective clinical trials of intracranial low-grade glioma in adult and children, *Neuro-Oncol.* 5: 153-160.

Shibamoto, Y., Kitakabu, Y., Takahashi, M., Yamashita, J., Oda, Y., Kikuchi, H., and Abe, M., 1993, Supratentorial low-grade astrocytoma: correlation of computed tomography findings with effect of radiation therapy and prognostic variables, *Cancer* 72: 190-195.

Shih, A.H., and Holland, A.C., 2004, Develpmental neurobiology and the origin of brain tumors, *J. Neuro-Oncol.* 70: 125-136.

Shin, W.-Y., Laufer, H,, Lee, Y.-C., Aftalion, B., Hirano, A., and Zimmerman, H.M., 1978, Fine structure of a cerebellar neuroblastoma, *Acta Neuropathol.* 42: 11-13.

Shin, J.H., Lee, H.K., Lee, J.K., Khang, S.K., Choi, C.G., and Suh, D.C., 2002, MR imaging and histopathologic findings of a case of cerebral ganglioneurocytoma, *Kor. J. Radiol.* 3: 214-217.

Shutz, B.R., Scheurlen, W., Krauss, J., du Manoir, S., Joos, S., Bentz, M., and Lichter, P., 1996, Mapping of chromosomal gains and losses in primitive neuroectodermal tumors by comparative genomic hybridization, *Genes Chromosomes Cancer* 16:196-203.

Sidransky, D., Mikkelsen, T., Schwechheimer, K., Rosenblum, M,L., Cavenee, W.K., and Vogelstein, B., 1992, Clonal expansion of p53 mutant cells is associated with brain tumor progression, *Nature* 355: 846-847.

Sieg, D.J., Hauck, C.R., Ilic, D., Klingbell, C.K., Schaefer, E., Damsky, C.H., and Schlaepfer, D.D., 2000, FAK integrates growth-factor and integrin signals to promote cell migration, *Nat. Cell Biol.* 2: 249-256.

Siegers, H.P., Zuber, P., Hamou, M.F., van Melle, G.D., and de Tribolet, N., 1989, The implications of the heterogeneous distribution of Ki-67 labelled cells in meningiomas, *Br. J. Neurosurg.* 3: 101-117.

Silber, J.R., Blanck, A., Bobola, M.S., Mueller, B.A., Kofstoe, D.D., Ojemann, G.A., and Berger, M.S., 1996, Lack of the DNA repair protein O^6-methylguanine-DNA-methyltransferase in histologically normal brain adjacent to primary human brain tumors, *PNAS USA* 93: 6941-6946.

Simmons, M.L., Lamborn, K.R., Takahashi, M., Chen, P., Israel, M.A., Berger, M.S., Godfrey, T., Nigro, J., Prados, M., Chang, S., Barker, F.G. 2nd, and Aldape, K., 2003, Analysis of complex relationships between age, p53, epidermal growth factor receptor, and survival in glioblastoma patients, *Cancer Res.* 61: 1122-1128.

Simon, M., Park, T.W., Koster, G., Mahlberg, R., Hackenbroch, M., Boström, J., Loning, T., and Schramm, I., 2001, Alterations of INK4a(p16-p14ARF)/ INK4b(p15) expression and telomerase activation in meningioma progression, *J. Neuro-Oncol.* 55: 149-158.

Singh, S.K., Clarke. I.D., Terasaki, M., Bonn, V.E., Hawkins, C., Squire, J., and Dirks, P.B., 2003, Identification of a cancer stem cell in human brain tumors, *Cancer Res.* 63: 5821-5828.

Singh, S.K., Clarke, I.D., Hide, T., and Dirks, P.B., 2004, Cancer stem cells in nervous system tumors, *Oncogene* 23: 7267-7273.

Singh, S.K., Hawkins, C., Clarke, I.D., Squire, J.A., Bayani, J., Hide, T., Henkelman, R.M.,Cusimano, M.D., and Dirks, P.B., 2004, Identification of human brain tumour initiating cells, *Nature* 42: 396-400.

Slowik, F., Jellinger, K., Gaszo, L., and Fischer, J., 1985, Gliosarcomas: histological, immunohistochermical, ultrastructural, and tissue cultures studies, *Acta Neuropathol.* 67: 201-210.

Small, R.K., Riddle, P., Noble, M., 1987, Evidence for migration of oligodendrocyte-type 2 astrocyte progenitor cells into developing rat optic nerve, *Nature* 328: 155-157.

Smith, M.T., Ludwig, C.L., Godfrey, A.D., and Ambrustmacher, W., 1983, Grading of oligodendrogliomas, *Cancer* 52: 2107-2114.

Smith, T.W., and Davidson, R.I., 1984, Medullomyoblastoma: A histologic, immunohistochemical and ultrastructural study, *Cancer* 54: 323-332.

Smith, J.S., Alderete, B., and Minn, Y., 1999, Localization of common deletion regions on 1p and 19q in human gliomas and their association with histological subtypes, *Oncogene* 18: 4144-4152.

Smith, J.S., Perry, A., Borell, T.J., Lee, H.K., O'Fallon, J., Hosek, S.M., Kimmel, D., Yates, A., Burger, P.C., Scheithauer, B.W., and Jenkins, R.B., 2000, Alterations of chromosome arms 1p and 19q as predictors of survival in oligodendrogliomas, astrocytomas and mixed oligoastrocytomas, *J. Clin. Oncol.* 18: 636-645.

Smith, J.S., Tachibana, I., Passe, S.M., Huntley, B.K., Borell, T.J., Iturria, N., O'Fallon, J.R., Schaefer, P.L., Scheithauer, B.W., James, C.D., Buckner, J.C., Jenkins, R.B., 2001, PTEN mutation, EGFR amplification, and outcome in patients with anaplastic astrocytoma and glioblastoma multiforme, *J. Natl. Cancer Inst.* 93: 1246-1256.

Song, J.H., Song, D.K., Pyrzynska, B., Petruk, K.C., Van Meir, E.G., and Hao, C., 2003, TRAIL triggers apoptosis in human malignant glioma cells through extrinsic and intrinsic pathways, *Brain Pathol.* 13: 539-553.

Sonoda, Y., Ozawa, T., Aldape, K.D., Deen, D.F., Berger, M.S., and Pieper, R.O., 2002, Akt pathway activation converts anaplastic astrocytoma to glioblastoma multiforme in a human astrocyte model of glioma, *Cancer Res.* 61: 6674-6678.

Soylemezoglu, F., Kleihues, P., Esteve, J., and Scheithauer, B.W., 1997, A typical central neurocytoma, *J. Neuropathol. Exp. Neurol.* 56: 551-556.

Soylemezoglu, F., Onder, S., Tezel, G.G., and Berker, M., 2003, Neuronal nuclear antigen (NeuN): a new tool in the diagnosis of central neurocytoma, *Pathol. Res. Pract.* 199: 463-468.

Stafford, S.L., Perry, A., Suman, V.J., Meyer, F.B., Scheithauer, B.W., Lohse, C.M., and Shaw, E,G., 1998, Primarily resected meningiomas: outcome and prognostic factors in 581 Mayo Clinic patients, 1978 through 1988, *Mayo Clin. Proc.* 73: 936-942.

Stafford SL, Pollock BE, Foore RL, Link MJ, Gorman DA, Schomberg PJ, Leavitt JA. 2001, Meningioma radiosurgery: tumor control, outcomes, and complications among 190 consecutive patients. *Neurosurg* 49: 1029-1037.

Steinbach JP, Supra P, Huang HJ, Cavenee WK, Weller M. 2002, CD-95-mediated apoptosis of human glioma cells: modulation by epidermal growth factor. *Brain Pathol* 12: 12-20.

Steinbach, J.P., and Weller, M., 2002, Mechanisms of apoptosis in central nervous system tumors: application to theory, *Curr. Neurol. and Nerosci. Rep.* 2: 246-253.

Steindler, D.A., and Laywell, E.D., 2003, Astrocytes as stem cells: nomenclature, phenotype, and translation, *Glia* 43: 62-69.

Stohwasser, R., Giesebrecht, J., Kraft, R., Muller, E.C., Hausler, K.G., Kettenmann, H., Hanisch, U.K., and Kloetzel, P.M., 2000, Biochemical analysis of proteasomes from mouse microglia: induction of immunoproteasomes by interferon-γ and lipopolysaccharide, *Glia* 29: 355-365.

Streffer, J., Schabet, M., Bamberg, M., Grote, E.H., Meyermann, R., Voigt, K., and Dichgans, J., and Weller, M., 2000, A role for pre-irradiation PCV chemotherapy for oligodendroglial brain tumors, *J. Neurol.* 247: 297-302.

Strik, H., Deininger, M., Streffer, J., Grote, E., Wickboldt, J., Dichgans, J., Weller, M., and Metermann, R., 1999, BCL-2 family protein expression in initial and recurrent glioblastomas: modulation by radiochemotherapy, *J. Neurol. Neurosurg. Psychiatry* 67: 763-768.

Stroebe, H., 1895, Uber Entstehung und Bau der Hirngliome, *Beitr. Path. Anat. (Jena)* 18: 405-486.

Stupp, R., Janzer, R.C., Hegi, M.E., Villemure, J.G., and Mirimanoff, R.O., 2003, Prognostic factors for low-grade gliomas, *Semin. Oncol.* 3: 23-28.

Stupp, R., Mason. W.P., Van den Bent, M.J., Weller, M., Fischer, B., Taphoorn, M.J., Belanger, K., Brandes, A.A., Marosi, C., Bogdahn, U., Curschmann, J., Janzer, R.C., Ludwin, S.K., Gorlia, T., Allgeier, A., Lacombe, D., Cairncross, J.G., Eisenhauer, E., Mirimanoff, R.O., 2005, Radiotherapy plus concomitant and adjuvant temozolomide for glioblastoma, *N. Engl. J. Med.* 352: 1036-1038.

Sugawara, K.-I., Kurihara, H., Negishi, M., Saito, N., Nakazato, Y., Sasaki, T., Takeuchi, T., 2002, Nestin as a marker in proliferative endothelium in gliomas, *Lab. Invest.* 82: 345-351.

Sugita, Y., Shigemori, M., Okamoto, K., Morimatsu, M., Arakawa, M., and Nakayama K., 2000, Clinicopathological study of pleomorphic xanthoastrocytoma: correlation between histological features and prognosis, *Pathol. Intern.* 50: 703-708.

Sultana, S., Zhou, R., Sadagopan, M.S., and Skalli, O., 1998, Effects of growth factors and basement membrane proteins on the U-373 MG glioblastoma cells as determined by the expression of intermediate filament proteins, *Am. J. Pathol.* 153: 1157-1168.

Swanson, C.R., Bridge, C., Murray, J.D., and Alvord, A.C. Jr. 2003, Virtual and real brain tumors: using mathematical modeling to quantify glioma growth and invasion, *J. Neurol. Sci.* 216: 1-10.

Swarz, J.R., and Del Cerro, M., 1977, Lack of evidence for glia cells originating from the external granular layer in mouse cerebellum, *J. Neurocytol.* 6: 241-250.

Székessy, D.P., and Stoltenburg-Didinger, G., 2001, Differentiation, proliferation and apoptosis in primary and recurrent primitive neuroectodermal tumors of childhood, *Child's Nerv. Syst.* 17: 320-327.

Tachibana, O., Lampe, J., Kleihues, P., and Ohgaki, H., 1996, Preferential expression of Fas/APO1 (CD95) and apoptotic cell death in perinecrotic cells of the glioblastoma multiforme, *Acta Neuropathol.* 92: 431-434.

Taphoorn, M.J.B., 2003, Neurocognitive sequelae in the treatment of low-grade gliomas, *Sem. Oncol.* 30: 45-48.

Takao, H., Nakagawa, K., and Ohtomo, K., 2003, Central neurocytoma with craniospinal diseemination, *J. Neuro-Oncol.* 61: 255-259.

Takekawa, Y., Sawada, T., and Sakurai, I., 1999, Expression of apoptosis and its related protein in astrocytic tumors, *Brain Tumor. Pathol.* 16: 11-21.

Takeuchi, H., Kubota, T., Sato, Y., Yao, Y., Kitai, R., and Arishima, H., 2003, Atypical neuronal-glial tumors of cerebral hemispheres on adults with PNET-like component: clinicopathological featuresof 5 cases, *Clin. Neuropathol.* 22: 47-56.

Takizawa, T., Nakashima, K., Namihira, M., Ochiai, W., Uemura, A., Yanagisawa, M., Fujita, N., Nakao, M., and Taga, T., 2001, DNA methylation is a critical cell-intrinsic determinant of astrocytic differentiation in the fetal brain, *Dev. Cell* 1: 749-758.

Tani, E., Kitagawa, H., Ikemoto, H., and Matsumoto, T., 2001, Proteasome inhibitors induce Fas-mediated apoptosis by c-Myc accumulation and subsequent induction of FasL message in human glioma cells, *FEBS Lett.* 504: 53-58.

Taratuto, A.L., Monges, J., Lylyk, P., and Leiguarda, L., 1984, Superficial cerebral astrocytoma attached to dura. Report of six cases in infants, *Cancer* 54: 2505-2512.

Taylor M.D., Liu, L., Raffel, C., Hui, C.C., Mainprize, T.G., Zhang, X., Agatep, R., Chiappa, S., Gao, L., Lowrance, A., Hao, A., Goldstein, A.M., Stavrou, T., Scherer, S.W., Dura, W.T., Wainwright, B., Squire, J.A. Rutka, J.T. and Hogg, D., 2002, Mutations in SUFU predispose to medulloblastoma, *Nat. Genet.* 31: 306-310.

Teo, C., Nakaji, P., Symons, P., Tobias, V., Cohn, R., and Smee, R., 2003, Ependymoma, *Child' Nerv. Syst.* 19: 270-285.

Tews, D.S., 1999, Cell death and oxidative stress in gliomas, *Neuropathol. Appl. Neurobiol.* 25: 272-284.

Thiessen, B., Finlay, J., Kulkarni, R., and Rosenblum, M.K., 1998, Astroblastoma:does histology predict biologic behavior ? *J. Neuro-Oncol.* 40: 59-65.

Tihan, T., Davis, R., Elowits, E., Di Costanzo, D., and Moll, U., 2000, Practical value of Ki-67 and p53 labeling indexes on stereotactic biopsies of diffuse and pilocytic astrocytomas, *Arch. Pathol. Lab. Med.* 124: 108-113.

Tohma, Y., Gratas, C., Van Meir, E.G., Desbaillets, I., Tennan, M., Tachibana, O., Kleihues, P., and Ohgaki, H., 1998, Necrogenesis and Fas/APO-1 (CD95) expression in primary (de novo) and secondary glioblastomas, *J. Neuropathol. Exp. Neurol.* 57: 239-245.

Tohyama, T., Lee, V.M., Rorke, L.B., Marvin, M., McKay, R.D., and Trojanowski, J.Q., 1996, Nestin expression in embryonic human neuroepitheliumn and in human neuroepithelial tumor cells, *Lab. Invest.* 66: 303-313.

Tomlinson, F.H., Scheithauer, B.W., and Hayostek, C.J: 1994, The significance of atypia and histologic malignancy in pilocytic astrocytoma of the cerebellum: A clinico-pathologic and flow cytometric study, *J. Child Neurol.* 9: 301-310.

Tomlinson, J.P.M., Bodmer, W.F., 1995, Failure of programmed cell death and differentiation as causes of tumors: some simple mathematical models, *PNAS, USA* 92: 11130-11134.

Tomura, N., Hirano, H., Watanabe, O., Watari, J., Itoh, Y., Mineura, K., and Kowada, M., 1997, Central neurocytoma with clinically malignant behavior, *AJNR* 18: 1175-1178.

Tong, C.Y., Ng, H.K., Pang, J.C., Hu, J., Hui, A.B., Poon, W.S., 2000, Central neurocytomas are genetically distinct from oligodendrogliomas and neuroblastomas, *Histopathology* 37: 160-165.

Tsai, R.Y., and McKay, R.D., 2000, Cell contact regulates fate choice by cortical stem cells, *J. Neurosci.* 20:3725-3735.

Tynninen, O., Aronen, H.J., Ruhala, M., Paetau, A., Von Boguslawski, K., Salonen, O., Jaaskelainen, J., and Paavonen T., 1999, MRI enhancement and microvascular density in gliomas. Correlation with tumor cell proliferation, *Invest. Radiol.* 34: 427-434.

Tysnes, B.B., and Mahesparan, R., 2001, Biological mechanisms of glioma invasion and potential therapeutic targets, *J. Neuro-Oncol.* 53: 129-147.

Uchida, N., Buck, D., He, D., Reitsma, M., Masek, M., Phan, T., Tsukamoto, A., Gage, F., Weissman, I., 2000, Direct isolation of human central nervous system stem cells, *PNAS USA* 97: 14720-14725.

Uchida, K., Mukai, M., Okano, H., and Kawase, T., 2004, Possible oncogenicity of subventricular zone neural stem cells: case report. *Neurosurgery* 55: 977-987.

Ueki, K., Wen-Bin, C., Narita, Y., Nasai, A., and Kirino, T., 1999, Tight association of loss of merlin expression with loss of heterozigosity at chromosome 22q in sporadic meningiomas. *Cancer Res*, 59: 5595-5598.

Ueki, K., Nishikawa, R., Nakazato, Y., Hirose, T., Hirato, J., Funada, N., Fujimaki, T., Hojo, S., Kubo, O., Ide, T., Usui, M., Ochiai, C., Ito, S., Takahashi, H., Mukasa, A., Asai, A., and Kirino, T., 2002, Correlation of histology and molecular genetic analysis of 1p, 19q, 10q, TP53, EGFR, CDK4 and CDKN2A in 91 astrocytic and oligodendroglial tumors, *Clin. Cancer Res.* 8: 196-201.

Uematsu, Y., Rojas-Corona, R.R., Llena, J.F., and Hirano, A., 1989, Distribution of epithelial membrane antigen in normal and neoplastic human ependyma, *Acta Neuropathol.* 78: 325-328.

Uhlmann, K., Rohde, K., Zeller, C., Szymas, J., Vogel, S., Marczinek, K., Thiel, G., Nurnberg, P., and Laird, P.W., 2003, Distinct methylation profiles of glioma subtypes, *Int. J. Cancer* 106: 52-59.

Uhrbom, L., Dai, C., Celestino, J., Rosenblum, M. K., Fuller, G.N., and Holland, E.C., 2002, Ink4a-Arf loss cooperates with Kras activation in astrocytes and and neural progenitors to generate glioblastomas of various morphologies depending on activated Akt, *Cancer Res.* 62: 551-5558.

Vajtai, I., Varga, Z., and Aguzzi, A., 1997, Pleomorphic xanthoastrocytoma with gangliogliomatous component, *Pathol. Res. Pract.* 193: 617-621.

Vajkoczy, P., Goldbrunner, R., Farhadi, M., Vince, G., Schilling, L., Tonn, J.C., Schmiedek, P., and Menger, M.D., 1999, Glioma cell migration is associated with glioma-induced angiogenesis in vivo, *Int. J. Dev. Neurosci.* 17: 557-563.

Van Den Bent, M.J., 2003, Can chemotherapy replace radiotherapy in low-grade gliomas ? Time for randomized studies, *Semin. Oncol.* 30: 39-44.

Van den Bent, M.J., Looijenga, L.H., Langenberg, K., Dinjens, W., Graveland, W., Uytdewillingen, L., Sillevis Smitt, P.A., Jenkins, R.B., and Kros, J.M., 2003, Chromosomal anomalies in oligodendroglial tumors are correlated with clinical features, *Cancer* 97: 1276-1284.

Van Tilborg, A.A.G., Al Allak, B., Velthuizen, S.C.J.M, de Vries, A., Kros, J.M., Avezaat, C.J.J., de Klein, A., Baverloo, H.B., and Zwarthoff, E.C., 2005, Chromosomal instability in meningiomas, *J. Neuropathol. Exp. Neurol.* 64: 312-322.

VandenBerg, S.R., May, E.E., Rubinstein, L.J., Herman, M.M., Perentes, E., Vinores, S.A., Collins, V.P., and Park, T.S., 1987, Desmoplastic supratentorial neuroepithelial tumors of infancy with divergent differentiation ("desmoplastic infatile gangliogliomas"). Report on 11 cases of a distinctive embryonal tumor with favorable prognosis, *J. Neurosurg.* 66: 58-71.

VandenBerg, S.R., 1993, Desmoplastic infantile ganglioglioma and desmoplastic cerebral astrocytoma of infancy, *Brain Pathol.* 3: 275-281.

Vaquero, J., Martinez, R., and Manrique, M., 2000, Stereotactic biopsy for brain tumors: is it always necessary? *Surg. Neurol.* 53: 432-437.

Vaquero, J., Zurita, M., Morales, C., Oya, S., abd Coca, S., 2000, Prognostic significance of endothelial surface score and MIB-1 labeling index in glioblastoma, *J. Neuro-Oncol.* 46: 11-16.

Varlet, P., Guillamo, J.S., Nataf, F., Koziak, M., Beuvon, F., and Daumas-Duport, C., 2000, Vascular endothelial growth factor expression in oligodendrogliomas: a correlative study with Saint-Anne malignancy grade, growth fraction and patient survival, *Neuropathol. Appl. Neurobiol.* 26: 379-389.

Vogelgesang, S., Kunert-Keil, C., Cascorbi, I., Mosyagin, I., Schröder, E., Runge, U., Jedlitschky, G., Kroemer, H.K., Oertel, J., Gaab, M.R., Pahnkee, J., Walker, L.C., and Warzok, R.W., 2004, Expression of multidrug transporters in dysembryoplastic neuroepithelial tumors causing intractable epilepsy, *Clin. Neuropathol.* 23: 223-231.

Von Deimling, A., Janzer, R., Kleihues, P., and Wiestler, O.D., 1990, Patterns of differentiation in central neurocytoma. An immunohistochemical study of eleven biopsies, *Acta Neuropathol.* 79: 473-479.

Von Deimling, A,, Kleihues, P., Saremaslani, P., Yasargil, M.G.,Spoerri, O., Sudhof, T.C., and Wiestler, O.D., 1991, Histogenesis and differentiation potential of central neurocytoma, *Lab. Invest.* 64: 585-591.

Von Deimling, A., von Ammon, K., Schoenfeld, D., Wiestler, O.D., Seizinger, B.R., and Louis, D.N., 1993, Subsets of glioblastoma multiforme defined by molecular genetic analysis, *Brain Pathol.* 3: 19-26.

Von Deimling A, Larson J, Wellenreuther R, Stangl AP, van Velthopven V, Warnick, R., Tew, J. Jr, Balko, G., and Menon, A.G., 1999, Clonal origin of recurrent meningiomas, *Brain Pathol.* 9: 645-650.

Von Deimling, A., Fimmers, R., Schmidt, M.C., Bender, B., Fassbender, F., Nagel, J., Jahnke, R., Kaskel, P., Duerr, E.M., Koopmann, J., Steinbec k, S., Wick, W., Platten, M., Muller, D.J., Przkora, R., Waha, A., Blumcke, B., Wellenreuther, R., Meyer-Puttlitz, B., Schmidt, O., Mollenhauer, J., Poustka, A., Stangl, A.P.,Lenartz, D.,and von Ammon, K., 2000, Comprehensive allelotype and genetic analysis of 466 human nervous system tumors, *J. Neuropathol. Exp. Neurol.* 59: 544-558.

Wagenknecht, B., Hermisson, M., Eitel, K., and Weller, M., 1999, Proteasome inhibitors induce p53/p21-independent apoptosis in human glioma cells, *Cell Physiol. Biochem.* 3: 117-125.

Wagenknecht, B., Hermisson, M., Groscurth, P., Liston, P., Krammer, P.H., and Weller, M., 2000, Proteasome inhibitor-induced apoptosis of glioma cells involves the processing of multiple caspases and cytochrome c release, *J. Neurochem.* 75: 2288-2297.

Waha, A., Waha, A., Koch, A., Meyer-Puttlitz, B., Wegen, S., Sorensen, N., Tonn, J.C., Albrecht, S.W., Goodyer, C.G., Berthold, F., Wiestler, O.D., and Pietsch, T., 2003, Epigenetic silencing of the HIC-1 gene in human medulloblastomas, *J. Neuropathol. Exp. Neurol.* 62: 1192-1201.

Wakimoto, H., Aoyagi, M., Nakayama, T., Nagashima, G., Yamamoto, S., Tamaki, M., and Hirakawa, K., 1996, Prognostic significance of Ki-67 labeling indices obtained using MIB-1 monoclonal antibody in patients with supratentorial astrocytomas, *Cancer* 77: 3763-380.

Wang, J.L., Zhang, Z.J., Hartman, M., Smits, A., Westermark, B., Muhr, C., and Nister, M., 1995, Detection of *TP53* mutation in human meningiomas: a study using immunohistochemistry, polymerase chain reaction/single-strand conformation polymorphism and DNA sequencing techniques on paraffin-embedded samples, *Int. J. Cancer* 22: 223-228.

Wang, X., Ohnishi, K., Takahashi, A., and Ohnishi, T., 1998, Poly(ADP-ribosylation is required for p53-dependent signal transduction induced by radiation, *Oncogene* 17: 2819-2825.

Watanabe, K., Sato, K., Biernat, W., Tachibana, O., von Ammon, K., Ogata, N., Yonekawa, Y., Kleihues, P., and Ohgaki, H., 1997, Incidence and timing of p53 mutations during astrocytoma progression in patients with multiple biopsies, *Clin. Cancer Res.* 3: 523-530.

Watanabe, K., Yokoo, H., Yokoo, M., Yonekawa, Y,. Kleihues, P., and Ohgaki, H., 2001, Concurrent inactivation of RB1 and TP53 pathways in anaplastic oligodendrogliomas, *J. Neuropathol. Exp. Neurol.* 60: 1181-1189.

Watanabe, T., Nakamura, M., Kros, J.M., Burkhard, C., Yonekawa, Y., Kleihues, P., and Ohgaki, H., 2002, Phenotype versus genotype correlation in oligodendrogliomas and low-grade diffuse astrocytomas, *Acta Neuropathol.* 103: 267-275.

Watanabe, T., Hirota, Y., Arakawa, Y., Fujisawa, H., Tachibana, O., Hasegawa, M., Yamashita, J., and Hayashi, Y., 2003, Frequent LOH at chromosome 12q22-23 and Apaf-1 inactivation in glioblastoma, *Brain Pathol.* 13: 431-439.

Watson, M.A., Perry, A., Budhjara, V., Hicks, C., Shannon, W.D., and Rich, K.M., 2001, Gene expression profiling with oligonucleotide microarrays distinguishes world health organization grade of oligodendrogliomas, *Cancer Res.* 61:1825-1829.

Watson, M.A., Gutmann. D.H., Peterson, K., Chicoine, M.R., Kleinschmidt-DeMasters, B.K., Brown, H.G., and Perry, A., 2002, Molecular characterization of human meningiomas by gene expression profiling using high-density oligonucleotide microarrays, *Am. J. Pathol.* 161: 665-672.

Weber, R.G., Sommer, C., Albert, F.K., Kiessling, M., and Cremer, T., 1996, Clinically distinct subgroups of glioblastoma multiforme studied by comparative genomic hybridazation, *Lab. Invest.* 74: 108-119.

Weber, R.G., Boström, J., Wolter, M., Baudis, M., Collins, V.P., Reifenberger, G., and Lichter, P., 1997, Analysis of genomic alterations in benign, atypical, and anaplastic meningiomas: Toward a genetic model of meningioma progression, *PNAS USA* 26: 14719-14724.

Wellenreuther, R., Kraus, J.A., Lenartz, D., Menon, A.G., Schramm, J., Louis, D.N., Ramesh, V., Gusella, J.F., Wiestler, O.D., and von Deimling, A., 1995, Analysis of the neurofibromatosis 2 gene reveals molecular variants of meningioma, *Am. J. Pathol.* 146: 827-832.

Weller, R.O., Foy, M., and Cox, S., 1977, The development and ultrastructure of the microvasculature of gliomas, *Neuropathol. Appl. Neurobiol.* 3: 303-322.

Wesseling, P., Van deer Laak, J., De Leeuw, H., Ruiter, D.J., and Burger, P., 1994, Quantitative immunohistological analysis of the microvasculature in untreated human glioblastoma multiforme, *J. Neurosurg.* 81: 902-909.

Wesseling, P., Ruiter, D.J., and Burger, P.C., 1997, Angiogenesis in brain tumors; pathobiological and clinical aspects, *J. Neuro-Oncol.* 32: 253-265.

Wesseling, P., Van der Laak, J., Link, M., Teepen, H., and Ruiter, D.J., 1998, Quantitative analysis of microvascular changes in diffuse astrocytic neoplasms with increasing grade of malignancy, *Human Pathol.* 29: 352-358.

Wessels, P.H., Hopman, A.H.N., Ummelen, M.I.J., Krijne-Kubat, B., Ramaekers, F.C.S., and Twiinstra, A., 2001, Differentiation between reactive gliosis and diffuse astrocytoma by in situ hybridization, *Neurology* 56: 1224-1227.

West, C.R., Bruce, D.A., and Duffner, P.K., 1985, Ependymomas: factors in clinical and diagnostic staging, *Cancer* 56: 1812-1816.

Westphal, M., Hänsel, M., Müller, D., Laas, R., Kunzmann, R., Rohde, E., König, A., Hölzel, F., and Herrmann, D.-D., 1988, Biological and karyotypic characterization of a new cell line derived from a human gliosarcoma, *Cancer Res.* 48: 731-740.

Westphal, M., Nausch, H., and Herrmann, H.D., 1990, Antigenic staining patterns of human gliomas cultures: primary cultures, long-term cultures and cell lines, *J. Neurocytol.* 19: 466-477.

Wharton, S.B., Chan, K.K., Hamilton, F.A., and Anderson, J.R., 1998, Expression of neuronal markers in oligodendrogliomas: an immunohistochemical study, *Neuropathol. Appl. Neurobiol.* 24: 302-308.

Wharton, S.B., Chan, K.K., Anderson, J.R., Stoeber, K., and Williams, G.H., 2001, Replicative Mcm2 protein as a novel proliferation marker in oligodendrogliomas and its relationship to Ki67 labelling index, histological grade and prognosis, *Neuropathol. Appl. Neurobiol.* 27: 305-313.

Whittle, I.R., 2004, The dilemma of low grade glioma, *J. Neurol. Neurosurg. Psychiatry,* (Suppl II) ii31-ii36.

Wiestler, O.D., Schiffer, D., Coons, S.W., Prayson, R.A., and Rosenblum, M.K., 2000, Ependymoma. *Pathology and genetics of tumours of the nervous system.*, P. Kleihues and W.K. Cavenee Eds, 2000, IARC, Lyon.

Wiestler, O.D., and Schiffer, D., 2000, Subpendymoma. In: *Pathology and genetics of tumours of the nervous system,* P. Kleihues and W.K. Cavenee Eds IARC, Lyon.

Wijsman, J.H., Jonker, R.R., Keijzer, R., van der Velde, C.J.H., Cornelisse, C.J., and van Dierendonck, J.H., 1993, A new method to detect apoptosis in paraffin sections: in situ end-labeling of fragmented DNA, *J. Histochem. Cytochem.* 41: 7-12.

Wild-Bode, C., Weller, M., Rimner, A., Dichgans, J., and Wick, W., 2001, Sublethal irradiation promotes migration and invasiveness of glioma cells: implications for radiotherapy of human glioblatoma, *Cancer Res.* 61: 2744-2750.

Willert, J., Epping, M., Pollack, J.R., Brown, P.O., and Nusse, R., 2002, A transcriptional response to Wnt protein in human embryonic carcinoma cells, *BMC Dev Biol* 2: 8.

Williams, G., and Stoeber, K., 1999, Clinical applications of a novel mammalian cell-free DNA replication system, *Br. J. Cancer* 80 Suppl 1: 20-24.

Wilson, C., 1994, Meningiomas: genetics, malignancy, and the role of radiation in induction and treatment, *J. Neurosurg.* 81: 666-675.

Wolf, H.K., Müller, M.B., Spänle, M., Zentner, J., Schramm, J., and Wiestler, O.D., 1994, Ganglioglioma: a detailed histopathological and immunohistochemical analysis of 61 cases, *Acta Neuropathol.* 88: 166-173.

Wolf, H.K., Wellmer, J., Muller, M.B., Wiestler, O.D., Hufnagel, A., and Pietsch, T., 1995, Glioneuronal malformative lesions and dysembryoplastic neuro-epitheloial tumors in patients with chronic pharmacoresistant epilepsies, *J. Neuropathol. Exp. Neurol.* 54: 245-254.

Wolf, H.K., Buslei, R., Blumke, I., Wiestler, O.D., and Pietsch, T., 1997, Neural antigens in oligodendrogliomas and dysembryoplastic neuroepithelial tumors, *Acta Neuropathol.* 94: 436-443.

Wolter, A., Reifenberger, J., Blaschke, B., Ichimura, K., Schmidt, E.E., Collins, V.P., and Reifenberger, G., 2001, Oligodendroglial tumors frequently demonstrate hypermethylation of the CDKN2A (MTS1, p16 ^{INK4a}), p14^{ARF}, and CDKN2B (MTS2, p15^{INK4b}) tumor suppressor genes, *J. Neuropathol. Exp. Neurol.* 60: 1170-1180.

Wong, A.J., Bigner, S.H., Bigner, D.D., Kinzler, K.W., Hamilton, S.R., and Vogelstein, B., 1987, Increased expression of the epidermal growth factor receptor gene in malignant gliomas is invariably associated with gene amplification, *PNAS USA* 84: 6899-6903.

Wong, K., Gyure, K.A., Prayson, R.A., Morrison, A.L., Le, T.Q., and Armstrong, R.C., 1999, Dysembryoplastic neuroepithelial tumor: in situ hybridization of proteolipid protein (PLP) messenger ribonucleic acid (mRNA), *J. Neuropathol. Exp. Neurol.* 58: 542-542.

Wrensch, M., Minn, Y., Chew, T., Bondy, M., and Berger, M.S., 2002, Epidemiology of primary brain tumors: current concepts are review of the literature, *Neuro-Oncology* 4: 278-299.

Wu, G., Chai, J., Suber, T.L., Wu, J.W., Du, C., Wang, X., Shi, Y., 2000, Structural basis of IAP recognition bt Smac/DIABLO, *Nature* 408: 408-1008-1012.

Wyllie, A.H., Bellamy, C.O.C., Bubb, V.J., Clarke, A.R., Corbet, S., Curtis, L., Harrison, D., Hooper, M.L., Toft, N., and Webb, S., 1999, Apoptosis and carcinogenesis, *Br. J. Cancer* 80 (suppl 1), 34-37.

Xao, K.C., Komata, T., Kondo, Y., Kanzawa, T., Kondo, S., and Germano, I.M., 2003, Molecular response of human glioblastoma multiforme cells to ionizing radiation: cell cycle arrest, modulation of the expression of cyclin-dependent kinase inhibitors, and autophagy, *J. Neurosurg.* 98: 378-384.

Xia, C.L., Du, Z.W., Liu, Z.Y., Huang, Q., and Chan, W.Y., 2003, A2B5 lineages of human astrocytic tumors and their recurrence, *Int. J. Oncol.* 23: 353-361.

Xiao, C., Yang, B.F., Asadi, N., Beguinot, F., and Hao, C., 2002, Tumor necrosis factor-related apoptosis-inducing ligand-induced death-inducing signaling complex and its modulation by c-FLIP and PED/PEA-15 in glioma cells, *J. Biol. Chem.* 277: 25020-25025.

Yagishita, S., Itoh, Y., Chiba, Y., and Kuvana, N., 1982, Morphological investigations on cerebellar "neuroblastoma" group, *Acta Neuropathol.* 56: 22-28.

Yamada, Y., Kuroiwa, T., Nakagawa, T., Kajimoto, Y., Dohi, T., Azuma, H., Tsuji, M., Kami, K., and Miyatake, S., 2003, Transcriptional expression of survivin and its splice variants in brain tumors in humans, *J. Neurosurg.* 99: 738-745.

Yamasaki, F., Yoshioka, H., Hama, S., Sugiyama, K., Arita, K., and Kurisu, K., 2000, Recurrence of meningiomas, Cancer 89: 1102-1110.

Yancopoulous, G.D., Savis, S., Gale, N.W., Rudge, J.S., Wiegand, S.J., and Holash, J., 2000, Vascular-specific growth factors and blood vessel formation, *Nature* 407: 242-248.

Yang, S., Wetzel, S., and Cha, S., 2002, Dynamic contrast-enhanced T2*-weighted MR imaging of gliomatosis cerebri, *Am. J. Neuroradiol*, 23: 350-355,

Yang, B.F., Xiao, C., Roa, W.H., Krammer, P.H., and Hao, C., 2003, Calcium/Calmodulin-dependent protein kinase II regulation of c-FLIP expression and phosphorylation in modulation of Fas-mediated signaling in malignant glioma cells, *J. Biol. Chem.* 278: 7043-7050.

Yin, X.L., Pang, J.C., Hui, A.B., and Ng, H.K., 2000, Detection of chromosomal imbalances in central neurocytomas by using comparative genomic hybridization, *J. Neurosurg.* 93: 77-81.

Yin, D., Zhou, H., Kumagai, T., Liu. G., Ong, J.M., Black, K.L., Kopeffler, H.P., 2005, Proteasome inhibitor PS-341 causes cell growth arrest and apoptosis in human glioblastoma multiforme (GBM), *Oncogene* 24: 344-354.

Yokota, N., Nishizawa, S., Ohta, S., Date, H., Sugimura, H., Namba, H., and Maekawa, M., 2002, Role of Wnt pathway in medulloblastoma oncogenesis, *Int. J. Cancer* 101: 198-201.

Zagzag, D., Friedlander, D.L., Miller, D.C., Dosik, J., Cangiarella, J., Kostianovsky, M., Cohen, H., Grumet, M., and Gredo, M.A., 1995, Tenascin expression in astrocytomas correlates with angiogenesis, *Cancer Res.* 55: 907-914.

Zagzag, D., Hooper, A., Friedlander, D.R., Chan, W., Holasch, J., Wiegand, S.J., Yancopoulos, G.D., and Grumet, M., 1999, *In situ* expression of angiopoietins in astrocytomas identifies angiopoietin-2 as an early marker of tumor angiogenesis, *Exp. Neurol.* 159: 391-400.

Zagzag, D., Zhong, H., Scalzitti, J.M., Laughner, E., Simons, J.W., and Semenza, B.L., 2000, Expression of hypoxia-inducible factor-1 in brain tumors. Association with angiogenesis, invasion and progression, *Cancer* 88: 2606-2618.

Zagzag, D., Amimovoin, R., Greco, M.A., Yee, H., Holasch, J., Wiegand, S.J., Zabski, S., Yancopoulos, G.D., and Grumet, M., 2000, Vascular apoptosis and involution in gliomas precede neovascularization: a novel concept for glioma growth and angiogenesis, *Lab. Invest.* 80: 837-849.

Zagzag, D., and Capo, V., 2002, Angiogenesis in the central nervous system: a role for vascular endothelial growth factor/vascular permeability factor and tenascin-C. Common molecular effectors in cerebral neoplastic and non-neoplastic "angiogenetic diseases", *Histol Histopathol.* 17: 301-321.

Zakeri, Z., Bursch, W., Tenniswood, M., and Lockshin, R.A., 1995, Cell death: programmed apoptosis, necrosis, or other ? *Cell Death Differ.* 2: 87-96.

Zassler, B., Ingolf, E., Humpel, B., and Humpel, C., 2005, Protein delivery of caspase-3 induces cell death in malignant C6 glioma, primary astrocytes and immortalized and primary brain capillary endothelial cells, *J. Neuro-Oncol.* 71: 127-134.

Zentner, J., Wolf, H.K., Ostertum ,B., Hufnagel, A., Campos, M.G., Solymos, L., and Schramm, J., 1994, Gangliogliomas: clinical, radiological and histopathological findings in 51 patients, *J. Neurol. Neurosurg. Psychiatry*. 57:1497-1502.

Zhang, Z., Vuori, K., Wang, H.-G., Reed, J., and Ruoslathi, E., 1996, Integrin activation by R-ras, *Cell* 85: 61-69.

Zhivotosky, B., Burgess, D.H., Vanags, D.M., and Orennius, S., 1997, Involvement of cellular proteolytic machinery in apoptosis, *Biochem. Biophys. Res. Commun.* 230: 481-888.

Zhou, Q., Wang, S., and Anderson, D.J., 2000, Identification of a novel family of oligodendrocyte lineage-specific basichelix-loop-helix transcription factors. *Neuron* 25: 331-343.

Zhou, Q., and Anderson, D.J., 2002, The bHLH transcription factors OLIG2 and OLIG1 couple neuronal and glial subtype specification, *Cell* 109: 61-73.

Zhou, R., and Skalli, O., 2000, TGf-α differentially regulates GFAP, vimentin, and nestin gene expression in U-373 MG glioblastoma cells: correlation with cell shape and motility, *Exp. Cell Res.* 254: 269-278.

Zhu, J.J., Maruyama, T., Jacoby, L.B., Herman, J.G., Gusella, J.F., Black, P.M., and Wu, J.K., 1999, Clonal analysis of a case of multiple meningiomas using multiple molecular genetic approaches: pathology case report, *Neurosurgery* 45: 409-416.

Zülch, K.J., 1956, Biologie und Pathologie der Hirngeschwülste. *Pathologische Anatomie der raumbeengenden intrakranielle Prozesse*, K.J. Zülch and E. Christensen Eds, Springer, Berlin.

Zundel, W., Schindler, C., Haas-Kogan, D., Koong, A., Kaper, F., Chen E., Gottschalk, A.R., Ryan, H.E., Johnson, R.S., Jefferson, A.B., Stokoe, D., Giaccia, A.J., 2000, Loss of PTEN facilitates HIF-1- E mediated gene expression, *Genes Dev*. 14: 391-396.

Zuzak, T.J., Steinhoff, D.F., Sutton, L.N., Phillips, P.C., Aggert, A., and Grotzer, M.A., 2002, Loss of caspase-8 mRNA expression is common in childhood primitive neuroectodermal brain tumour/medulloblastoma, *Eur. J. Cancer* 38: 83-91.

INDEX

α-sm-actin 76, 80, 193, 195, 198